D. C Christner

The International Encyclopedia of Scientific Tailor Principles, for All Kinds and Styles of Garment-Making

D. C Christner

The International Encyclopedia of Scientific Tailor Principles, for All Kinds and Styles of Garment-Making

ISBN/EAN: 9783337224103

Printed in Europe, USA, Canada, Australia, Japan

Cover: Foto ©berggeist007 / pixelio.de

More available books at **www.hansebooks.com**

THE INTERNATIONAL ENCYCLOPEDIA

—OF—

SCIENTIFIC TAILOR PRINCIPLES,

—FOR—

ALL KINDS AND STYLES OF GARMENT-MAKING,

—AND—

Drafting the Actual Measures or Indices

—The Absolute Mathematical Mould or Perfect Fit
of the Human Form—

TRUE TO ANY DESIGN OR FASHION PLATE,

By Scientific Square Measurement, either

DIRECTLY OR INDIRECTLY UPON THE GOODS,

WITH ANY MATHEMATICAL INCH-RULE AND TAPE MEASURE ALONE,

—ALSO—

DESIGNING, COPYING, BASTING, SEAMING, STITCHING, TAILOR FINISHINGS, TAILOR BUTTONHOLES, TRIMMING, DRAPING, STAMPING, EMBROIDERY, CROCHETING, KNITTING, WORSTED WORK, FANCY AND ARTISTIC NEEDLE WORK, LACE WORK, BEAD WORK, MILLINERY, HAIR WORK, FEATHER WORK, WAX FLOWERS AND WAX WORK, SHELL FLOWERS AND SHELL WORK, TRANSFER WORK, TRANSPARENCIES, LEAF WORK, LEATHER WORK, FANCY DYEING, WATER-COLOR PAINTING, CHINA AND POTTERY PAINTING, GRECIAN OIL PAINTING, LUSTRE-PAINTING, LINCRUSTA PAINTING, KENSINGTON PAINTING, UTILITY AND DECORATIVE ARTS.

IN ONE VOLUME OF FOUR PARTS,

ILLUSTRATED AND SELF-INSTRUCTIVE,

—BY—

PROF. D. C. CHRISTNER, V. D. M.

PHILADELPHIA:
1888.

To the millions in the Work-room and the Home, who are striving for excellence —in its province—and who would and may educate, refine, and ennoble humanity, by giving the most beautiful, impressive, and perfect expression to the best, most appropriate, and perfect conceptions of thought, through the characters and language of UTILITY and DECORATIVE ART—this book is respectfully dedicated, by

THE AUTHOR.

DIVISIONAL CONTENTS.

PART   I.—All branches of *Ladies'* and *Children's Dress Cutting* and *Basting*.

PART   II.—All branches of *Ladies'* and *Children's Cloak* and *Wrap Cutting* and *Basting*.

PART   III.—All branches of *Designing Ladies'* and *Children's Garments, Copying, Basting, Seaming, Stitching, Tailor Finishings, Tailor Buttonholes, Trimming, Draping, Stamping, Embroidery, Crocheting, Knitting, Worsted Work, Fancy* and *Artistic Needlework, Lace Work, Bead Work, Millinery, Hair Work, Feather Work, Wax Flowers and Wax Work, Shell Flowers* and *Shell Work, Transfer Work, Transparencies, Leaf Work, Leather Work, Fancy Dyeing, Water-Color Painting, China* and *Pottery Painting, Grecian Oil Painting, Lustre Painting-Lincrusta Painting, Kensington Painting, Utility* and *Decorative Art*.

PART   IV.—All branches of *Gentlemen's* and *Boys' Coat, Vest, Pantaloon*, and *Overcoat Cutting*.

PRICE-LIST---AND EXPLANATION·

ENCYCLOPEDIA, printed on extra fine heavy paper, elegantly bound in fine cloth; stamped in gold, with heavy gilt edges; price, $5.00.

"*The Scientific Tailor Principles of Garment-drafting*" alone, as contained in *The Encyclopedia*, have an actual and positive standard commercial value—among the professional connoisseurs and leading tailors throughout Europe and America—ranging in price, all the way from $50 to $100; and in order to obtain a knowledge of each and all of the various other Arts and Sciences, fully and clearly taught and beautifully illustrated in the work—if learned in the usual and ordinary way and manner, and at the usual and standard prices, which range from $2.50 to $10.00, for each special department—a still greater expense would necessarily be incurred thereby! It will thus be seen at once, by all thoughtful and fair-minded persons, that through *The Encyclopedia*, a complete, scientific, and exhaustive manual of instruction is now furnished to the public, in all of these valuable, costly, and eagerly sought arts and sciences at an exceedingly and surprisingly small cost, when we fix the price at only $5.00, for what would otherwise cost the student or public—through the channel of the conservative and exorbitant prices of the former and once popular, but now rapidly waning, "Oral school" of Teachers—several hundred dollars! All of which is at least, more thoroughly and accurately taught, in theory; and certainly by means of our lucid and exhaustive system of diagrams, equally as well if not better, in practice. If we are asked: "Do you mean to convey the idea—accepting the principles of your system of drafting as being absolutely correct, that all persons can, through it, be taught to measure, and to use it, absolutely correct?" If the question is asked, as pertaining to such persons who are capable of receiving, retaining, and practically using any knowledge whatever—either orally or from books—we ask in return, "why not?—through careful study and experiment, coupled with a determined care for accuracy, subject to the same mistakes only, that occur in other mechanical operations!"

Is it not strange, yea, passing strange, that such a question, should even be asked by an intelligent mind, in this age? Show us any principle, either in Science or Philosophy, which is correct in the abstract, and we will show you one which can be reduced to practice! And as touching the demonstration and proof, in this case, we verify the same, in the given references—from those whose experience and testimony should certainly be a satisfactory answer to any intelligent mind. Of this we are certain: If we could not assure a practical knowledge of these Arts and Sciences to any ambitious, industrious, and patient student—through the printed and illustrated instructions of *The Encyclopedia*—certainly we could not conscientiously promise and assure a practical knowledge of the same to such, in any other way! Hence logically, practically, and by demonstrated proof we sustain our claim, of furnishing to the public—through *The Encyclopedia*, the same practical knowledge and commercial value, and that too in a more exhaustive and permanent form, for the mere nominal price of $5.00, which would, through any and every other available avenue, cost the public at least, several hundred dollars! No one, therefore—not even the most obtuse mind—capable of receiving, retaining, and practically using any knowledge either orally or from any book or books whatever, and who is industrious and patient, need have any fears but that—by devoting the spare hours of each day, or the evenings of a few weeks, to the study and practice of the given problems, and the instructions and illustrations for their solution, and in their natural and given order —all can be easily and thoroughly comprehended and perfectly reduced to practice, without being necessitated to expend either time or money on oral instructions. For the oral instructions would of necessity be simply the oral repetition of the printed and illustrated instructions—all being able to comprehend these, if they but will: the question therefore, simply resolving itself into this: "Whether or not the student is willing to exercise a sufficient degree of industry, patience, accuracy, and care?" The time, in which said knowledge is attained by each, depending entirely upon the degree of said properties possessed and exercised, by said student! But as a provision for those who can not, dare not, or will not trust their own industry and patience, to comprehend and practically master the printed and illustrated instructions, and who wish or insist upon the additional help of extra and oral instructions; in any part or parts of the work; and who are willing to pay for said time and service—such can arrange for said instructions with the agent from whom *The Encyclopedia* is purchased, at such times and places as may be mutually agreed upon; and in any department of the work that may be desired by said purchaser, and that may be mutually decided upon. Furthermore, arrangements can be made thereafter, and at any time, with any manager or agent of any of our metropolitan offices nearest and most convenient to and for said purchaser, for any number of lessons in any department of *The Encyclopedia*, as may be mutually agreed upon. The terms for said lessons or instructions, to be uniform in price throughout the United States, and not to exceed nor to be less in rate than fifty cents per hour; said lessons to be taken at the offices of said agents, unless otherwise and especially arranged for.

Believing in and practicing the principles of "live and let live," and "one and the same price to all," and having spared no cost—either in time, labor or money— in the preparation, arrangement and completion of a work for the small sum of $5.00, which contains and represents a positive and standard commercial value of at least several hundred dollars to the public, we solicit the patronage and influence of the profession, and an intelligent and fair-minded public everywhere—and especially that of our patrons, in its introduction; which will insure to us a constant, permanent, and a grateful patronage thereafter.

Squares, inch-rules, tape-measures, tracing wheels, etc., are things entirely separate and distinct from *The Encyclopedia*, or book of instructions—as much so, as the sewing machine, the plaiter, the dress-form, the scissors, the crochet-hook, or needles, etc.; and must each and all be provided for, and arranged and paid for, separately. The system of drafting is not founded on any particular kind or form of the square, or inch-rule, or any special form of the "inch-instrument" whatever. Nor yet is it based as some systems are—simply and solely upon twelfths or sixteenths, etc., as integral parts of the mathematical inch; thus requiring a special square or inch-rule, upon which said divisions are especially graded—along with or without various other fixed proportions, numbers, divisions, etc.; but the system is in every way, entirely, and in a preëminent sense independent of all such unscientific and deceptive proportional gradings, whatever. In all its elements of construction and principles of application, it is purely, simply, and positively, actual measurement, alone; and it is not based upon any special form of the square or inch-rule, but upon the mathematical inch itself—and upon this alone! Since the greater necessarily includes the lesser—and the whole its parts—we of course include all its scientific, mathematical, and natural divisions; and nothing more, and nothing less!

It will readily be seen, therefore, by any intelligent mind, that any form of the inch instrument, by means of which straight lines can be drawn of sufficient length, and at right angles with each other, will satisfy all demands—in connection with the tape-measure—as the instrument of execution, in drafting. Any square or inch-rule therefore, even a school child's penny inch-rule, or otherwise a YARD STICK, will be sufficient, in the hands of the experienced student! But as a matter of the highest convenience, preference, and general adaptability, for all kinds of work, under all circumstances, the author both recommends and personally uses, the common, plain, hardwood, mathematical inch square, having THE INCH and its mathematical divisions—down to sixteeths—repeated twelve times on the short arm, and twenty-four times on the long arm;—"sixteenths" being a sufficiently close division of the inch, for all ordinary practical purposes.

The inside of the end of the long arm should be neatly and evenly curved, commencing from a point four inches from the end and curving to the depth of one-half inch, at the end, so as to give the ordinary and standard curves for the curved lines in the darts, and also at and below the waist line, and as these same lines are illustrated in the engravings. But any one having a mathematical inch-rule of any kind —as given above—need not necessarily be at any extra expense in the purchasing of new or different instruments, in accepting and adopting this system of drafting. But if straight rules are used, the same curve should be given to the inside of one of the ends—as described above, on the plain, common square. Students can purchase Rules and Squares, all the way from fifty cents to five dollars—according to style, quality, convenience, and finish. *The Encyclopedia* simply teaches the student the scientific and practical use of any mathematical inch-rule and tape-measure—alone. It teaches the science of garment-drafting, by actual mathematical inch measurement! Great care must however be taken in so drafting that the lines will be exactly at right angles with each other—where called for in the instructions—if a straight inch-rule is used. We furnish through our agents, or at our offices, all kinds and grades of Rules, Squares, Tape-Measures, Tracing-Wheels, Tailors' and Dressmakers' Furnishing Supplies, and indeed all supplies or necessaries pertaining to the arts and sciences taught in *The Encyclopedia*, at a very small margin of profit; and will always furnish our customers the best, and the most that is possible, for their money, promptly, and according to order. We do a strictly cash business, on business principles, only. Agents pay cash for all goods received, and our customers must expect the same terms only, either from them or from us. Agents are strictly confined to tabulated prices. ☞ Agents found guilty of deception, falsehood, dishonesty, or immorality, will immediately be discharged, regardless of any sacrifice whatever to ourselves, in a financial view. ☜ Hence the public can rely implicitly upon the statements, business transactions, and character of our Agents; and we bespeak for them the courtesies due a lady or gentleman, unless they should prove themselves other than such, in which case, the reporting of their names and conduct at our office will be accepted as a favor by THE AUTHOR.

SUPPLEMENTARY PRICE-LIST.

1 Plain Straight Inch-Rule, (18 inches long),	$0 50
1 Plain Straight Inch-Rule, (21 inches long),	0 75
1 Plain Straight Inch-Rule, (24 inches long),	1 00
1 Plain Straight Folding Inch-Rule, (24 inches long),	1 25
1 Plain Hardwood Square, (24x12 inches),	1 50
1 Plain Hardwood Square, with Brace, (24x12),	1 75
1 Plain Hardwood Square, highly polished, with Hinges for folding,	2 50
1 Fine Sateen Tape Measure, (60 inches),	25
1 Best Sateen Tape Measure, with "Spring-fold,"	1 00
1 Best Steel Tape Measure, with "Spring-fold," (60 inches),	2 25
1 Measure Book,	50
1 Good Single Tracing Wheel,	35
1 Best Quality Single Tracing Wheel,	65
1 Good Double Tracing Wheel,	1 00
1 Best Quality Double Tracing Wheel,	1 25
1 Pair Best Steel Scissors, 8½ inches,	1 75
1 Pair Best Steel Scissors, 9 inches,	2 00
1 Pair Best Steel Scissors, 10 inches,	2 25
1 Pair Best Steel Scissors, 11 inches,	2 50
1 Pair Best Steel Scissors, 12 inches,	3 50
1 Pair Best Steel Tailors' Shears, 14 inches,	5 50
1 Plain Wire Draping Form,	5 00
1 Patent Adjustable Draping Form, changeable neck, bust, waist and skirt, from 32 to 44 inches at the bust, and 20 to 32 inches at the waist,	12 50
1 Best and Latest Plaiter,	5 00
1 Plain Gauge for Bias Cutting,	2 00
1 Jointed Gauge for Bias Cutting,	2 50
1 Beautiful Stamped Leather or Morocco "Portfolio" for carrying book, folding-square, fashion journal, measure book, scissors, and sewing outfit—on going out to sew in families,	4 50
1 Quire good Manilla Drafting Paper	25
1 Quire best Manilla Drafting Paper,	35
1 Box "Tailor's Chalk," (three colors)	25
1 Box Best "Crayons".	35
1 "Test Lining," cut and basted to measures	1 00
1 Cut Paper Pattern of Waist, for lady or child, to measures.	50
1 Cut Paper Pattern of waist and sleeves, for lady or child, to measures,	75
1 Cut Paper Pattern of Sleeve, for lady or child, separately, to measures	35
1 Cut Paper Pattern of Basque for lady or child, to any Fashion Plate chosen, and to measures,	1 00
1 Cut Paper Pattern of Skirt and Over-Skirt for lady or child, to any Fashion Plate chosen, and to measures.	1 00
1 Cut Paper Pattern of Skirt, separately, for lady or child, to any Fashion Plate chosen, and to measures	50
1 Cut Paper Pattern of ladies or misses Train-skirt, to any Fashion Plate chosen, and to measures,	1 00
1 Cut Paper Pattern for lady's or misses' polonaise, to any chosen Fashion Plate and to measures	1 25
1 Cut Paper Pattern of ladies' or child's Short Jacket or Coat, to any Fashion Plate, and to measures,	1 00
1 Cut Paper Pattern of lady's or child's long Coat or Ulster, to any chosen Fashion Plate, and to measures,	1 50
1 Cut Paper Pattern of lady's or misses' Dolman-Sleeved Wrap, to any chosen Fashion Plate, and to measures,	1 75
1 Cut Paper Pattern of lady's or child's Circular, to any Fashion Plate, and to measures	1 00
1 Cut Paper Pattern of entire suit (complete) for lady or child, to any chosen Fashion Plate, and to measures	2 50
1 Cut Tissue Paper Pattern for lady or child (complete suit) to any Fashion Plate, and to measures, neatly pinned into position, with draperies, etc., fixed and indicated,	3 00
"SHOW PATTERNS," in fine French Tissue Paper—indicating the styles and combinations of colors, etc., thoroughly and securely made according to taste or order, *for the Trade*, ranging in price per suit, from $2.50 to	20 00
1 Cut Paper Pattern of gent's and boys Shirts, to measures,	50
1 Cut Paper Pattern of gents' or boys' entire Suit to measures, from $3.00 to	5 00
1 Cut Paper Pattern of gents' or boys' Overcoat, to measures	2 00

N. B.—Any of the aforesaid articles along with any other articles needed, in any of the arts and sciences taught in *The Encyclopedia*, not tabulated, sent, prepaid by mail or express, to any address in the United States, on receipt of the price, by cash, draft, or money order.

Any fashion journal in the world furnished by mail, on receipt of 20 per cent. above cost and carriage—where foreign correspondence is required—otherwise, at cost and carriage.

Dress goods, woolens, silks, trimmings, and all dressmakers', cloakmakers' and merchant tailors' supplies, promptly furnished through mail or by express, according to sample and order, on receipt of 5 per cent. above actual cost and carriage, by cash, draft, or money order.

Where inquiries are made requiring answers through mail, postage must invariably be inclosed to insure an answer. Do not forget, in ordering anything, to be brief, and very explicit and plain in description, so that it is not possible to misunderstand your wishes. No goods sent C. O. D. unless cost of carriage and expressage, both ways, is paid in advance. Write very plainly your full name, number, street, post-office, county, and State; and address all orders and letters very plainly.

PROF. D. C. CHRISTNER,

PHILADELPHIA, PA.

IMPORTANT NOTICE TO THE PUBLIC.

"*The International Encyclopedia of French, Prussian, English and American Scientific Tailor Principles, for all kinds and styles of garment-making, and drafting the actual measures or indices—the actual mathematical mould or perfect fit of the human form—true to any design or fashion plate, by scientific square measurement, either directly or indirectly upon the goods, with any mathematical square or inch rule and tape measure alone: Also, designing, copying, basting, seaming, stitching, tailor finishings, tailor buttonholes, trimming, draping, stamping, embroidery, crocheting, knitting, worsted work, fancy and artistic needle-work, lace work, bead work, millinery, hair work, leather work, wax flowers and wax work, shell flowers and shell work, transfer work, transparencies, leaf work, leather work, fancy dyeing, water-color painting, china and pottery painting, Grecian oil painting, lustre painting, lincrusta painting, Kensington painting, utility and decorative arts—in one volume of four parts, illustrative and self-instructive*—has been in all its parts and contents copyrighted in full, with all rights reserved; and has been entered in the office of the Librarian of Congress, at Washington, D. C., in full and strict compliance with the late United States copyright laws. And a printed certificate showing the purchase of said Encyclopedia, and the right of personal use and benefit of its contents, will be given to each purchaser of said Encyclopedia—having the author's signature attached thereto; or, otherwise, the signature of his lawfully authorized representative or agent, bearing printed and dated certificate, authorizing him or her as such,

as well as that of being the only rightful and lawful agent for the sale of said work in the territory, where said sale is made, and at the time said sale is made; said certificate of said agent, having the author's signature attached thereto. Any person using or selling said Encyclopedia, or any part or parts thereof, for personal interest otherwise—or in any way whatever infringing by appropriation, or plagiarism, on the author's copyrights thereon, will be prosecuted and will receive the full penalty of the law.   D. C. CHRISTNER, Author and Proprietor,
PHILADELPHIA, PA.

IMPORTANT NOTICE PERTAINING TO AGENTS AND AGENCIES.

Letters of application for terms and agencies must invariably contain a two-cent postage stamp, with full name and P. O. address of "given reference," touching moral character, honesty, and general mental and business qualifications, as well as the stated amount of cash capital that could or would be invested in the business; provided the terms and conditions offered proved sufficiently inducive and were accepted—otherwise they will remain unanswered. The author is too busy, and his time too valuable to waste in either talking or writing (in his office and on business matters) to any one who is merely curious and does not strictly mean business ; or who does not propose business on business principles—and to whom, in return, he cannot propose business in a direct and business-like way, on a definite and ready capital and on stated goods and territory.

Letters with the above qualifications will be treated *bona fide*—and their authors will, by return mail, receive confidential papers, circulars, propositions, etc., which are submitted strictly for personal and private knowledge, and for private and personal use, only—and for a limited time only. And should their contents and propositions not be accepted, said papers, circulars etc., must—at the request or demand of their author, and on receipt of return postage for the same—be returned immediately, thereupon. We ask or expect no one to accept our propositions— only voluntarily, and from self interest and purely selfish motives alone ; neither do we deal dishonestly or make better propositions on the same capital, stock, and territory, to one than to another ; neither have we any business secrets, other than such as are just, honest, and fair to all, or the publicity of which could reflect upon us in the least, or that would injure us legally or morally—nor which we in any way fear or dread. But from the past teachings of an extensive experience in this particular line of business, and for the best of reasons—which are best known to ourselves—as a protection to our agents, and against the impositions of envious, intriguing, financially interested, and unprincipled persons, who under the false pretence of business, seek information only for the purpose of effecting possible financial injury thereby ; and also for other, and the very best of reasons, we have a "decided preference" not to gratify either idle curiosity, envy, or selfish deceit ; and, therefore, the above prerogative, touching "confidential papers," will in all cases be strictly and positively reserved—and when made necessary, by the circumstances in the case, will be fully enforced by law. Immoral and dishonest persons will neither be accepted knowingly, nor retained if accepted through deception and by mistake—regardless of any financial loss or sacrifice whatever, to the author. Although very select and choice in the securing and retaining of agents, nevertheless true moral worth and intelligence, wedded to honesty and industry, will always find a most hearty welcome; and such will receive, as a token of appreciation, a certain and rapid advancement to the highest, most lucrative, and most promising positions—and with exclusive benefits, under legal protection! We do not sell territory, but give it away—to be possessed and controled, legally—to the extent and degree of value, expressed in the ratio and proportion of money invested in stock (consisting of Encyclopedias, Supplies, etc.,) when entering into the business. To the right parties, and in the ratio of their investment, unparalleled inducements are offered and given, for a certain and rapid accumulation of wealth ; and that too, in an easy, polite, legitimate, useful, and a thoroughly exclusive and protected business. An opportunity is thereby furnished for combining the pleasure of travel, with the pleasure of accumulating wealth, and that of "doing good," at one and the same time! The right persons will find places awaiting them as agents in villages, townships, county seats, counties, towns, cities, and states ; and as organizers, principals, and managers of schools and conservatories in the various metropolitan cities throughout the United States, Canada, and foreign countries. All our agents were most happily and astonishingly surprised, beyond even their highest and most brilliant anticipations and hopes, in the degree of success that attended their immediate efforts, when they first began to sell the work—which success is increasing, in the ratio of their experience. Agents are ecstatic everywhere over its wonderful sale, and are wild in their anxiety to secure and control more territory. It stands preëminently without a rival, in its record in sales, and in the satisfaction it renders, and the profits it yields to those who are introducing it. The universal verdict of our agents is: "It sells at sight!" Those who have been in the habit of thinking it a difficult matter to sell a book—have found out that it all depends upon the kind, the quality, and the demand! The contents of "The Encyclopedia" reduced to practice, bring gold to the purchasers; hence, they anxiously and gladly give gold for that which brings them more. It is the old doctrine of human selfishness—a selfish reciprocity! The book is a grand compendium of new ideas, and new principles; and it stands preëminently by itself—in its line, as well as its success—and has long and anxiously been waited for by thousands, who have painfully felt the need of just such a work ; and who only need to be shown its title page and index, to become anxious patrons, at any price. The author challenges the "world of agencies" to produce its equal, in the ease and rapidity of its sale, the price it brings and for which it is so eagerly sought, and the rich profits it yields to those whose fortune it has been to chance upon its sale and introduction, under exclusive rights, in special and protected territory.

Write full name, number, street, P. O. Address, County and State, with nearest express office, as qualified above, and address all letters very plainly to

PROF. D. C. CHRISTNER,

PHILADELPHIA, PA.

AGENTS CAN BEGIN IMMEDIATELY!

They will however be expected to become industrious students.

Any lady or gentleman of mere ordinary intelligence and ability—however inexperienced in the business—can at once proceed, and succeed in a very successful canvass and sale of the work; and by simply devoting the evenings of the first eight or ten days to a careful reading, study, and practice of the various parts, and the illustrated examples and solutions of the given problems—beginning at "the very beginning" and mastering everything on every page, and in the order given—any agent can also prepare himself or herself in advance, for any and every emergency; in case they should be requested to give, or aid the purchaser through, oral instructions. But you might possibly sell the work successfully for many months, without ever even being solicited or requested to give a single lesson of instruction. And yet, it would certainly be to the agent's interest, in a financial sense, as well as to the interest of the author and the patron, that the qualification existed in the agent; and that the giving of lessons—by sound argument and correct reasoning—was encouraged. For many persons, especially in measurement and garment-drafting, tracing, basting, etc., through a lack of sufficient industry, accuracy, and patience, most probably would obtain as a result, in their first effort—a failure, which would dishearten them; and which would most probably discourage all further effort, in the attainment of its true teaching and practice; and which would in some cases, through this ignorance and disappointment, create a prejudice and opposition against the work—to the extent and in the exact ratio of said influence. All of which evil, under more careful and wise management, might not only have been avoided, but the very opposite result might have been obtained—and that too, at a profit to the agent! Here as elsewhere, therefore, "honesty is the best policy," and "merit brings its reward!"

While it is possible, therefore, to be very successful without even knowing anything about the contents—thus representing "The Automatic Agent"—nevertheless, that success is increased in the exact ratio of the knowledge possessed of the work and its contents, by the agent. And in justice to himself or herself financially, to the reputation of the work, and to the financial interest of the author—who furnishes profitable employment thereby—no agent accepting the opportunities and benefits of its sale, should rest content until the entire work is thoroughly mastered and practically understood; but which is by no means necessary at first, and for all of which ample time and opportunity is afforded, by devoting one hour each evening to its study and practice—while selling it, and making money rapidly at the same time.

To any intelligent mind, the evidence as given in our printed references—who are living, and can be addressed or seen—is entirely sufficient, being unimpeachable both in quality and kind; and which is the highest possible evidence that can be given.

It is simply infallible and exhaustive, to any fair-minded and intelligent person. Those who ask that the system should be proved over and over, ten thousand times repeated again and again, and to each individual to whom it is presented for sale—make an unjust and unreasonable request or demand—unless they are willing to pay for said time and service.

Generally speaking, we do not expect it of agents, neither do we advise it, as being wisest and best under many circumstances—many unprincipled persons in this manner, seeking only the principles and secrets of measurement, etc., at no cost to themselves, under the pretence of testing the system.

The agent's business, therefore, is less proving the system than selling it—upon its own merits and intrinsic worth, and the given and demonstrated proofs and references within the work itself.

Neither is it fair to hold either the work or the author responsible for the lack of knowledge, the inexperience, the inabilities, or even the failures of any agent or agents whatever. Nor yet does the reputation of the system depend either upon the success or failure of the agent, in case an attempt is made to demonstrate the accuracy of the system. A thing once proved and demonstrated openly and before the eyes of the best and most disinterested critics—under the severest and most crucial tests possible—ends the controversy, if their verdict is in its favor! Such are the proofs given and offered within *The Encyclopedia*.

Personally, we hold ourselves in readiness to place our system of measurement, drafting, tracing, and basting (in garment-making) on test, against any respectable, honorable, and well-known authors' system, and according to the printed instructions; or against any system or systems of the world—separately or collectively—and that too, in the most public place and manner possible.

But we are not now engaged in the business of proving" our system ; that is in the past with us. We are now "selling" the system, to an intelligent, discriminating and appreciative public, upon given proofs!

And we beg for ourselves and our agents, to be excused from paying any attention whatever, to the vulgar and cowardly boastings and so-called "challenges" of such scientific (?) system-makers and system-venders, who can neither write a legible challenge nor repeat the multiplication table, and who have neither system, reference, reputation, character, or principle; and who only wish to provoke notice, through controversy, and gain free advertisement and public notoriety through the reputation and influence of others.

But there are circumstances—which must be discerned by the wisdom and discretion of the agent—where an ocular demonstration and proof of its claims would not only effect a sale, but win in its favor large and extensive influences, in the way of its introduction, both in cities, towns, and communities; and where the investigation is honestly sought. Such opportunities no intelligent, right-thinking and honest agent, would fail to improve and utilize, both for his or her own interest, as well as those of the author and proprietor—who furnishes profitable employment thereby.

Any agent, therefore, who, having had a fair opportunity and a reasonable amount of time in its sale and introduction, and who, under the above or similar circumstances, lacks sufficient practical knowledge of its principles, or confidence, either in himself or the system, to place it on test against anything and everything in its line—without any exception whatever—said agent is requested to immediately abandon his or her position. It is due the author, the reputation of the work, and to the public, that such an agent should voluntarily relinquish all claims to his or her place, in order that it might be filled by a more industrious, honorable, worthy, and able representative—otherwise we will be compelled to insist upon their so-doing! In the above, and in similar cases, results only, should be submitted in the proofs given—the process of drafting, basting, etc., not being given or performed in the presence of those who have not yet paid for the knowledge of said principles! There are, therefore, circumstances under which the public has a right to ask for proof and evidence—in results obtained; but there can be no circumstances under which the public has any right whatever to see the process—until the price for the same has been paid! But while we expect the above qualifications and abilities in agents, after extensive opportunity and experience, we neither expect it nor need it in those just having entered upon its sale; for *The Encyclopedia* might be very successfully sold for six months or a year, without a similar circumstance presenting itself. The business of the agent is to sell the work, and aid those who wish additional oral instruction, in practically understanding it—for which service he or she is paid at the rate of five dollars per day—independent and extra of all commissions on *The Encyclopedia* and furnishing supplies! To qualify for this requires but a few hours' preparation, with any common and ordinarily intelligent person, however inexperienced they may be in this line.

Furthermore, all agents investing fifty dollars in goods, on entering upon the business of its sale and introduction, are entitled to five lessons—free of charge—either at our offices, or from the agent from whom the purchase is made. This will doubly secure success. Most tailors will purchase the work merely on its presentation, without wishing any oral instructions—as they will readily comprehend its principles and its value at a mere glance; and lady agents are not prohibited from soliciting tailors and selling *The Encyclopedia* to them, where the mere presentation of the work induces them to purchase the work. But gentlemen can only receive extra or oral instructions from our gentlemen agents or teachers; or at our offices! And while we expect our gentlemen agents to qualify and become efficient instructors to both ladies and gentlemen, at least in garment-drafting, it is not expected of our lady agents to necessarily qualify themselves in part fourth. Nor are they allowed to give oral instructions in said department of the work—to gentlemen tailors. Which ruling is based upon sound principles of mental, moral, and business philosophy, and practical experience and wisdom—as well as the general public common sense of propriety—and will be found to be most wise and practical in all cases, and under all circumstances; and to result in the highest and best financial interests of the author and proprietor, and for all parties concerned—and it will therefore be strictly observed and enforced. Furthermore, no agent selling this work, is permitted during said engagement, to give oral instructions on any art or science treated in *The Encyclopedia*, to any person or persons not having first purchased said *Encyclopedia*. Any agent violating any of the above rules, thereby forfeits any and all claims and rights to the sale of said *Encyclopedia*—and will be dismissed.

Respectfully,

THE AUTHOR.

PREFACE.

The author neither feels inclined nor deems it apropos, to do homage to the false and stereotyped forms of an apologetic preface. No apologies were either offered or expected when Harvey gave us the discovery and law of the circulation of the blood, and Guttenberg that of metal type and the press ; or when Galileo, and Howe, and Watts, gave us the telescope, the sewing machine, and the power of steam ; or when Franklin brought electricity from the clouds, and Morse annihilated distance thereby, making it the fleet messenger of the world! When light is given for darkness, truth for error, and liberty for slavery, apologies are uncalled for and are out of place ; and when made under such circumstances reflect seriously, either upon the head or the heart of their author, or otherwise upon the offering he would thereby introduce to the public. The title-page of the work in the reader's hands, is a sufficient index to its province and character.

It was undertaken and completed in compliance with the wishes and oft-repeated requests of many, who now occupy and adorn the most eminent positions in the profession throughout the various metropolitan cities of America, as well as many of the authors numerous customers, at his former establishments on Girard avenue, and Market and Chestnut Streets, of Philadelphia ; and he feels it will at least be appreciated by those. In combining with *The Science of Garment-making*, what the title-page indicates, he feels that he simply furnished its necessary counterpart ; and that a deep and long-felt want is thereby supplied, hitherto almost entirely overlooked and neglected.

The *System of Garment-Drafting* is based upon the world-renowned and universally acknowledged and received principles of measurement and drafting, honored and common to all the most eminent and learned French and Prussian, as well as English and American tailors and connoisseurs in the art and science of garment-cutting. And they are no longer problematic, but have been mathematically proved by scientific square-measurement, to correct scientific bone and circumference measures of the human form ; and they furthermore represent the science, experience, and practical wisdom of centuries. They have long been practically tested and used —as given—by the author, through an extensive and varied experience ; and they have been found to invariably produce the same absolutely accurate result under all circumstances, and under every conceivable complication of form and garment possible. The author has therefore no hesitancy in announcing to the public, that the principles of garment-drafting, as systemized and herein taught, are invariably, absolutely, and mathematically perfect—in the broadest and deepest sense of that term—when rightly understood and applied to absolutely accurate measures, and as given and taught. And he challenges the world to a successful scientific demonstration and practical contradiction of these claims. In the plan and simplification of the department for gentlemen's garments, the author was influenced by the most advanced thought and practice of Parisian, Berlin, and London tailors, as foreign factors ; and those of Philadelphia and New York, as American authorities, prominent among which in the latter field, we place the name of Lynthicum, the scholarly and learned teacher of New York. In the department for ladies' and children's garments are given in principle, those shades of thought common to science, and the practice of such cutters as Pingat, Durnzean, Felix, Hantenaâr, Martini, Tainturier, Meyer, Sara, Lipperheide, and Pacaude; with those of other foreign as well as American authorities of equal or even greater ability, but of less notoriety. But the measures and system are, entirely and in a preëminent sense, new and original.

Most cheerfully accepting scientific truth wherever found, the author has nevertheless chosen and adopted only such principles which proved themselves absolutely true, in a purely mathematical sense, and from an independent and purely scientific standpoint; and he has founded, prosecuted, and completed the work under the influence of a thorough conviction that any method, not equal to, or commensurate in variation with that of fashion and form, is of necessity deficient to that extent; and is therefore utterly inadequate to the necessities and emergencies of the cutter. *Principia non homines* has been the motto of the author, in all his investigations and researches; and *multum in parvo secundum artem*, in the mode of construction. He has studiously and conscientiously sought "the truth, the whole truth, and nothing but the truth," on this, to many, seemingly vexatious problem; and he has aimed to give expression to the practical use of its solution, in the most simple and easily comprehended form possible. Accuracy, simplicity, economy of time, labor, material, and money—wedded to ease of comprehension—have been the grand watchwords of the author.

Those having thought absolute accuracy in garment-cutting from actual measurement alone, as necessarily requiring a very complicated system of details, will be happily surprised at the exceeding simplicity of science, as taught and applied in the work. The system is truly and mathematically founded, in all its elements and principles of application, in measurement and drafting—from the neck to the floor—upon actual measurement, and nothing but, and all possible and necessary actual measurement. It is unique, in that it is the first and only system in the world founded on the practical common sense and scientific principle of the mathematical inch rule and its mathematical divisions only, independent of any particular form of the mathematical inch instrument whatever; which is the only true and proper instrument assigned by science, philosophy, and mathematics, to the cutter, in transferring the mathematical mould of the human form, as expressed by the measures or indices, upon the goods—whether directly or indirectly, It is also unique, in that it is the first and only system in the world based strictly upon scientific mathematical principles alone, absolutely independent of any fixed proportional forms or seams; thus making fashion its perpetual slave! And it is for this very reason, and by virtue of its purely scientific principles of construction, and the philosophy of their execution through the only instruments assigned to the province of the cutter, by mathematics, the first and only system in the world that is fully and truly commensurate with all possible variation of form and fashion, whether past, present, or future. Provincial to, and founded in its axiomatic principles of construction, and in its instruments of execution, on the immovable rock of pure mathematics, and commensurate therewith in variation of application and result, it is of necessity as immutable in its laws, as those which govern that science, when delineating the orbits of the planets, or in measuring the cycles of the ages.

Deep down underneath the agitated and angry waters of the sea of confusion, contradiction, and bitter strife—along the shores of its province, we have carefully laid its foundation upon this rock; and out of its elements and principles, we have reared the system, as a strong lighthouse, towering far above the loud roar and confusing spray of the angry billows—of a bitter controversial strife; and we cannot but feel that, not only in the present, but in the far off and distant future, its light will only shine with increased brilliancy and splendor, as its principles are assailed and tested; and that in each successive and retreating wave—which, having spent its force against its immovable base, will only be more clearly revealed and proclaimed the utter weakness and folly of its assailants, and its own inherent and perpetual strength and utility.

We venture to say to all, that the work—in all its departments, will be found replete with a spring-like newness in principle and method, which in a peculiar and a preëminent sense, is its own. And from the intrinsic merit and utility of its principles and contents alone, it is truly believed and hoped—at least, that it will be hailed with gladness by an intelligent and appreciative public; and that it will accomplish its intended mission of usefulness, upon which it is now sent—and to which it is hereby sacredly dedicated by

THE AUTHOR.

INTRODUCTION.

Designing, draping, ornamental finishing, etc., in garment-making, is the work and ministration of art. In this province she is supreme lawgiver. Her scepter gives command in all things, until the province of fitting is reached. Here she is powerless, and must appeal to the inexorable laws of the science of mathematics! Drafting the actual mathematical mould or the measure upon the goods, true to a certain style or fashion plate, and so combining all the measure and various parts that, when cut and completed, the garment will conform truly and mathematically to the shape and contour of the person measured; fitting the form, independent of the degree of tightness or looseness, without wrinkles or distortion, and embodying at one and the same time comfort, symmetry, grace, and perfection in style and fit—this is the province of mathematics, and is a science, as much so as music, surveying, or astronomy. And any method not commensurate with the laws of mathematics governing that science—thereby constituted adequate to the above emergencies, whether on perfect or imperfect forms—is to the science of garment-making, what discord is to music! Tailors having been the first to reduce these mathematical principles to a "System," this science naturally received the name—"The Tailor System." A correct definition, therefore, of the expression "The Tailor System," is this, viz.: "The Tailor System" is the science of mathematics which teaches perfect garment-drafting and making, from the perfect indices or measures of the human form, with the instruments assigned to that province by mathematics! In other words, it is that system of principles in measurement and drafting, each and all of which in all their elements of construction, as well as the instruments of their execution, are now, and in all ages have alone been, universally honored and practiced in common by the vast majority of the world's most eminent, scholarly, and successful tailors; and which principles, in like manner have been, and are now, universally accepted by the civilized and scientific world, as the voice and teachings of pure mathematics alone, in the science of garment-drafting and making—in the same manner as those universally received principles that constitute the science of surveying, or any other science of mathematics, "A Tailor System" at most, is but the best thought and experience of one individual tailor's principles systematized—and may be only the system of a very erroneous and unsatisfactory tailor—having nothing in common with either the true mathematical science of garment-drafting, or that of any other individual tailor. Nor yet is "The Taylor System" anything more at most, than the best thought and experience of one individual person, whose name chanced to be Taylor. And although these terms sound alike, and much confusion and deception has been caused thereby, the thoughtful reader will detect a vast difference between the provincial meanings of the terms "The Taylor System," "A Tailor System," and "The Tailor System," as well as in their real and commercial values, relatively and respectively considered! And when the proposition, "Whether or not there is or can be more than one genuine science of geometry, trigonometry, or surveying," will be proposed, entertained, and discussed by the intelligent public, we will feel under obligations, and will hold ourselves in readiness to entertain and discuss the proposition, "Whether or not there is or can be more than one genuine scientific tailor system, or science of garment-cutting—and not until then! And furthermore, when the proposition "Whether or not the instruments assigned by mathematics, to the provinces of geometry, trigonometry, or surveying, as the means of applying their principles in the elimination of truth, in each of these respective sciences, are either imperfect, insufficient, superfluous, or can be substituted by proxy—or whether or not absolute mathematical truth ever was, is, or can be obtained in the science of surveying, by any other and different instruments than the compass and chain," will be proposed, entertained, and discussed by the intelligent public, we will feel under obligations, and will hold ourselves in readiness to entertain and discuss the proposition, "Whether or not the mathematical inch-rule and tape-measure as used, and assigned by mathematics to the province of the cutter, are either imperfect, insufficient, superfluous, or can be substituted by proxy; or whether or not absolute mathematical truth ever was, is, or can be obtained in the science of garment-cutting, by and through any other instruments, in transferring the indices and reproducing the mathematical mould or perfect fit of the human form upon the goods, by anything other, different, more or less, than the mathematical inch-rule and tape-measure"—and not until then!

There is an eternal fitness as well as that of fixedness, in all the appointments of mathematics, both as to the principles and the instruments of their execution—in the elimination of truth—in each and all of her provincial sciences; and "the truth, the whole truth, and nothing but the truth," in any department of her provinces, can only be attained by full and perfect obedience to all her imperial behests—under which circumstances the decision voiced by her court, is supreme! The majesty and immutability of her laws are constantly proclaimed by millions of revolving systems of suns, and are written in characters of fire—in the burning orbits of the flying comets. And whether the question involved is, fixing the time of an eclipse, by astronomy; or locating the boundary lines between individuals or nations—involving at times, the property, lives, and destinies of millions of human beings—by the compass and the chain of the surveyor; or whether otherwise, it is simply that of guiding the hand of the artist, in delineating the mathematical mould of the human form—true to the mandate of fashion and taste—upon the rich and costly fabrics prepared for its comfort and adornment—by means of the inch-rule and tape-measure of the cutter; the decision of the court of mathematics, announced through her only authorized and lawfully appointed instruments of interpretation, application, and execution of her laws, is the voice of the Supreme Court; and is the end of all appeals, and ends the controversy!

Recognizing both the fitness, fixedness, and majesty of her appointments, and having cheerfully accepted them in our province, the question "Why she has placed into the hands of the architect, the square and the compasses; into those of the surveyor, the compass and the chain; and into those of the cutter, the square or inch-rule and tapemeasure—rather than something different—is a question which the author deems of less importance, than the attainment of a correct knowledge of their use; and the correct knowledge of the laws which they interpret, apply, and execute: and which question, will be left to be discussed and disputed with her, by those who are in controversy with her touching our subject—and whose teaching and practice is at variance and in direct antagonism with the voice and teachings of the world's practical experience and sound philosophy, as well as with that of the science of mathematics! To feign contempt for her appointments, whether as pertaining to principles or instruments, in any of her provinces, is the superlative degree of both folly and ignorance—no matter in what garb or form of dress it comes! Quickly as a flash of lightning, science brings millions of worlds, millions of miles distant, within the compass of man's naturally limited vision: but she brings them through the telescope, or pieces of glass! Again, instantly she brings us into direct communication with the nations beyond the outstretched and stormy waters of the vast ocean —with whom we can converse, as if they were our guests and seated with us in our parlors; but she has seen fit to vouchsafe this privilege to us, through the means of the telegraph—or a piece of wire! Those who quarrel with science and refuse to look through glass, will have a limited vision! And those who refuse to hear through wire, have even a much more limited circle of sound!

And the cutter, who out of either a lack of industry, or ignorance, conceit or prejudice, refuses to be guided by science through the mathematical inch-rule and tapemeasure—and these alone—will always find every problem, "a tangled skein!"

Our first "inventors" in this line, treated the world to "the one pattern for all." Then to patterns innumerable, fitting none! Then came the swarms of "charts" and "models," and even "machines"—like the plagues of Egypt, whose name is legion! then followed "scales"; and as these dropped from off the cutter's eyes, he began to see the faint outlines of the square or inch-rule in picture-form, around the edges of the various paste-boards, and so-called "scales"; and which, finally appeared in wood—underneath a confusing shower of "numbers"! But from "the one pattern for all" down to the various so-called "systems" of squares or inch-rules with "numbers," one and the same erroneous and deceptive principle of proportion prevails in and through all. There is not a single element of sound reason, philosophy, science, or actual measurement, in any or all of them, separately or collectly. And there is no more excuse or reason for abandoning "the mathematical inch," in the science of garment-cutting for those nonsensical, erroneous, and fixed proportions represented by "scales" or "numbers" than there would be in substituting them for the mathematical inch in mechanics, architecture, or surveying! The inconsistency is so palpable and so great, that we are astonished to think that it ever received even a passing recognition, from the profession; but we are not surprised, however, on the other hand, that those who are financially interested, and therefore the advocates of "scales" and "numbers," that such, when closely questioned by the intelligent public and asked to give a reason for their abandonment of the mathematical inch for such a supreme absurdity—we are not surprised, we say, that these, under such circumstances, should at once proceed to "throw dust in the air," and attempt to confuse and over-awe the general public into a certain degree of reverence for these said mysterious "scales" and "numbers," claiming for them both a hoary age and the honor of a heritage handed down to us, from the ancient Egyptians! But where, when, and by whom this important discovery was made always was, is now, and always will remain, a secret—known only to those who assert it! The author nevertheless holds himself in readiness to prove from all the records extant touching this subject, that such statements are not only without any foundation whatever in fact, or even based upon anything that has the faintest shadow of inference in their favor, but that the very opposite is true; and can be substantiated beyond even a shadow of doubt, to any intelligent and unbiased mind. But alas ! for the Egyptians ; they must suffer in profound silence, this insolent reflection upon their hitherto unsullied name and reputation, both for wisdom and science ; only as they have engraven their testimony in tables of stone against it, in the towering pyramids that perpetually proclaim their knowledge, use, recognition, and honor, of the "mathematical inch," and this alone as a unit, in computing height, and depth, and length, and breadth ! But "scales" on the contrary, are of recent date ; and had it not been for the "scales" on these inventors eyes, "scales" would never have had an existence, in the science of garment cutting—no more than in architecture, mechanics, or surveying !

We ask and demand in the name of pure mathematics, correct science, sound philosophy, and practical experience, for "the mathematical inch and its mathematical divisions"—and this alone—in the art and science of garment cutting!

Furthermore, the half-filled chalice of wine, when filled with water, ceases chemically to be either wine or water ; and in like manner the decoy-square, or inch-rule, printed or fastened upon the edges of these so-called "scales," or "charts" and "models," is not the square or inch-rule of mathematics, either in form or principle ; neither does this "shadow of the mathematical inch" falling upon such "systems", heal their deformities nor change their delusive principles of proportion into those of actual measurement. The instant the square or inch-rule receives the impress of division marks, other than those of the mathematical inch and its divisions repeated, whether in the form of so-called "neck numbers," "chest numbers," "bust numbers," "waist numbers," etc., either stamped upon it or upon "scales", or otherwise arranged in "tables," and used either separately or jointly in connection with the mathematical inch rule, as means of obtaining results, at that instant it becomes a medium of proportion ; and its identity, both in form and principle, with that of the mathematical inch is lost—and ceases *de facto* ; and at that very instant, such a process drops out of the mathematical domain of actual measurement into the domain of proportion, and that of proportion alone ; which principle in garment cutting, is no less inconsistent in its province than the process of measuring the trunk of a tree and attempting to base thereon, as a "scale, " a correct estimate of its true height, or the number of cubic feet contained in its branches ; basing said estimate, on the exact ratio of the expression of said circumference! Or, again, to measure the circumference of a lady's waist, to compute the circumference of a ring that will fit her finger. Nor is it any more reasonable and consistent, as a process, than that of an artist who would attempt from the portrait of some perfect and beautiful face, to truthfully paint the faces of all ! And notwithstanding the frequently brilliant, but brief and local newspaper reputation and notoriety—great or small, in the exact ratio of the money and business sagacity invested—which is sometimes attained by systems founded on such principles and instruments ; or even if attained from the seeming degree of success sometimes approximated in their use, by those who may have adopted them ; in both cases, the facts admit of an easy solution, viz: the first is simply the natural order of things, and only proves that by shrewd and extensive newspaper advertisement the most worthless nostrum can, for a time, be popularized as a princely medicine. But reputation built upon such a foundation, is short-lived. The second fact is invariably due, to the superabundance of good "common sense" practical knowledge and experience—inherent in the cutter using it—in knowing when, where, and how, to make the necessary changes; and just how, to supply the natural defects in each and every particular case. Therefore, whatever success might possibly fall from such hands upon said "system" or "systems," it is wholly or entirely due to the cutter; and not to the inventor of said "system," nor because of any merit contained in any such "system" whatever.

Had said cutter lived in the halcyon days of yore, he or she might have reflected the same credit upon the "one pattern for all system!"

Again, it will not suffice for these inventors with such conceptions of the science of garment-cutting, when driven from behind this gauzy veil of "numbers," "scales," "forms," "curvatures," "curvalinears," etc., used as adjuncts, at least, in drafting, to hold up a square or inch rule, and claim that their systems are inch measurement. Indeed! Why not then consign these false and deceptive things to the flames, and let the winds scatter their ashes far and wide, that their very existence may be forgotten? for they have only proved themselves a sore and grievous plague to all who ever attempted to use such systems; and they have discouraged thousands from even hoping to ever understand the science of garment-cutting, and have hopelessly ruined cloths, silks, and velvets, to the amount of millions of dollars! Furthermore, it will not better the case by throwing even all of these away, and adopting the inch rule alone, and then basing a system, simply and alone, either on twelfths or sixteenths, as integral parts of the inch. This is simply building a system on "scales" again, the scales being based either only on the twelfth of the inch, or otherwise on the sixteenth, thus ignoring all other mathematical divisions of the inch, and insisting upon using the twelfth, or otherwise the sixteenth of an inch where the facts of the case, mathematically, might demand neither of these; but instead of them some other division, called for, and demanded by the actual measure, in the measure book.

Such systems and methods are anything but actual measurement.

Neither does the sole and actual use of the inch and its mathematical divisions alone, prove anything whatever; for the methods of drafting might be unscientific and inadequate to cover every possible emergency.

Many persons, doubtless, have plain simple mathematical squares or inch rules, who know not the first elements of their scientific use. And in like manner, teachers whose principles are both erroneous and utterly worthless might—and such most likely will, in the future—in order to weaken criticism and to cover the deformity of their "systems," take refuge behind the inch rule alone. But the possession, or even the fact of going through the process of using the compass and chain, does not by any means prove that the possessor has any true and correct scientific knowledge of surveying whatever. Again, the drafting might even be correct, while the fatal deficiency might be in the method of measurement of the human form. Not only has mathematical law fixed the principles and instruments to be used in the science of garment-drafting, but also the method of measurement, or the mathematical indices, which alone can possibly express the correct and absolute mathematical mould or perfect fit of the human form. Under the same law and with equal certainty that light reflects and reproduces, through the camera, the exact image of the object before it, that "mathematical mould" which is expressed by the given indices or measures, whether those measures express either truth or error, or a mixture of both, will be mathematically reflected by the draft upon the goods—provided the correct scientific instruments are used in the correct scientific method in constructing said draft from said measures.

It is, therefore, all-important that the method of reading the measures from the human form, is mathematically correct and truly scientific, as well as the instruments used both in measuring and drafting ; and the method or process, in each. However perfect, therefore, the instruments, the system of measuring and that of drafting may be, if the measures are wrong the draft will be wrong to the same extent—and in the exact same ratio. The tape-measure, the measures or indices, and the mathematical square or inch-rule, are to the cutter, what the chain, the field-notes, and the compass are to the surveyor; and in the province of either, the draft is an expression of mathematical truth in the exact ratio that the indices are the mathematical expression of the truth sought.

Hence in this province, as everywhere in her dominion, mathematics has fixed under the imperial seal of an immutable decree :

1. The Measurement or Indices.

2. The Instruments, or the means of obtaining, conveying, and interpreting the Indices.

3. The Law of Government in each and over all.

And until twice one ceases to be two ; until we can trample upon the law of gravitation by walking upon the air ; until the microscope can be substituted for the telescope, and discord becomes enchanting music ; until the law of light produces darkness; until motion is self-productive, and effect is produced without cause ; until style can effect the needle of the compass, or turn the countless systems of stars and their planets out of the orbits described for them, on the face of "The Mathematical Clock of the Heavens;"—until then, upon the throne erected by mathematics, in the royal palace within her temple, and resting upon these pillars, Fashion, arrayed in the princely gifts of the world's choicest looms, and crowned with the earth's richest minerals and most precious stones, sits, the queen of beauty, the adored empress of the world.

Respectfully,

THE AUTHOR,

Philadelphia, Pa.

Introductory from the Profession,
— OR —
THE VOICE OF OUR PATRONS,
AND
Our Neighbors.

"To the Public:

We join in the highest praise of Prof. Christner's International System of Garment-Cutting, with any square and tape alone. He has given the most convincing proofs of his system fitting the most difficult forms perfectly; in waists, sleeves, coats, wraps, etc., purely, and absolutely from the measures alone, drafted with any square and tape alone. It is absolutely accurate and perfect, in the ratio that the measures are taken accurate, and perfect and true—all of which is easily learned. Time, fashion, and form cannot affect its principles, which are magnificently perfect and exhaustive. Its province embraces economy of time, labor, and money, to any dressmaker. We would not part with it for many times its cost; and can most cordially recommend both the system and its urbane author as worthy of the fullest confidence of all who are interested in learning the true science and art of garment making.

Very Respectfully,
MRS. M. WOOD and MISS M. WOOD, Modists,
No. 1320 Poplar St., Philadelphia, Pa."

"To the general public and professional cutters :

I have learned and used Prof. Christner's International Scientific Tailor System of Garment-Cutting, with any square and tape-measure alone; and having thoroughly tested it, I pronounce it absolutely perfect and graceful in result, to all styles; and very simple and easy to learn.

MISS J. MILLER, Dressmaker,
No. 2032 Warnock St., Philadelphia, Pa."

"To all who are interested :

I learned and used quite a large number of 'systems;' and have been an extensive and practical dressmaker for twenty years. I know what is a perfect fit. I learned Prof. Christner's International Tailor System, after seeing it thoroughly tested. I can conscientiously recommend it to be all the author claims for it. Ladies will find Prof. Christner a master cutter, and a thorough and honest teacher. The price is money well spent.

MRS. A. A. FISH, Modiste,
No. 1228 Poplar St., Philadelphia, Pa."

"To all whom it concerns :

I learned Prof. Christner's International Scientific Tailor System of Garment-Cutting. It is simple and easy to learn, and is perfect. It makes most exquisitely beautiful fits, from the measures alone, and that too, in about ten minutes, with any square and tape alone. I cannot say too much for it; and I would not part with it for many times its cost. Mr. Christner is an honorable and fair dealing gentleman ; and ladies can rely upon what he claims, and that he will do and as he agrees.

Respectfully,
MISS CHRISTINE POEHNER, Dressmaker,
No. 1219 Girard Ave., Philadelphia, Pa."

" No. 600 Parrish St., Philadelphia, Pa.

Having learned Prof. Christner's International Scientific Tailor System of Garment-Cutting, with any square and tape alone, I wish to say : I saw it tested, and tested it myself, both in body, sleeve, and coat-cutting; and testify to its marvelous accuracy and perfection. I cheerfully recommend both the author and the system, as worthy of the highest confidence of the profession and public. I learned it in four lessons.

MISS M. SCHLESINGER, Dressmaker."

"To whom it may concern :

Believing in the motto, 'Honor to whom honor is due,' I wish to say I learned Prof. Christner's International System of Garment-Cutting with any square and tape alone, after an expensive experience and disappointment, in other pretentious and so-called 'systems.' I found Mr. Christner's system a panacea for all my former losses, and my disappointments. His instructions have forever simplified that seemingly difficult subject of wrap-cutting. I now feel equal to any emergency of form or fashion in Dolman-sleeved wraps, as well as plain cutting, along with all those difficult 'little things,' such as capes, collars, cuffs, circulars, etc. Ladies can feel assured that Prof. Christner's system is all he claims for it; and will find his dealings honest and fair, in every way.

MISS ROSE McELWEE, Dressmaker,
No. 705 N. 18th St., Philadelphia, Pa."

"No. 218 Girard Ave. Philadelphia, Pa.
Prof. D. C. Christner took my measures and cut me a waist and sleeves, which fit exquisitely, without change. I know what is a fit. I have learned his International Tailor System, and unhesitatingly pronounce it the most perfect system ever invented. The sleeves are exquisitely beautiful and grand. Prof. Christner is a perfect Christian gentleman; and ladies can rely upon what he advertises and agrees to do—and even more. Any one can more fully convince themselves by either calling upon, or writing to me.
MRS. M. J. WARD, Dressmaker,
No. 218 Girard Ave., Philadelphia, Pa."

"To whom it may concern:
I learned the International System of Garment-Cutting of Prof. Christner, the inventor, in five lessons. I saw it tested, and tested it myself. It is very easy to learn, quick and exceedingly accurate; and is simply perfect. I speak from actual knowledge. The sleeves are par excellence. All styles are easily and quickly drafted to any form. Too much cannot be said for the author and the system. No lady should be without it for ten times its cost.
Very Respectfully,
MRS. S. McCARTY, Dressmaker,
No. 1521 Ogden St., Philadelphia, Pa."

"To whom it may concern:
Prof. Christner cut a waist and sleeves for us, from the measures alone, which were charmingly beautiful and perfect in design and fit. We join in recommending his International System above anything we ever saw in its line; and having learned it of the inventor, we can now heartily recommend the system to all, and as being all its accomplished author claims. Ladies will find Mr. Christner a thorough gentleman, and can rely upon his business agreements. The system is easy, simple, comprehensive, quick, and absolutely perfect, if the measures are taken accurately and perfectly.
MISS M. M. WILEY, }
MISS A. RICHARDETTE, } Modists,
No. 817 N. 16th St., Philadelphia, Pa."

"No. 416 Girard Ave., Philadelphia, Pa.
To all who may be interested:
I learned Prof. Christner's International System of Drafting and Cutting; and I wish to say it is easy to learn, and exceedingly accurate and perfect, as he claims. I can cut a basque in ten minutes. The sleeves are exquisite and grand. The author is a true gentleman, and does what he agrees, and claims, and promises. I would not part with his system for many times its price; and I recommend it above anything and everything in the market—in its line—and to all.
MISS MARY M. PHILIPBAR,
Corset and Dressmaker."

"To all who are interested:
This is to certify that I have learned the International Scientific Tailor System of Garment-Cutting, of Prof. D. C. Christner, and pronounce it by far the best system ever offered, and as being equal to all forms and styles of garments; and having a grace and beauty of seams and shape, peculiarly its own. Mr. Christner is an honorable gentleman, and ladies can rest assured he will do as he agrees.
MRS. E. B. DURR, Dressmaker,
No. 1518 Stiles St., Philadelphia, Pa."

"To the public:
I learned Prof. Christner's International System of Garment-Cutting, and pronounce it perfect. It is very simple and yet comprehensive. It all depends on the accuracy of the measures, and accuracy in the cutter. The system is certain to be perfect, every time; and ladies can depend on this system as being what it is represented, viz: a thoroughly exhaustive, scientific, practical, and perfect system; which teaches you to cut all kinds and styles of garments. I advise all who wish such a system, to investigate Prof. Christner's Foreign and American Tailor Methods, and they will find The System of the World; and that which is certainly destined to become 'The Universal Standard.'
MRS. F. STEVENSON, Modiste,
No. 1302 Brown St., Philadelphia, Pa."

"To all whom it may concern:
I learned the International Scientific Tailor System with any square and tape alone, of Prof. D. C. Christner the inventor. My first effort was a polonaise; which, was charmingly beautiful and perfect. It is a marvel of perfection and simplicity; and must become notedly, as it is in reality, 'The Standard of Standards:' and 'The King of all Systems.' Having carefully investigated all the many systems in Philadelphia, I became convinced of the superiority of this great system in every way, over all the others; and having learned it, I am delighted with it; and I advise every lady to convince herself by comparing all with it, and judging for themselves. Prof. Christner is a thorough and master teacher, and a gentleman of honor—who will do just as he agrees.
MRS. M. A. MARONEY, Dressmaker,
No. 1227 N. 16th St., Philadelphia, Pa."

Philadelphia, April 28, '85.
"To all interested in systems of cutting, I can truly say, Prof. D. C. Christner's system of drafting is exceedingly accurate, and has a matchless grace and style; and that it is superior to anything I ever saw or used. The sleeves are most beautiful; and the system moreover, is not difficult. Ladies will find its author a thorough Cutter, and an honest and courteous gentleman. Am delighted with the system in every way.
Respectfully,
MRS. IDA M. BROWN, Modiste and Draper,
No. 1323 Davis St., Philadelphia, Pa.

"No. 912 Ontario St., Philadelphia, Pa.
Prof. Christner cut a basque-body and sleeves for me, from the measures alone; which, were artistic in design, and fit perfectly without change. I learned his International System with any square and tape alone, in five lessons. Am delighted with it; and I consider it the solution of that vexatious problem, of a system of fitting whch is commensurate with, and equal to all the demands in irregularity of form and the fluctuations of style.
MRS. J. MOSS, Modiste."

"To whom it may concern:
I learned Prof. Christner's International Tailor System of Garment-Cutting, after seeing it thoroughly tested. His system is perfect, and quite simple and easy to learn. Ladies will find Mr. Christner a thorough Christian gentleman. He does as he agrees, and what he claims.
MRS. A. H. KNIGHT, Dressmaker.
No. 1122 N. 4th St., Philadelphia, Pa."

"Philadelphia, Pa.:

Prof. D. C. Christner:

Dear Sir—I have been in business for a number of years, and learned several systems. I find your system by far ahead of all others I ever used or saw—and very perfect and accurate.

Yours, Respectfully,
MRS. A. DOWNS, Modiste,
No. 1118 Brown St., Phila."

"I learned Prof. Christner's International Scientific Tailor System with any square and tape alone; and cut my first waist and sleeves, in fifteen minutes, which were elegant and very perfect. It is the most perfect system I ever saw, and I am perfectly delighted with it. I can say truthfully, that it is all that is claimed for it by its author, who is a gentleman both of ability and honor; and who deals 'on the square,' as accurately and carefully, as he teaches 'on the square.'

MRS. JULIA WACHTER, Dressmaker,
No. 1333 Myrt e St., Philadelphia, Pa."

"To any who are desirous of learning Garment-Cutting:

Having learned Prof. Christner's International Tailor System of Drafting by square and tape alone, and having seen Mr. Christner test it thoroughly, and having tested it myself with like results, I can say truthfully to all, that it is the simplest, quickest, and most perfect system taught. It is grand! Equally perfect to all forms, and to all styles. It is refreshing to find as courteous and honorable a gentleman in that line of business as its author, whose word can be depended upon, in all that he agrees and promises, as well as his system.

Very Respectfully,
MRS. E. V. MAXWELL, Modiste,
No. 908 N. 12th St., Philadelphia."

No. 127 N. Howard St., Baltimore, Md.
"To all whom it may concern:

I learned Prof. D. C. Christner's International Scientific Tailor System of Garment-Cutting by square and tape alone. I pronounce it one of the grandest achievements of the age, in its province. It is simple, comprehensive, and grandly perfect. It is destined to become universal; and will receive a glad welcome from the profession, which has long waited for just such a work. Am more than delighted with it.

MRS. H. H. WARDD, Modiste and Draper."

"To all who are interested in learning the art of Garment-Cutting:

Common sense taught me long since that ladies' garments could not be cut perfectly-fitting—and as the scientific tailor cuts—until the same scientific principles and instruments were used, to the exclusion of all others; such as patterns, charts, models, machines, etc. I am the wife of a first-class tailor, and have been a practical dressmaker for twenty-five years, in experience and time. I have watched and studied with exceeding great care and anxiety, the multitude of so-called "systems," "tailor systems," "Taylor system," and "charts," and "models," and "machines," innumerable—as were offered, and imposed upon the ladies of Philadelphia; as well as in New York, and other cities. I investigated, studied and proved all the latest and most generally received systems, both here and elsewhere; and must if I wish to be truthful, confess that I found them anything but what they professed to be; and in the main, utterly worthless and deceptive, and a libel on the name " The Tailor System." Not even knowing the meaning of the term, nor the use of the tailors' instruments, all sorts of devices have been resorted to, in order to ensnare and confuse the public; and in many instances, unsuspecting and poor hard-working girls have been literally robbed of their hard earnings, in being induced to buy or learn these so-called "tailor systems." I therefore feel it a duty to say to all who are seeking to learn ladies' garment-cutting, by the same instruments and in the same manner that the scientific tailor does, to avail themselves of Prof. Christner's great International System. It is the standard among standards, and cannot be too highly praised. He has done a great work for the profession; and in such a thorough masterly way, that he should receive the aid and gratitude of all who love truth, justice, and right. It should be published free, and far, and wide! It gives the same principles, both for ladies and gentlemen; and with the same instruments. I learned the "ladies' department" in three lessons; and can cut any garment perfectly with it, in from ten to fifteen minutes. I would not part with it for ten times its cost. It is simply the end of this whole matter, and the end of all perfection. It saves more time, labor and money, than many times its cost, every month, whether in the establishment, or in the home. In its accuracy, comprehensiveness, and simplicity, with economy of time, material, labor, and money, along with a matchless excellency and grace of result, it outshines all other systems ever offered in America, as the sun outshines the stars.

MRS. S. C. BAITENGER, Modiste and Draper.
No. 331 W. Thompson St., Philadelphia, Pa."

We have above 500 more references throughout this and other American cities, similar and even more expressive than some of the above, and could fill a volume with references alone. But we must beg to be excused from granting any more space to references. One's character, ability, and reputation, are best known to his neighbors and patrons; and in the city and place where he has his home and business. We have therefore chosen the above few from among the many—not because they are more flattering in their praise of the work, but because of the peculiar weight of influence they have, or should have, being residents of the same or adjacent wards, and in the same line of business. We have many similar references from the highest and most influential and leading tailors both in this and in other cities, as well as from the leading commercial houses. We have their letters from their own hands, and they can be seen by those who are sufficiently interested, along with all the others and those given above; whom can be seen or addressed, if postage is enclosed, as given above. And while unwilling to offer a sufficient inducement to tempt any one to do wrong; nevertheless, to show that we are not offering manufactured reference, we hereby challenge anyone to investigate for themselves; and we offer to forfeit $100 for any reference given above, the original of which we cannot produce, and as written in their own hand over their own signatures; or if ever lost, that we will not substantiate and testify to, under oath.

Very Respectfully,
THE AUTHOR.

INDEX TO CONTENTS.

Agents and Agencies, Important Notice to . . . 15, 16
Agents, can begin Immediately 17, 19
Art, The Coming or Future 220, 221
Ball Dresses and Low Bodices 112, 149, 150
Basting the Linings, Rules for 116, 117
Basting the Materials, Rules for 117, 118, 138
Basting the Sleeves, Rules for 118
Basque, How to cut Drapery for 112
Basques, Rules for Cutting 112
Basques, Rules for Cutting Plain Round 112
Basques, Rules for Cutting Pointed 112
Basques, Rules for Cutting Vests for 112
Bead Work, Rules for making Bracelets and Chains
. . . . 194, 195
Bead Work, Rules for making leaves, flowers, buds, 195
Bead Work, How to make French Card-Case in
Gold Thread and Steel Beads 194
Bead Work, Rules for making Toilet-Cushion, . . . 195
Berthae, Rules for making the 150
Binding 147, 148
Bias Cutting 115
Bias Cutting on Twill 115
Bias, Rules for making the Fourfold 165
Bodices and Ball Dresses, Low Combination 150
Bodices and Ball Dresses, Low, Pointed Basque 150
Bodices and Ball Dresses, Low, Round Basque . . . 150
Bodices and Ball Dresses, Rules for Low 150
Bows, Principles of Making 166
Boys Pantaloons, International Tailor rules for Cutting Vest and Over-coat Knee-pants, Coat, 71, 76, 222,
. . . . 251, 152
Bouillion, Rules for the gathered 145, 146
Buttonholes 152, 155
Buttonholes, Rules for making Herring-bone 155
Buttonholes, Rules for making Piped 155
Buttonholes, Rules for making Plain 154
Buttonholes, Rules for making Pointed 155
Buttonholes, Rules for making Round 154
Buttonholes, Rules for making Double-Stitch . . . 154
Buttonholes, Rules for making Tatted 154, 155
Buttonholes, American Tailor, Prussian Tailor,
French Tailor, and English Tailor 152, 153
Buttonholes, Rules for making Twisted Stitch . . . 155
Buttons, Rules for sewing-on Covered 150, 151
Buttons, Rules for sewing-on Linen 151
Buttons, Rules for sewing-on Metal-shanked 151
Buttons, Rules for sewing-on Woven 151

Buttons, Rules for sewing-on, With Eyes 151
Capes, Rules and principles for cutting 123, 126
Cape, Rules for cutting the Dolman 125
Casings, Rules for making Whalebone 148
Children, Arranging the seams, styles and degrees
of fit in drafting for 71
Children's Dresses, open at the back or at the front . 71
Children, Franco-Prussian method of drafting for . 68, 76
Children, Instructions to draft all kinds and styles
of costumes for 68, 76
Children, Measures for 68, 69
Circulars, Rules and principles for cutting 126
Cloaks, Rules for Cutting 127, 131
Coats, Rules for cutting 127, 131
Coats, Rules for increasing the measures 128, 129
Coats, To cut with one dart 129
Coat, The English or Ulster Skirt 131
Collar, method and rule for drafting "The Berlin,". 101
Collar, method and rule for drafting, "The Byron,". 102
Collar, method and rule for drafting "The Circular," 104
Collar, method and rule for drafting "The English
Rolling," 103, 104
Collar, method and rule for drafting "The French
Coat," 104, 105
Collar, "The Rolling Choker," Method and rule for
drafting 101, 102
Collar, "The Sailor," method and rule for drafting . 105
Color, Harmony of Complexion with 200, 201
Computation of required material 41
Copying, Principles and rules for 121
Copyright notices 7, 14, 15, 48
Cording and Binding, Rules for 146, 148
Cording, Rules for making Double . . . 146, 147, 162, 163
Cording, Rules for making Single 147
Cording, Treble and Quadruple 146, 148
Crazy Patch-Work, Principles and rules for making
the 178, 179
Crazy Patch-Work, Stitches used in 179
Crazy Patch-Work Stitch, "The Border." 179
Crazy Patch-Work Stitch, "The Brier," 179
Crazy Patch-Work Stitch, "The Chain Satin." . . . 179
Crazy Patch-Work Stitch, "The Coral," 179
Crazy Patch-Work Stitch, "The Cross," 179
Crazy Patch-Work Stitch, "The Crow's-foot," . . . 179
Crazy Patch-Work Stitch, "The Diagonal," 179
Crazy Patch-Work Stitch, "The Double Feather," . 179
Crazy Patch-Work Stitch, "The Feather," 179

UTILITY AND DECORATIVE ART.   33

Crazy Patch-Work Stitch, "The Knot,"   179
Crazy Patch-Work Stitch, "The Loop,"   179
Crazy Patch-Work Stitch, "The Point Russe," . . . 179
Crocheting, Principles and rules for 179, 186
Crocheting, Rules for Chain   179
Crocheting, Rules for Double   179
Crocheting, Rules for Double-Long   180
Crocheting, Rules for Long   179
Crocheting, Materials and instruments used for   . . 179
Crocheting, Rules for Open   180
Crocheting, Rules for Single   179
Crocheting, The Stitches used in, and how to make them   179
Crocheting, Rules for Treble-Long   180
Crochet Work, Rules and regulations for Clover-Leaf Stitch   181
Crochet Work, Rules for American Tidy   181
Crochet Work, Rules and explanations for the Forget-me-not Flowers   184
Crochet Work, Prussian Edging   180
Crochet Work, Russian Carriage Rug   180
Crochet Work, the Louise-Collar   180
Crochet Work, Crazy-Stitches   183
Crochet Work, in Wavy-Braid   184
Crochet Work, Rules and explanations for the Laburnum Flower   184
Crochet Work, Rules and explanations for Geometrical or Honey-Combed Netting   186
Crochet Work, Rules and explanations for Grecian Netting   185
Crochet Work, Rules and explanations for the Snow-Drop, Knitted Flowers etc   185
Crochet Work, Rules for Knitted Spanish Wheat-ear Edging 181, 182
Crochet Work, Rules and explanations for Mignardise-Work   181
Crochet Work, Rules and explanations for Narrow Scotch Edging   180
Crochet Work, Rules and explanations for Grecian Netted-Border   185
Crochet Work, Rules and explanations for Oriental or Persian Netting   186
Crochet Work, Rules for American Raspberry Stitch, 182
Crochet Work, Rules and explanations for the New Berlin-Stitch   180
Cuff, rule for drafting the French Tailor   105
Curves, Indications for the proper Method of basting 63, 71
Cutting Gores, Rules for   116
Cutting Gores of Train Skirts, Rules for   116
Cutting Material, Rules for   115
Cutting Skirts, Rules for 92, 95
Cutting Train Skirts, Rules for 94, 95, 116
Dart, Rules for Drafting with one   81
Dart, Rules for the back, or the arms eye   88
Decorative Arts-Utilito, Rules and principles of 212, 221
Decorative Art, Electro Glass-Engraving   213
Decorative Art, Paper Flowers, Art of making . . . 216
Decorative Art, Gold-Liquid in   214
Decorative Art, To Stain Wood Black, Marble, and Mahogany   214
Decorative Art, Pendrawing 218, 216
Decorative Art, on Glass 217, 218
Decorative Art, on Metal   219
Decorative Art, on Linen   219
Decorative Art, Decorating Christmas Cards   . . . 216
Decorative Art, Method of Preservation of natural flowers   217

Decorative Art, for decorating Lamp Shades 217
Decorative Art, Method of Japanese Decorations . . 215
Decorative Art, Table Covers in   215
Decorative Art, in Embroidered Chair Backs 219
Decorative Art, in Satin and Plush Portieres . . . 220
Dedication   5
Designing, Principles and rules for 121, 122
Divisional Contents   8
Diagonals, Rules for Drafting and Cutting   81
Double-breast, Rules for drafting the   81
Dolmans, Rules for drafting and cutting   . . . 132, 138
Dolman, Rules for drafting the Standard English. 132, 136
Dolman-Sleeve, Rules for drafting the Princess Louise 136, 137
Dolman-Sleeve, Rules for cutting the Joined-Back . 138
Drafting, Rules for English and American direct body, front and side-front 78, 84
Drafting, Rules for English and American direct body, the back 84, 86
Drafting, Rules for English and American direct body, the side-back 86, 88
Drafting, Requisites and preparations for 55, 56
Drafting, Direct and Indirect 77, 91, 57, 75
Drafting, French and Prussian Compound Body . 57, 58
Drafting, Instructions for French and Prussian Compound 58, 63
Drafting, All Kinds of Styles of Ladies Costumes . 42, 138
Drapery, Rules and principles of Back 169, 170
Drapery, Rules and principles of Bouffante   . . . 170
Deapery, Rules and principles of En Biais   170
Drapery, Rules and principles of Front 169, 170
Drapery, Rules and principles of Cascade   170
Drapery, Rules and principles of En Chale   170
Drapery, Rules and principles of En Coquille . . . 170
Drapery, Rules and principles of En Eventail . . . 170
Drapery, Rules and principles of Flots Coques . . . 170
Drapery, Rules and principles of High 169, 170
Drapery, Rules and principles of Low 169, 170
Drapery, Rules and principles of Medium   . . . 169, 170
Drapery, Rules and principles of Pannier   170
Drapery, Rules and principles of Side 169, 170
Draping, Rules and principles for 168, 171
Dress Fabrics and Table of widths 40, 41
Dress Linings, Table of   41
Dressmaking, the principal stitches used in . . . 142, 143
Dressmaking, The Principal Seams used in . . . 144, 145
Dressmaking, Stitches used in 142, 143
Dressmaking, Terms used in 37, 40
Dyeing Fancy Blue, Black, Gray, Crimson, Golden-Yellow, Imperial-Blue, Sky-Blue Claret, Fawn-Drab, Flesh-Color, Salmon, Mauve, Violet, Magenta, Plum, Peagreen, Grey, Lilac, Pink, Purple, Scarlet, Red, Nankeen, Orange, Lavender, and Blue-Black 201, 203
Embroidery, Rules and principles for   . . . 173, 179
Embroidery, Fabrics used for   173
Embroidery, Object and aim of   173
Embroidery, The various kinds   173
Embroidery, The Instruments used in   173
Embroidery, Method for the stamens and veins . . . 175
Embroidery, Tacking down the edges in   175
Embroidery, Rules for making Stitches used in . 175, 177
Embroidery-Stitch, Rules for making the Skeleton . 175
Embroidery-Stitch, The Tracing   175
Embroidery-Stitch, Rules for making the French-Knot   175
Embroidery-Stitch, Rules for making Darning . 175, 176
Embroidery-Stitch, Rules for making the Weaving . 176

3

34   THE ENCYCLOPEDIA OF GARMENT-MAKING,

Embroidery-Stitch, Rules for making the Plush . . 176
Embroidery-Stitch, Rules for making the Kensington 176
Embroidery-Stitch, The Outline 176
Embroidery-Stitch, The Satin 176
Embroidery-Stitch, The Appliqued 176
Embroidery-Stitch, Rules for making the French Knot 174
Embroidery-Stitch, the Arasene 176
Embroidery-Stitch, The Grecian 176
Embroidery-Stitch, The Persian 176
Embroidery-Stitch, The Tent 176
Embroidery-Stitch, The Star 176
Embroidery-Stitch, The Feather 176
Embroidery-Work, Rules for making Rick-Rack-Daisy 178
Embroidery-Work, Rules for making the Wild Rose 178
Embroidery-Work, Rules for making the Snow Ball 178
Embroidery-Work, Rules for making Wild Clematis 178
Embroidery-Work, Rules for making the Transparent-Tidy 178
Engravings, To transfer 213
Explanations for eliminating the various parts of the draft 113, 114, 94, 95, 99, 100, 136, 138
Extensions, Rules and pointed lines for finishing the 63, 71
Extensions, Rules for Plaiting and Drapery . . . 100, 111
Fabrics, Table of widths 40, 41
Fancy and Artistic Needle Work, Principles and Instructions for 193
Feather Work, The art of Making Flowers in . 193, 194
Feathers, To clean 193, 194
Feathers, To curl 193, 194
Feathers, To dye: Blue, Green, Lilac, Red, Yellow, and Scarlet, etc 193, 194
Finishings, Rules for Tailor 157, 158
Flounce, Rules for making the Plain Gathered . . . 161
Flounce-Heading 165
Fold, Rules for making the French 163
Fringe, Rules for making Raveled 166
Fringe, Rules for making Berlin 166
French Seam 144
Garments, How to draft Gents' and Boys' . 222, 251
Garments, Half-fitting 81
Garments, Two-third-fitting 81
Garments, Three-fourth-fitting 81
Garniture, For Sleeves and Bodice 159, 160
Gathering, Rules for the Bouillion 145, 146
Gathers, Rules for Sewing in the 146
Gauging or Gathering, Explanations for 145, 146
Gauging, Rules for Double 146
Gold Liquid, Receipt for making 214
Goods, How to proceed for very narrow 115
Goods, Having Geometric Patterns, How to proceed with 115, 116
Gores Two under arm—Rules for, in large forms . 83, 84
Habits, How to cut Riding 131
Hair Work, Principles and Instructions for 193
Hem, The False 145
Hem, The French half 145
Hem, The Stitched 144
Hemming, Rules for 160, 165
Hoods, Rules and methods of drafting all kinds and styles 105, 106
Hose, Fancy pattern, for open worked French . . . 191
Introductions 23, 28, 120, 141, 223
Introductory, From the Public and Profession . . 29, 31
Ivory and Bone, To stain, Black, Blue, Green, and Purple, Receipt for 214

Jackets, Rules for cutting 119, 131
Kaleidoscope Work, Rules and explanations for 179
Lace Work, Its origin, object, aim, and principles for making 186, 190
Lace, Description and rules for Brussels 186, 187
Lace, Rules for making Point d' Alencon 187
Lace, Rules for making Brussels Point 187
Lace, Rules for making Scotch 187
Lace, Rules for making Berlin Insertion 188
Lace, Rules for making Venetian 188
Lace, Rules for making Sorento 188
Lace, Rules for making Torchon 188
Lace, Rules for making Spanish Point 188
Lace, Rules for making English Point 188
Lace, Rules for making Open English Point 188
Lace, Rules for making Spanish Bar 188
Lace, Rules for making French Rosette 188
Lace, Rules for making Portuguese 188
Lace, Rules for making Saw Teeth Edging 189
Lace, Rules for making Double Oak-Leaf 189
Lace, Rules for making Clydesdale 189
Lace, Rules for making wide Normandy 189
Lace, Rules to make Collar in Waved-Braid . . 189, 190
Leaf-Work, Rules and principles for 196, 197
Leather-Work 194
Leather, To clean 194
Leather, To join 194
Leather, To make water-proof 194
Linings, Table of 41
Loops, Rules for 155
Making-up the dress 142, 194
Mantelets, Rules for drafting and cutting . . . 132, 138
Mantaux, Rules for drafting and cutting 132, 138
Mantilla, Rules for drafting the Standard French 137, 138
Mantles, Rules for cutting 115, 133, 138
Material, How to fold Double width 115
Material, Method for computing the number of yards 41
Measures, Introductory to taking the 44, 46
Measures, Rules for taking the 47, 53
Measures, Rules for taking the Sleeve 53, 54
Measures, The Extra 52, 53, 54
Measures, Methods of proving the 52
Measures, The 42, 43
Measures, The International System of 47, 48
Measure-Book, plan of, for Gentlemen's Measures . 251
Measures for practice 56, 69, 128
Measure-Book, plan of, Ladies' Measures 244
Millinery, Rules and Principles for 197, 200
Mourning, Dresses for 167, 118
Narrow Fabrics, mode of adding to the width . 115
Notice, Important to the public 14, 15
Notice, Pertaining to Agencies and Agents 15, 19
Nomenclature, Of the Art and Science of Dressmaking 37, 40
New-Markets, Rules for drafting 99, 131
Neck, Rules for Binding the 148, 149
Needlework, Rules and instructions for Fancy and Artistic 193
Needlework, Rules for working Table Covers . . . 215
Overskirts, Rules for cutting Drapery of 115
Painting, China and Pottery 205
Painting, Embossed Pottery 205
Painting, Embroidery 204
Painting, on Glass 212, 213
Painting, Grecian Oil, ... 205, 207
Painting, In Gold and Silver 208
Painting, on Ivory 218

UTILITY AND DECORATIVE ART. 35

Painting, Lace in Water-Colors 204
Painting, Lustre 209, 210
Painting, Kensington 211, 212
Painting, Lincrusta 210, 211
Painting, Oil Photograph 208, 209
Paletots, Rules for cutting 127, 131
Paletots, Rules for increasing the measures 128
Painting, Tapestry 207, 208
Painting, in Water-Colors 203, 204
Pencil and Chalk Drawings—To fix 214
Pelisses, Rules for drafting and cutting 132, 138
Pelisses, Rules for drafting the Standard Romanoff. 138
Peplumes, How to cut 115
Piping, Rules for making Bias 161
Piping, Rules for making Double-materialed 162
Piping, Rules for making Double 162
Plait, Rules for the Greek 150
Plaiting, Rules for making Kilt 169
Plaiting, Knife 167
Plate, Showing Embroidery and Fancy Needlework 174
Plate, Showing Skirt Trimmings 164
Plate, Showing Sleeve and Bodice Trimmings . . . 159
Polonaises, Rules for drafting all kinds and styles
of 107, 111
Polonaise, Rules for drafting Special darts for . . 111
Polonaise, Rules for drafting Front drapery of . . 111
Polonaise, Rules for drafting Back drapery of . . 111
Polonaise, Rules for drafting Side drapery of . . 111
Polonaise, Rules for drafting Vest Fronts of . . . 111
Polonaise, Methods of draping for remodeling . . 111
Polonaise, Rules for Pleating and drapery of . . . 111
Prefaces 20, 22, 120, 140, 223
Price-List and Explanation 9, 12
Princess, the Franco-Prussian Method of Drafting . 95, 99
Princess, Front of the Promenade 98, 99
Princess, Side-back of the Promenade 98, 99
Princess, the Back of the promenade 98, 99
Princess, Rules for the Demi-train 100
Princess, Rules for the Train 100
Principles, not rules, System built upon . . . 21, 23, 28
. . . . 45, 55, 81, 223
Riding Habits, How to cut 131
Roses, How to improve their color 217
Rouleaux 165
Ruffle, Rules for making the plain gathered . . . 161
Ruffles, The French and English 161
Ruffle, Rules for making the Corded 162
Running-Stitch, The 142
Sacques, Rules for cutting 115, 127, 138
Seams, The principal 144, 145
Seams, To change the 55
Seams, To the shoulder in front 63
Seams and Spacings, How to govern, not fixed . . 67
Seams, Rule for allowing the 63
Seams, The kinds, their names and descriptions 144, 145
Seam, The plain 144
Seam, The flat 144
Seam, The French 144
Seam, The Invisible cloth 144
Seam, The double 144
Seam, Corded 144
Seam, The stitched-hem 144
Seam, The pocket-slit 145
Shell Flowers and Shell Work, Rules and principles
for 195, 196
Shirts, Rules for Cutting 75, 76
Shirts, Drafting and rules for cutting Gents' and
Boys' 75, 76

Shoulder, Curved line for the 63
Skirts, How to trace the 94, 113
Skirts, Rules for cutting Walking or promenade . 92, 94
Skirts, Rules for cutting Demi-promenade 94
Skirts, Rules for cutting Demi-train 94
Skirts, Rules for cutting Full train 94
Skirts, Rules for cutting Square-train 94, 95
Skirts, Rules for cutting their widths 95
Skirts, Cutting out the goods 116
Skirts, Trimmings and Finishing 160, 167
Skirts, Train 94, 116
Skirt, Sleeve, ... 116
Skirts, Rules for Gathered 146
Skirts, Franco-Prussian Method of drafting all kinds
and styles 92, 95
Sleeve, Methods of Drafting French and Prussian
Compound 64, 67
Sleeves, Methods of making 148, 149
Sleeves, The Coat 66, 67
Sleeve, The close 64, 66, 88, 91
Sleeves, The open 149
Sleeves, How to cut Draped 149
Sleeves, Rules for drafting 64, 67, 88, 91
Sleeves, Rules for Binding the 148
Sleeves, Rules for Opening 149
Sleeves, Oriental 149
Sleeve, Mediaeval 149
Sleeves, Greek 149
Sleeves, Roman 149
Sleeves, Cassock 149
Sleeves, The Judge 149
Sleeves, The Religious 149
Sleeves, The Leg of Mutton 149
Sleeves, The Engageantes 149
Sleeves, Rules to trim open 148, 149
Sleeves, for children Dress and Coat 71
Spatter-Work, Rules for 213, 214
Stamping, Rules and principles of 171, 172
Stitching and Basting Back 117, 118
Stitches, The principal dressmaking 142, 143
Stitch, The Running 142
Stitch, The Back 142
Stitch, The Overcast 142
Stitch, The French Invisible 142
Stitch, The Chain 142
Stitch, The Buttonhole 142
Stitch, The Embroidery 142
Stitching, The grand secrets of 142
Supplementary Price-List 13, 14
Tabliers, Rules for cutting 115
Tailor-Finishings 156, 158
Tapes, Rules for Sewing-on 165, 166
Title Page 3, 6
Tracing, Introductory rules for 113
Tracing, How and what seams to allow in 113
Tracing, Rules for the front 113
Tracing, Rules for the back 113
Tracing, Rules for the side-back 113
Tracing, Rules for the side-front 113
Tracing, Rules for the Princess-back and side-back . 113
Tracing, Rules for the vest-front 114
Tracing, Rules for the French sleeve 114
Tracing, Rules for the English sleeve 114
Tracing, Rules for the Children's draft, Gabrielle . 114
Tracing, Rules for the Children's draft, Gabrielle
side-front 114
Tracing, Rules for the Children's draft, Gabrielle,
the back 114

Tracing, Rules for the Children's draft, Gabrielle, the side-back	114
Tracing, Rules for the Children's draft, Gabrielle, the English Back	114
Tracing, Rules for the Children's draft, Gabrielle, the French back	114
Tracing, Rules for the Children's draft. All kinds and styles of Garments	114
Trains, Principles and Methods of cutting	94
Trains, Lengths of	116
Transparencies, Rules for	196
Transfer-Work, Rules for	196
Trousseaux, Rules for Wedding	149
Trimmings, Rules for making the	160, 167
Trimmings, The proper choice of materials for the	160, 161
Trimmings, Rules of good taste for	160, 167
Tunics, Rules for cutting	115
Ulsters, Rules for drafting	99, 127, 131
Ulsters, Rules for drafting Demi	99, 127, 131
Ulsters, Rules for increasing the measures of	128
Underwear, Rules for cutting and making ladies' and children's	75
Vests, Lines and rules for tracing and cutting ..	63
Wadding, Rules for	184
Waterproofs, Rules for drafting	99
Wax Flowers, Rules and principles for making ...	195
Whalesbones, ...	148
Wraps, Rules to cut from double fold material ...	115
Wraps, Rules for drafting and cutting Dolman-sleeved	132, 138
Wrappers, Method and rules for drafting	99
Wedding Trousseaux	149
Worsted Work, Rules and principles for	190, 193
Worsted Work Mosaic Knitting, Rules for making .	190
Worsted Work, Rules for making "Berlin"	190
Worsted Work, Rules for working fancy mittens ..	191
Worsted Work, Rules for knitting Spanish fringe .	191
Worsted Work, The Violet	190
Worsted Work, Saxony Edging	191
Worsted Work, Forest-Moss-Border	191
Worsted Work, French Half-Hose	191
Worsted Work, Rules for knitting ladies' French hood	191
Worsted Work, Rules for knitting gentlemen's English scarf	191
Worsted Work, Rules for knitting "French open-hose"	191
Worsted Work, Rules for knitting wristlets	191
Worsted Work, Rules for knitting Persian afghans .	192
Yokes, How to draft and cut	112

THE NOMENCLATURE OF THE ART AND SCIENCE OF DRESSMAKING.

No science or art is without a language peculiarly its own, which must be mastered by the student who would excel. The nomenclatures of chemistry and botany have their sources in the dead languages, the Greek and Latin; while musical terms are chiefly Italian. The connoisseur of the kitchen unlocks the secrets of his profession and art, in modern times, with a French key. And to France, as the author and founder of modern Fashion's throne and empire, we turn for instruction in the alphabet and nomenclature of dressmaking.

We call attention to such terms only, as have become anglicised, and are universally accepted and understood by the profession. We may instance the word, *Bouillion*, as one of these terms, which it is almost impossible to correctly translate into English, and which is nevertheless in universal use by professional and artistic dressmakers, modistes, and drapers.

In the following alphabetical list we give: 1st, the exact or literal meaning of the word; 2d, the explanation when necessary; and 3d, an example of the application also, when necessary. There are others of less importance and much less frequently used, which are omitted. Novelties constantly arise, for which new terms are coined, and which soon become universal. The following are therefore the principal and universally adopted technical terms used by the profession, at the present time.

LIST OF FRENCH TERMS USED IN DRESSMAKING.

Agraffe.—A clasp; applied also to gimp fastening.
Apprêt.—1. Finish; the dressing put into calicoes, etc. Ex.: *Percale sans apprêt,* undressed cambric. 2. Also, the trimming at the back of a bonnet, either a lace lappet, or any finish to a head-dress.
A rez de chausee.—Even with the ground.
Au fond.—To the bottom.
Aumonière.—Alms-bag; a small bag hanging from the waist.
Baleine.—Whalebone.
Bandeaux.—Bands; applied also to bands of hair.
Bas.—1. The lower edge. 2. Stockings.
Basques.—Applied to the ends of a jacket or bodice falling below the line of the waist.
Biais.—1. Bias, on the cross. 2. Crossways.
Bombé.—Rounded or puffed.
Bordé.—Round; edged with.
Bordé à cheval.—Binding of equal depth on both sides.

Bourré.—Wadded or stuffed; a term often applied to quilted articles.
Boutez en avant.—Push forward.
Calotte.—Crown; the crown of a cap or a bonnet.
Camisole.—A loose jacket; applied to dressing and morning jackets.
Capitoné.—Drawn in like the seat of a chair or sofa; buttoned down.
Capuchon.—A hood on a mantle.
Cascade.—A fall of lace; generally used in speaking of lace that is made to flow, with zigzag bends, like a river.
Ceinture.—Belt, waistband, or sash.
Chemise.—Shift; *chemise de jour,* day chemise; *chemise de nuit,* night dress; *chemise d'homme,* a night shirt.
Chiquetté.—Pinked out.
Clos.—Closed or fastened.
Coiffure.—A head-dress; manner of dressing the hair.
Coive.—Bonnet lining.
Confection.—A term applied to all kinds of made-up mantles, cloaks, and jackets, and all out-door garments.
Coques.—Looped bows of ribbon.
Cornet.—The cuff of a sleeve opening like the large end of a trumpet, larger at the wrist than above.
Corsage—Bodice.
Corset.—Stays.
Costume.—Complete dress.
Coulisse.—Small slip-stitched plaiting, sewn on the dress by slip-stitches.
Crénelé.—Crenelated; cut in square scallops, like battlements.
Dentelle.—Lace.
Dentellé.—Scalloped.
Dents.—Scallops; either pointed or square.
Dessous.—Underneath.
Dessus.—Above.
Devant.—Front.
Dos.—Back.
Droit et avant.—Right, and forward.
Echarpe.—A scarf; applied also to scarves tied round the hat.
Ecru.—The color of raw silk.
Effilé.—Fringe; generally a narrow one.
En biais.—On the cross.
Encolure.—The opening at the neck of a dress, or the arm's-eye.
En châle.—Resembling a shawl; applied to bodices and drapery.
En cœur.—Heart or V-shaped; applied to bodices.
En coquille.—Folded backwards and forwards in zigzags. Shell points.
En echelle.—Like a ladder.
En éventail.—Like a fan.
En tablier.—To look like or imitate a tablier.
Envers.—The wrong side.
Epaise.—Thick.
Epaisseur.—Thickness.

UTILITY AND DECORATIVE ART. 39

Essayez.—Try; attempt.
Farretière.—Garter.
Fendu.—Slashed, cut open; applied to a jacket.
Fichu.—A half-square, cut from corner to corner; any small covering for the shoulders.
Flots.—Quantities of lace or ribbon so arranged as to fall over each other like waves.
  Ex.—*Flots de dentelle*, rows of gathered lace falling one over the other.
Frange grillée.—A rather deep fringe, with an open heading, like net-work.
Fronces.—Gathers. *Froncé*; gathered.
Front-à-front.—Face to face;—basques, sleeves, etc.
Fupe.—Skirt.
Fupou.—Petticoat.
Grande parure.—Full dress.
Lingerie.—Collars and cuffs, made either of linen, cambric, or muslin and lace.
Lisere.—A narrow edging or binding.
Lisière.—Selvage; applied also to the colored edges of silks.
Manche.—Sleeve.
Manchette.—Cuff.
Manteau.—Cloak.
Mauvais goût.—Bad taste.
Nocud.—A bow or knot.
Noue.—Tied, or knotted.
Ombrelle.—Parasol.
Parement.—Cuff on the outside of a sleeve.
Parure.—A set of collars and cuffs; applied also to a set of jewelry or ornaments.
Passant.—Piping with a cord.
Passe.—The front of a bonnet or cap.
Peignoir.—Dressing-gown; dressing-jacket.
Pelerine.—A small mantle, rounded like a cape.
Petit côte.—Side-piece.
Plastron.—Breast-piece; a piece put on the front or back of a dress bodice, generally of a different color and material.
Pli.—Fold.
Plis.—Folds.
Plisse.—A plait or fold.
Plisses.—Plaits.
Ras-terre.—Just touching the ground.
Robe.—Dress.
Robe de chambre.—Dressing or morning-gown.
Rouleautes.—The same.
Rouleaux.—Rolled trimming made of crossway strips of material.
Ruches.—Gathered trimmings; called ruches.
Saut de lit.—Dressing-gown.
Simuler.—Simulate; to imitate.
Soulier.—Shoe.
Tourner casaque.—To turn the coat.

Tournure.—A bustle; also the general appearance of a dress, costume or person. Ex.: *Tournure distinguée*, lady-like appearance.
Tout ensemble.—The whole taken together.
Traine.—A train. *A traine.* With a train.
Tunique.—Tunic.
Tuyaux.—Fluted plaitings.
Tuyaux d'orgue.—Wide flutings, like the pipes of an organ.
Velours.—Velvet.
Veloute.—Soft, like velvet.
Vetement.—Garment.
Volant.—Flounce or frill.

DRESS FABRICS.

As materials for dresses vary so much in width, we submit a table of some of the standards; as a help to beginners. Fancy names are often given by drapers to certain fancy goods, each season; and, in many cases, the same fabric is sold under four or five different names. Special makes, too, in silks and cashmeres, present other difficulties in classification, needing special instruction. Yet there are certain time-honored dress materials, the widths of which are fixed by standard looms, and which are unchanged from year to year. And as they are also the most useful and durable in dressmaking, we will only tabulate these. Silk, poplin, merino, cashmere, alpaca, velveteen, muslin, print, and piqué form a sufficient amount of materials for the amateur; who would do well to make prints and muslins the fabrics upon which to experiment, in her first efforts.

TABLE OF FABRICS AND THE VARIOUS WIDTHS OF EACH.

	INCHES.		
Alpaca, 1st quality,	30	36	—
Alpaca, 2d quality,	24	36	—
Barathea,	42	—	—
Black Silk,	21	27	—
Beige,	25	28	—
Crape,	23	42	—
Crêpe de Chine,	24	—	—
Challé,	28	—	—
Cloth,	28	54	60
Cashmere,	23	46	—
Colored Silk,	22	26	—
Damask,	24	—	—
Foulard,	24	—	—
Grenadine,	18	26	—
India Silks,	32	34	—

	INCHES		
Janus Cord,	28	32	—
Merino,	45	46	—
Mousseline de Laine,	26	—	—
Muslin,	33	—	—
Ottoman and Turquoise Silks,	18	20	—
Paramatta,	42	—	—
Percales,	33	—	—
Piqués,	33	—	—
Prints,	33	—	—
Poplin,	30	32	—
Rep,	30	32	—
Satin,	18	27	—
Serge,	28	32	—
Tweed,	28	54	60
Velvet,	16	20	22
Velveteen,	27	28	—

DRESS LININGS.

It is money well invested to purchase the best quality of lining. It should combine the properties of strength, firmness, and pliability.

Among the standards we place,
1. The best drilling.
2. The firmest and best brands of Prussian silesia.
3. The double-faced silesias; and
4. Best quality of foulard silks and cambrics.

COMPUTATION OF MATERIAL REQUIRED.

The method of computing the number of yards, in plain garments, by founding the basis upon the width of the largest pieces, and the length required for each, according to the width of the goods, is so sensible and plain, that we deem it unnecessaay to occupy more space than to merely hint it. All draperies and trimmings being allowed extra, as a matter of course. Furthermore, it is a work of supererogation; since all the plates and designs have the amounts required, given under the same.

THE MEASURES.

The measures are all-important. First, as to the elements and principles of their foundation. Second, as to the accuracy with which they are taken and express the truth sought. Third, as to the accuracy with which they are transferred through the draft upon the goods, and reproduced in the completed garment.

The entire superstructure of the science of garment-making rests and depends upon the measurement, as its chief corner-stone. They must cover every possible emergency, contingency, and complication in the variations of form and fashion. Mathematics has fixed their location and number; and the instruments for obtaining, expressing, and transferring them.

To vary from her appointments, is suicidal to any possible hope of ever attaining to anything higher than mere guess-work, in the science of garment-making. To exceed her requirements, is only muddying a clear stream; and is an inconsistent work of supererogation. It is possible to found a very deficient "system of drafting" upon the correct measures; but it is certainly utterly impossible to found, develop, and complete the correct and truly scientific system of drafting, without—all and only—the necessary and truly scientific measure or indices. Furthermore, both the measures, the system of drafting, and the instruments may be mathematically and scientifically correct, and yet the garment may come forth a total failure, and as a misfit; the fault being wholly in the careless manner of taking the measures, drafting, basting, and stitching.

To the patient, accurate, and careful student, we promise a certain and sure result of perfection and grace of fit, that will abundantly repay all time and care expended; but to the impatient, slovenly, and careless student, we promise nothing—and expect nothing.

For instance, there are usually ten seams in a close-fitting garment, requiring twenty lines, in indication, on the draft; now, if by a blunt or illy-sharpened pencil, crayon, or chalk—or even with a sharp-pointed pencil—the lines are drawn one-eighth of an inch inside or outside of the points called for in the System, it will make the garment two and one-half inches too small or too large, according to the variation from correct drafting, in either direction, from the given points. And yet such students express astonishment that their efforts should end in complete failures; and then sometimes are loud in their declamation against the system, which they really think they followed. Because a careless and inexperienced marksman fails to hit the mark, holding the rifle far from the centre of the mark or object, should the gun receive censure and condemnation, when all the others hit the centre of the same mark every time, with the same gun or instrument? And what degree or extent of error might we not expect of such, when the same degree of utter recklessness and carelessness, that is here shown in drafting, only, is also carried into the work of taking the measures, tracing, basting, and stitching? In the same ratio of error, as explained and seen in the process of drafting, the garment would, when completed, be either twelve and one-half inches too small or too large.

We repeat, therefore, that to such persons we promise nothing, and of course expect nothing ; and the work was neither intended nor prepared for those who have neither the ambition to be perfect and excel, nor the required industry, patience, and willingness to be accurate. We commend such people to the school of charts, patterns, models, and machines.

Surely, if the world is willing to expend so much physical and mental labor, time, patience, care, and money in the work of manufacturing, acquiring, and possessing the various expensive cloths, silks, velvets, etc., it should not be considered a "waste of time" (?) for the cutter to spend from ten to twenty, or even sixty minutes, if necessary, on careful, thoughtful, and accurate work in cutting the same, so as to gracefully fit and adorn the form for which it was intended.

Science, in this province as elsewhere, crowns our efforts with an unerring degree of success or failure, in the exact ratio of the honor or dishonor with which we treat her laws. Neither science nor the author of this system is responsible for the inglorious achievements, in this province, by many former authors who claimed to represent science, and even praised her name, and then proceeded from the very first step to the last, through their entire teachings, to violate not only every known law of mathematics, but those of philosophy and common sense as well ; and ruthlessly trampled under foot all their most sacred precepts. We hold ourselves responsible for what we have taught and written—nothing more and nothing less. As to actual proof of its superior merit, we refer to results ; and as to the elements, principles, and the process of our method and system, we appeal to the court of science and mathematics, and are willing to abide by her decision.

Here as everywhere in mathematics, truth is much more simple, plain, beautiful, and far less intricate, than error ; and both the measures, the method of taking them, transferring them through the draft upon the goods, and reproducing them in the completed garment, is an exceedingly simple process, when the elements and principles are thoroughly mastered, understood, and honored with exceeding great care and accuracy—all of which takes less time, and gives much less trouble, than carelessness and in accurancy, in any thing ! Be studious ; be patient ; and be accurate ; and your efforts will be crowned with abundant success.

INTRODUCTORY TO TAKING THE MEASURES.

1. No instruments whatever are necessary, or should be used in obtaining, reading, and interpreting the measures or indices of the human form, other than those appointed by mathematics, viz.: the square or inch-rule and tape-measure.

2. Measures should only be taken over the corset intended to be worn with the garment, for which the measures are taken; including the desired wraps underneath; and over a close and neat-fitting body garment. In establishments, it is well to have a number of the principal average, with a few of the extreme sizes of thoroughly-stitched bodices of cheap but firm material, in readiness for those who come to leave their orders and measures, whose garments worn are loose and ill-fitting. Otherwise, use a leather or steel-jointed belt, drawing the garment well down on all sides, and then adjusting said belt in a horizontal position around the waist and fastening it firmly in that position by means of a buckle. It should be remembered, however, that said belt has nothing to do with taking the measures, other than in case of the above emergency, to firmly hold in a fixed position the loose covering of the form intended to be measured, in order to locate fixed points; from which and by means of which, the accurate measures or indices can be obtained, by the use of the tape-measure; and as given and illustrated in the engraving that illustrates the measures.

3. If the body or form to be measured is covered by heavy goods—even if close-fitting—the texture and weight of the same must be taken into consideration, as touching the degree of tightness or looseness with which the tape-measure is drawn, in comparison with the same, when measuring over lighter fabrics, before each particular measure is decided upon and placed upon the measure-book; as a basis and data from which and upon which to found and form the draft, at each respective and given point where each respective and given measure is taken. Industry, patience, and accurate care, with a little experience, will soon make the student completely master of all such circumstances.

4. Great care must be exercised by beginners to so take the circumference measures of the chest and bust that they shall express a graceful neatness, and yet be coupled with thorough ease and comfort; and so that any looseness or folds of the garment worn by the person being measured, and over which the measures are taken, will not deceive them when obtaining the exact and correct measures. Also, so that the position of the tape-measure on the form will not mislead them into error, in deciding upon said measures, for they are very important factors in the indices; and the inexperienced may easily be deceived as to what would actually express the correct measure representing grace of fit and comfort combined, on account of the often very ill-fitting garments worn by customers. All of which, indeed, applies with equal force to the waist and arm's eye circumferences, as well as to all the measures; but in a special manner to the above.

5. Too much care cannot be bestowed on the location of the points on the garment of the form to be measured, before measuring, as given and illustrated in the

engraving that shows how to take the measures; and also, in measuring from and to said points, and on the indicated lines passing from, to, and through said points; so that the expressions of the measures representing the various parts of the draft will always, when added to their opposites, produce the expression of the measure representing the whole. For instance: the sum of the Back-shoulder height plus the Front-shoulder height, must equal, or "balance," the entire measure from the position of the Chest circumference line at the back, up the back and along the side of the neck down the front, to the position of the Chest circumference line at the Front. And this principle of "balance proofs" should be practiced on all parts of the form where the measures are taken, by the beginner when taking, and before deciding upon and placing the measures in the measure-book; at least, until experience will teach a greater degree of direct accuracy, and a certain and well-founded confidence.

6. In the perfect form the blade width is equal to $\frac{1}{2}$ of the chest circumference; but always locate point 1 according to instructions, and directly under the centre of the junction of the arm with the body; and as illustrated in the engraving, showing the points and measures. The secret and philosophy of its location is the key that unlocks and unravels the cutter's greatest mysteries; and also makes it equally easy to fit both the stooped, round-shouldered, and even hump-backed form, with that of the erect and perfect form. For, the philosophy and principles upon which the measures and draft are founded, of which this is the "chief corner-stone," are such that the measures or indices are at one and the same time both "the proofs" and the "balance proofs." We challenge the scientific world in garment-cutting, to name and produce a system of measurement and drafting that even approaches this truth and discovery, and foundation of our system! It stands pre-eminently alone in this; and therein lies the secret of a large per cent. of its superlative worth and superiority. Our common sense, as well as our philosophy and practical experience, taught us long ago that if a measure is good to test the draft by, when completed, it must of necessity be equally good and useful to make, or help to form the draft by, directly and at once. Any principle of science and philosophy that is true in the mathematics of civil engineering, analytical and spherical geometry, and trigonometry, etc., can not be ignored in the science of garment-cutting, when the principles, the method of application, and all the contingencies are the same.

In the founding of the International System, we have therefore obviated all necessity for the once highly-honored and seemingly-necessary "balance proof measures," or "test measures."

The measures and system both "balance" and "test" the draft, in the natural process of its development and formation! And while we advise it, and while it is well for beginners to "test the work of their hands", before cutting, to see that no mistake has been made in applying the principles and the measures, according to the given instructions, and to correct any errors that they may have committed in applying the system, we can assure all that if the instructions, and measures, and principles, have all been applied as given and taught, it will be time lost and spent in vain, to "test the draft" to see if it expresses what the measures express!

7. All outside neck-adornments should be removed before locating points or measuring, so as to be ocessible to correct measurement.

In locating the given points and indicating the given lines on the form, use very neatly pointed crayon or tailor's pencil-chalk ; and lightly, but accurately and distinctly, indicate the points to a mathematical accurateness, upon the dress or garment worn over the form—all of which is easily and instantly erased by a mere touch of the hand or brush, when no longer needed or desired. Some prefer to locate the given and required points by pins; thoroughly adjusting them in such a manner and position, that when locating a point, the head of the pin when in its fastened and firmly fixed positon will be exactly upon said point.

8. Bearing the preceding instructions in mind, and having in readinesa, the crayon or tailor's pencil of chalk, otherwise pins, the square or inch-rule, the tape measure, the measure-book, and some one to write the measures as they are read and called, proceed according to the following given instructions for taking the measures.

THE
INTERNATIONAL SYSTEM

—OF—

FOREIGN AND AMERICAN

SCIENTIFIC TAILOR PRINCIPLES

—OF—

MEASUREMENT,

FOR OBTAINING AND READING THE INDICES

—OF—

THE MATHEMATICAL MOULD OR PERFECT FIT OF THE HUMAN FORM

—WITH ANY—

MATHEMATICAL SQUARE OR INCH-RULE

—AND—

TAPE-MEASURE ALONE.

THE INTERNATIONAL MEASURING PLATE.

THE BODY MEASURES.

PLATE XVII.—Figure LVII.

Instructions.

1. Locate point 1 on the garment of the form to be measured, on the exact centre point underneath the arm, on the junction line of the arm and the body ; and as illustrated in its correct position in the cut. Repeat point 1, in like manner, on the opposite side of the form, by the instructions for fixing its location on the first side, and as the same is illustrated in its first position, in the cut.

2. With the tape-measure, carefully and accurately take the exact circumference measure around the bare neck, on the junction line of the neck and the body ; or in other words, on the line of the proper junction of the collar with the high and close-fitting garment ; reading it, and placing or having it placed in the measure-book.

3. Adjust the collar of the garment worn by the form being measured, closely and securely around the junction of the neck and the body, by fastening it with a pin at the front ; turning it up, if not an upright collar. Then locate point 2 on the garment of the form being measured, on the exact centre point at the back of the form and at the neck, on the junction line of the neck and the body ; or, in other words, on the exact centre of the proper junction line, of the collar with the high and close-fitting garment, at the back of the form ; and as the same is illustrated in its correct position, on the cut.

4. Mentally, find one-eighth of the neck circumference, which has already been taken. Then locate point 3 on the garment of the form being measured, at the side of the neck and on the junction line of the neck and the body, or on the proper junction line of the collar with the high and close-fitting garment, at the distance of one-eighth of the neck circumference from point 2, towards the front of the form ; and as the same is illustrated in its correct position, on the cut. Repeat point 3, in like manner, and give it the like position at the opposite side of the form, by the structions for fixing its location at the first side ; and as the same is illustrated in its first and proper position, on the cut.

N. B.—The distances from point 2 to point 3 on the right side, and from point 2 to point 3 on the left side, must each be exactly one-eighth of the neck circumference ; and the distance from point 3 on the right side, over point 2, to point 3 on the left side, must exactly equal one-fourth of the neck circumference. Prove the correctness of the location of said point in this manner, before proceeding further ; and if wrong, correct your mistake.

5. Locate point 4 on the garment of the form being measured, on the exact centre point at the front of the form at the neck, and on the junction line of the neck—and the body ; or, in other words, on the exact centre of the proper junction-line, of the collar with the high and close-fitting garment, at the front of the form ; and as the same is illustrated in its correct position, on the cut.

6. Locate point 5 on the garment of the form being measured, on the centre point at the front of the form and at the waist, on the horizontal line with the actual depth of the waist, at the front; and as the same is illustrated in its correct position on the cut.

7. Locate point 6 on the garment of the form being measured, on the exact centre point at the side of the form, on the horizontal line with the actual depth of the waist at the side; and as the same is illustrated in its correct position, on the cut. Repeat point 6, in like manner, and give it the like position at the opposite side of the form, by the instructions for fixing its location at the first side; and as the same is illustrated in its first and proper position on the cut.

8. Locate point 7 on the garment of the form being measured, on the exact centre point at the back of the form, on the horizontal line with the actual depth of the waist at the back; and as the same is illustrated in its correct position on the cut.

9. Locate point 8 on the garment of the form being measured, on the exact centre point at the side of the form, on the horizontal line with the actual depth of the most prominent part of the hips, at the side; and as the same is illustrated in its correct position on the cut. Repeat point 8, in like manner, and give it the like position at the opposite side of the form, per instructions for fixing its first and proper location, at the first side; and as the same is illustrated in its first and proper position on the cut.

10. Placing the form of the person being measured in an easy, natural, and standing position; having the side towards your front and the arms down at the side of the form in their natural hanging position, and placing the square or inch-rule close up underneath, and firmly against the junction of the under side of the upper part of the arm and body; and having said part of the square, or otherwise the straight inch-rule, in a horizontal position with the form, then locate point 9 as follows, viz: locate point 9 on the garment of the form being measured, immediately in front of the arm, and even with the top of the inch-rule; and as the same is illustrated in its correct position on the cut.

Before changing the position of the inch-rule, locate point 10 immediately back of the arm, and even with the top of the inch-rule; and as the same is illustrated in its correct position on the cut: then reversing, or changing the position of the form being measured, so as to have the opposite side towards your front, repeat points 9 and 10 in like manner, and give them like positions, respectively, at the opposite side of the form, per instructions for fixing their first and proper positions, and as the same are illustrated in their first position, on the cut.

11. Place the inch-rule in a horizontal position across the chest at the front of the form, having the top edge exactly even with point 9 on the right side, and also even with point 9 on the left side; then locate point 11, on the garment of the form being measured, on the exact centre point at the front of the form, on a line with, or even with, the upper edge of the inch-rule; and as the same is illustrated in its correct position on the cut. Then locate point 12 on the garment of the form being measured, on a line even with the upper edge of the inch-rule, and at a point which is the distance of $\frac{1}{4}$ or the neck circumference from point 11 toward point 9, on the right side, at the front of the form; and as the same is illustrated in its correct position on the cut. Before changing the position of the inch-rule, repeat point 12 on

the garment of the form being measured, on the same line, at the upper edge of the inch-rule, and the same distance, or ¼ of the neck circumference from point 11, toward point 9, on the left or opposite side, at the front of the form ; and as point 12 is illustrated in its first and correct position on the cut.

12. Place the inch-rule in a horizontal position across the chest, at the back of the form being measured, having the top edge exactly even with point 10 on the right side, and also, even with point 10 on the left side ; then locate point 13 on the garment of the form being measured, on the exact centre-point at the back of the form on a line with, or even with, the upper edge of the inch-rule ; and as the same is illustrated in its correct position on the cut. Then locate point 14 on the garment of the form being measured, on a line even with the upper edge of the inch-rule, and at a point which is the distance of ⅙ of the neck circumference from point 13 toward point 10, on the right side, at the back of the form ; and as the same is illustrated in its correct position on the cut.

Before changing the position of the inch-rule, repeat point 14 on the garment of the form being measured, on the same line, or at the upper edge of the inch-rule, and the same distance, or ⅙ of the neck circumference from point 13 toward point 10, on the left side, at the back of the form ; and as point 14 is illustrated in its first and correct position on the cut.

13. Locate point 15 on the garment of the form being measured, at a point which is the distance of the actual or desired shoulder depth from point 3 toward the shoulder point, at the side of the form, and 1 inch from the top centre of the shoulder point, toward the back of the form ; and as the same is illustrated in its correct position on the cut.

14. Now having located all the necessary points, carefully and accurately take the following respective measures, over the following respective given lines and distances, as per instructions given for each; reading each measure as taken, and placing or causing it to be placed in the measure-book, and in the following order, viz :

a. Skirt Depth.—Measure closely on a perpendicular line, with the stretched tape measure, over the form, from point 6 toward the bottom, to the actual or desired depth of the skirt ; as correctly illustrated on the cut.

b. Blade Width.—Measure closely on a horizontal line, with the stretched tape-measure, over the back of the form, from point 1 on the left side to point 1 on the right side ; and as the same is correctly illustrated on the first side on the cut.

c. Chest Circumference.—Measure with a comfortable of degree closeness, on a circumference line, of the stretched tape measure around the chest of the form, on a horizontal line with the junction of the under side of the upper part of the arm and the body ; and as the same is correctly illustrated on the first side, on the cut.

d. Waist Depth.—Measure on a perpendicular line of the stretched tape measure, over the form, from point 1 to point 6 ; and as the same is correctly illustrated on the cut.

e. Back Waist Depth.—Measure on a perpendicular line of the streched tape measure, over the form, from point 13 to point 7 ; and as the same is correctly illustrated on the cut.

f. Back Height.—Measure on a perpendicular line of the stretched tape measure, over the form, from point 2 to point 13 : and as the same is correctly illustrated on the cut.

g. Front-Waist Depth.—Measure on a perpendicular line of the stretched tape-measure, over the form, from point 11 to point 5—and as the same is correctly illustrated on the cut.

h. Front Height.—Measure on a perpendicular line of the stretched tape measure, over the form, from point 4 to point 11 ; and as the same is correctly illustrated on the cut.

i. Arm's-Eye Circumference.—Measure on a circumference line of the stretched tape-measure, around the junction of the upper part of the arm and the body, over the shoulder, and very closely ; and as the same is correctly illustrated on the cut.

j. Back-Shoulder Height.—Measure closely on a perpendicular line of the stretched tape-measure, over the form, from point 3 to point 14 ; and as the same is correctly illustrated on the cut.

N. B.—Prove the correctness of your measure of the back-shoulder height and front-shoulder height, by observing whether the sum of the two, equals or "balances" the entire distance from point 12 at the front, over point 3 at the side of the neck, down to point 14, at the back ; if wrong, correct your mistake before entering either of the shoulder heights.

k. Front-Shoulder Height.—Measure closely on a perpendicular line of the stretched tape-measure, over the form, from point 3 to point 12 ; and as the same is correctly illustrated on the cut.

l. Shoulder Depth.—Measure on a line of the stretched tape-measure, held upon and over the form, from point 3 to point 15 ; and as the same is correctly illustrated, on the cut.

m. Bust Circumference.—Measure, with a degree of closeness representing both elegance of fit and an easy comfort, on a diagonal circumference line of the stretched tape-measure, around the fullest part of the form at the front, and over point 13 at the back ; and as the same is correctly illustrated on the cut,

n. Waist Circumference.—Measure *very* closely on a horizontal circumference line of the stretched tape-measure, around the form at the lower terminus of the natural waist ; and as the same is correctly illustrated, on the cut.

o. Hip Depth.—Measure on a line of the stretched tape-measure, held upon and over the form, from point 6 to point 8 ; and as the same is correctly illustrated on the cut.

p. Hip Circumference.—Measure, with a degree of closeness representing both elegance of fit and a *very* easy comfort, on a horizontal circumference line of the stretched tape-measure, around the fullest part of the form below the waist circumference, at the position of point 8 on the garment ; and as the same is correctly illustrated, on the cut.

N. B.—1. The extra "corset measure," as illustrated in its proper position on the cut, between the bust circumference and the waist circumference, is not positively necessary ; only in exceptionally depressed forms at that location, or for some styles of peculiarly-shaped corsets ; in which cases, it should be taken, and the surplus amount subtracted from the depths of the curves in the draft, and to the degree indicated by the actual measure at that part of said form ; before the draft is used

UTILITY AND DECORATIVE ART. 53

as a basis for completing the garment. Furthermore, the opposite, in principle and process, is true of the extra "hip circumference," illustrated in its proper position on the cut—between the waist and hip circumference. Where some peculiar forms, or otherwise peculiarly shaped corsets, will require this special and extra measure, in order that the exact amount of the required extra fullness may be certainly indicated thereby; which must in said cases be provided for by special indication on the draft, after it is completed according to the standard measure and per instructions given for general drafting, before transferring the indications of the same for the needle ; or before using the same as a basis for completing the garment.

2. The front-skirt depth and the back-skirt depth, are not positively necessary only in cases where the form is "stooped" or in cases of full *tournure*.

THE SLEEVE MEASURES.

PLATE XVII.—FIGURE LVIII.

Instructions.

1. Having the upper part of the arm down, along and in a line with the body, in an easy and natural position, and the lower part of the arm in a horizontal position with the body and at the front, carefully and accurately take the following respective measures over the respective given lines and distances, per instructions given for each respective measure ; reading each measure as taken, and placing or causing it to be placed in the measure book in the following order, viz :

a. Extreme Depth.—Locate point 16 on the top of the shoulder of the garment of the form being measured, at a point on the line of the junction of the upper and top part of the arm with that of the shoulder ; or, in other words, at the point on the true arm's eye circumference, which is one inch toward the front of the form, from point 15 ; and as the same is illustrated in its proper position on the cut. Then measure from point 16 on a direct line of the stretched tape measure around and over the outside point of the bone of the elbow, and along the outside of the arm to the most prominent bone at the outside of the wrist; and as the same is illustrated in its proper position on the cut.

b. Arm's Eye Circumference.—This is one and the same measure with that given in the measures and instructions for the "Body Measures ;" and the same measure being used, it need not, therefore, be re-taken. And if taken for sleeve-cutting only, it is taken per instructions for the same measure under the head of taking the measures for " Body Drafting ;" and as the same is illustrated in its correct position on the cut.

c. Elbow-Circumference.—Measure on a diagonal circumference line of the stretched tape-measure, around the arm, at the point of the deepest depression on the inside of the arm at the elbow, and over the extreme point of the bone at the outside of the elbow, to the degree of closeness desired ; and as the same is illustrated in its correct position on the cut.

d. Elbow Depth.—Locate point 17 on the sleeve of the form, at a point which is on the centre of the outside point of the outside bone of the elbow ; then measure on a line of the stretched tape-measure from point 16 to point 17 ; and as the same is illustrated in its correct position, on the cut.

e. Hand Circumference.—Measure on a circumference line of the stretched tape-measure, around the fullest part of the hand, over the junction of the thumb with the hand, to the degree of closeness desired ; and as the same is illustrated in its correct position, on the cut.

N. B.—If the sleeve is to fit closely, at the wrist, and is finished with buttons and button-holes, clasps, or a buckle, etc., the close wrist circumference is substituted for the hand circumference.

f. Inside Height.—Measure on a line of the stretched tape-measure pressed and held firmly on the point of the deepest depression at the inside of the elbow, at the inside of the arm, up along the proper position for the inside seam of the sleeve, close up to the junction of the front inside of the upper part of the arm with that of the body ; and as the same is illustrated in its correct position on the cut.

g. Arm Circumference.—Measure on a circumference line of the stretched tape-measure, around the arm, at the point which is $\frac{1}{4}$ of the inside height from the junction of the lower part of the arm with that of the body, toward the elbow, to the degree of closeness desired ; and as the same is illustrated in its correct position on the cut.

h. Outside Height.—Measure on a line of the stretched tape-measure from point 17, to the point on the true arm's-eye circumference, which is equidistant from point 1 and point 15, at the back of the arm ; or, in other words, at the proper point on the correct arm's-eye circumference, where the seam at the back of the sleeve joins with the garment of the form ; and as the same is illustrated in its correct position on the cut.

N. B.—1. The extra arm's circumference measures, illustrated in Figure lviii, equidistant in opposite directions from the elbow circumference measure, are not positively necessary, only in case of special depression of the arm at those points, and where a very close-fitting sleeve is desired ; under which circumstances they should be taken and recorded in the measure-book, for use when the draft is completed; and before tracing or transferring the impression of the fit, for the needle. It is easily and quickly done, by simply adding or subtracting the required amount in transferring the draft, either by the pencil or through the tracing-wheel.

2. When the sleeve is to be close-fitting at the wrist and is either buttoned, buckled, or clasped, substitute the wrist circumference for the hand circumference; and draft per instructions for the same, as given for the hand circumference, in sleeve drafting.

3. Although locating all the points necessary for "double measurement"—in measuring the form—it is not necessary to measure but one side for the shoulder-depth, back-height, front-height, arm's-eye circumference, waist-depth, and hip-depth, and that should be the right side, under all ordinary circumstances ; but in case the form differs sufficiently and noticeably, and it is plainly apparent that a double measurement is needed, the above measures should all be taken for each side, separately, and distinct from the opposite side, and should be recorded in the measure-book; and should be marked R and L, for right and left ; and the draft in that case

UTILITY AND DECORATIVE ART. 55

must be made only to one of the sides, per instructions for general drafting; and after being completed for the right side, the indications can quickly be placed thereon, true to the opposite side, and true to the measure expressing the fit of the opposite side; and then the lines indicating the fit of each respective side must be lettered R and L, and careful patterns traced from each respective draft, as the data upon which to base the work, and from which to complete the garment. In this way you will be able to fit such forms as perfectly as any other forms. It only requires industry, care, accuracy, and patience; and this you would be compelled to exercise in any other way, with a far less beautiful and perfect result.

4. If to gratify any special fancy or taste, it should be desired to throw the shoulder seam farther forward, either at the neck or at the arm's-eye, it can easily be done. But at the neck, it must under no circumstances exceed one-fourth of the neck circumference from point 2; and point 15 dare not exceed, in distance, two-fifths of the arm's-eye circumference from point 1; and when any other position is chosen for said points, than the ratios given, the respective allowances must be made in the draft, from the rules given, touching said points; *i. e.*, what is added to the "back" in the method or change of taking the measure, must be subtracted from the "front," in the draft; and *vice versa*. But the method of measurement given, is based upon sound logic, good reason, and best usage, and is standard; and it should not be changed or interfered with. For unless the student is thoroughly master of the principles, it would be easy to make a serious mistake in the draft, if the location of points 3 and 15 were not located per instructions given for the same. But we simply drop this hint to the "graduate" in the science; to show how easily it is to draft entirely independent of any fixed location of seams whatever.

N. B. This is the only system ever invented by which we are enabled to scientifically and accurately read the mathematical indices of the human form with any square or inch-rule, and tape measure alone; and interpret and transfer the same thereby, alone, either directly or indirectly upon the goods! It stands preeminently alone in this, and is a half century in advance of the age, in this great discovery alone!

REQUISITES, AND PREPARATIONS FOR DRAFTING.

1. A well jointed, strong, plain soft wood table; 10 by 4, the desired height.

2. In learning either the compound or direct methods, the student must learn drafting, upon good, strong manilla paper.

3. Tailor's pattern-paper should be used for patterns, if patterns are cut to lay upon the goods.

4. If the draft is made directly, either upon the linings or the material, the material used should be thoroughly pressed, carefully placed, and securely fastened at the edges, by means of the "double-tack", pins, or "weights". In compound drafting, draft upon good, strong, heavy manilla paper; place it upon the linings, carefully fas-

tening it by pins; and trace the pieces seperately, cutting each part before tracing another. This draft can then be kept for future use, for the same person. Good paper will answer all purposes, and is best.

5. For drafting upon the goods or linings, use the crayon pencil, or tailors' chalk, well pointed; and be exceedingly accurate.

6. To draft upon paper, use a medium soft lead pencil, carefully and neatly sharpened; draw a distinct but fine line; the points must simply be *points*, and very accurately located.

7. See therefore that you have the following, viz :
   a. A clean table.
   b. The material to draft upon, the paper or lining.
   c. The square or rule, and the tape measure.
   d. The sharpened pencil, crayon, or tailors' chalk.
   e. The system and diagrams, in the most easy and natural position, to guide you.

8. Then proceed according to instructions for drafting; using the measures given, for practice. Be careful from the very first instruction to the last, that you understand what is printed, and what is meant; and be sure you are right before locating a point, or drawing a line. Strive for accuracy and neatness, in constructing the draft. Excellency is attained by care, study, and accuracy in details.

THE INTERNATIONAL SYSTEM
OF
Measures for Practice.

		INCHES.
1.	Neck Circumference,	12
2.	Skirt-Depth,	10
3.	Blade-Width,	18
4.	Chest-Circumference,	36
5.	Waist-Depth,	8
6.	Back-Waist Depth,	8½
7.	Back-Height,	7½
8.	Front-Waist Depth,	9
9.	Front-Height,	5
10.	Arm's-Eye Circumference,	15
11.	Back-Shoulder Height,	7¾
12.	Front-Shoulder Height,	8½
13.	Shoulder-Depth,	6¼
14.	Bust-Circumference,	37
15.	Waist-Circumference	24
16.	Hip-Depth,	7
17.	Hip-Circumference,	48

18 Sleeve :			INCHES.
	1.	Extreme-Depth,	22
	2.	Arm's-Eye Circumference,	15
	3.	Elbow-Circumference	10
	4.	Elbow-Depth,	13
	5.	Hand-Circumference,	8
	6.	Inside-Height,	8
	7.	Arm-Circumference,	12
	8.	Outside-Height,	11

THE

INTERNATIONAL METHOD

—OF—

FRENCH AND PRUSSIAN

SCIENTIFIC TAILOR PRINCIPLES,

—FOR ALL—

KINDS AND STYLES

—OF—

COMPOUND "BODY-DRAFTING."

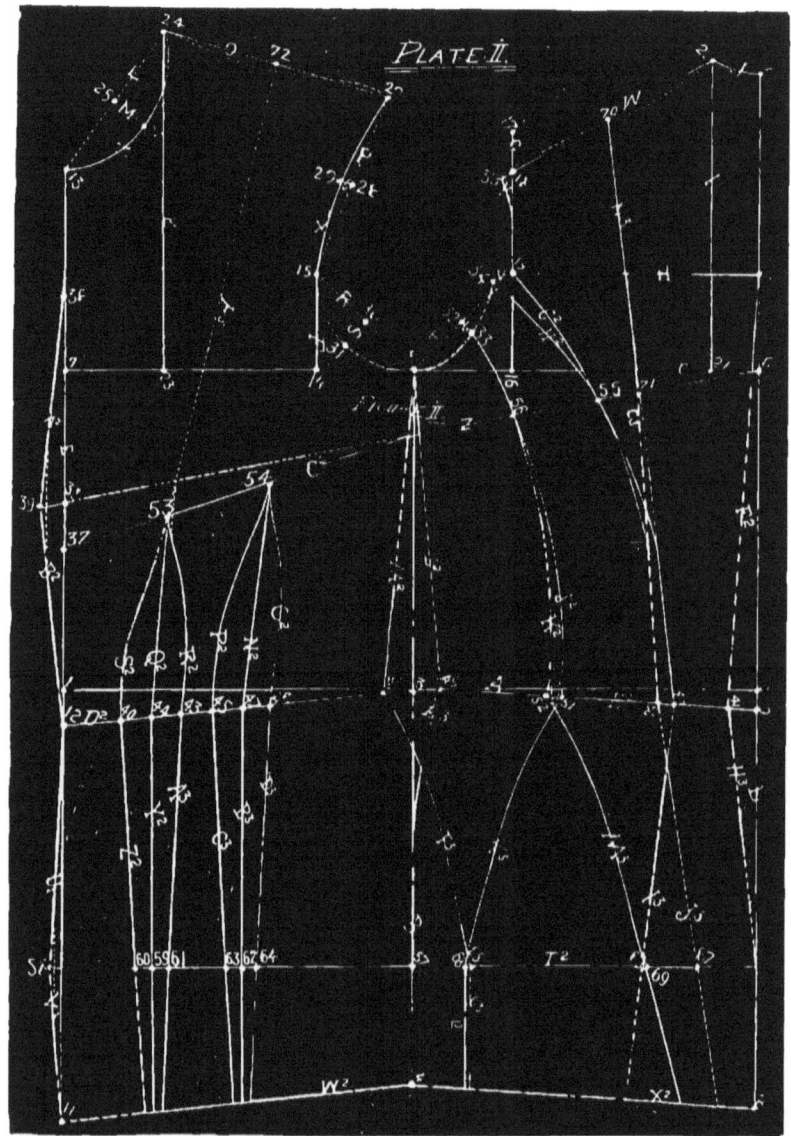

PLATE 1.—FIGURE 1.

INSTRUCTIONS.

1. Add 1½ inches to the skirt depth; locate point 1 at that distance from the bottom, or right-hand edge; and 1½ inches back, from the front edge of the paper or material.

UTILITY AND DECORATIVE ART.

2. Draw line A from point 1, back toward the opposite edge, at right angles with the front edge, to the distance of ½ of the chest circumference from point 1; at which point, locate point 2.

3. Locate point 3 on line A, ½ of the blade-width from point 2 toward point 1.

4. Draw line B from point 3 toward the bottom, at right angles with line A, to the distance of the skirt depth from point 3; at which point, locate point 4: then extend line B in the opposite direction, to the distance of the waist depth from point 3; at which point, locate point 5.

5. Draw line C from point 5, at right angles with line B, to the distance of ½ of the blade-width from point 5, toward the opposite edge; at which point, locate point 6. Subtract ½ of the blade-width from ½ of the chest circumference; extend line C, in the opposite direction, to the distance of that difference from point 5—at which point, locate point 7.

6. Add the back-waist depth to the skirt depth; draw line D from point 6 toward the bottom, at right angles with line C, to the distance of that sum of depths; at which point, locate point 8; also locate point 9 on line D, the distance of the back-waist depth from point 6, toward point 8. Then extend line D in the opposite direction, to the distance of the back height from point 6—at which point, locate point 10.

7. Add the front waist depth to the skirt depth; draw line E from point 7 toward the bottom at right angles with line C, to the distance of the sum of these depths from point 7, at which point locate point 11; also, locate point 12 on line E, the distance of the front waist depth from point 7, toward point 11; then extend line E in the opposite direction, to the distance of the front height from point 7, at which point, locate point 13.

8. Locate point 14 on line C, ⅙ of the arm's eye circumference from point 5, toward point 7.

9. Draw line F, at right angles with line C, toward the top from point 14, to the distance of ⅙ of arm's eye circumference from point 14; at which point locate point 15.

10. Locate point 16 on line C, ⅙ of the arm's eye circumference from point 5 toward point 6.

11. Draw line G at right angles with line C, from point 16 toward the top, to the distance of two-fifths of the arm's eye circumference from point 16; at which point locate point 17.

12. Locate point 18 on line G, at a point which is the distance of one-third of the arm's eye circumference from point 16 toward point 17.

13. Locate point 19 on line G, at a point equidistant from, or half way between, point 16 and point 18.

14. Draw line H, at right angles with line G, from point 19 to line D; and locate point 20, at the point of intersection of lines H and D.

15. Subtract ¼ inch from ⅙ of the neck circumference; locate point 21 on line C, that distance, from point 6 toward point 16.

16. Draw line I at right angles with line C, from point 21 toward the top, to the distance of the back shoulder height, from point 21; at which point locate point 22.

17. Draw curved line J, from point 22 to point 10; as illustrated in the cut.

18. Subtract ½ inch from ¼ of the neck circumference; locate point 23 on line C, that distance from point 7 toward point 14.

19. Draw line K at right angles with line C, from point 23 toward the top, to the distance of the front shoulder height from point 23; at which point locate point 24.

20. Draw pointed line L, from point 24 to point 13; as illustrated in the cut.

21. Locate point 25 on line L, equidistant from point 24 and point 13.

22. Draw pointed line M, from point 25 toward point 15, to the distance of one-twelfth of the neck circumference from point 25; at which point, locate point 26.

23. Draw curved line N, from point 24 through point 26, to point 13; as illustrated in the cut.
24. Draw line O, from point 24 toward point 17, to the distance of ⅜ of an inch less than the shoulder depth from point 24, at which point, locate point 27.
25. Draw pointed line P, from point 27 to point 15.
26. Locate point 28 on line P, equidistant from point 27 and point 15.
27. Draw pointed line Q, from point 28 toward point 25, to the distance of ¼ of the distance between point 17 and point 18 from point 28; at which point, locate point 29.
28. Draw pointed line R, from point 15 to point 5.
29. Locate point 30 on line R, equidistant from point 15 and point 5.
30. Draw pointed line S, from point 30 toward point 14, to a point equidistant from point 30 and point 14; at which point, locate point 31.
31. Draw pointed line T, from point 5 to point 19.
32. Located point 32 on line T, equidistant from point 5 and point 19.
33. Draw pointed line U from point 32 toward point 16, to that distance from point 32, which is one-seventh of the entire distance between point 32 and point 16; at which point locate point 33.
34. Draw pointed line V from point 19 toward point 14, to a point which is twice as far from point 19 as the distance between point 32 and point 33; at which point, locate point 34.
35. Draw line W from point 22 toward point 15, to the distance of the shoulder depth from point 22; at which point, locate point 35.
36. Draw curved line X, from point 27 through points 29, 15, 31, 5, and 33, to point 34; as illustrated in the cut.
37. Draw curved line Y, from point 35 to point 19; as illustrated in the cut.
38. Locate point 36 on line E, equidistant from point 13 and point 7.
39. Locate point 37 on line E, equidistant from point 7 and point 12.
40. Locate point 38 on line E, ¼ of the distance between point 7 and point 37, from point 37 toward point 7.
41. Draw line Z, from point 6, through point 38, to the distance of ½ inch more than ½ of the bust circumference, from point 6; at which point, locate point 39.
42. Draw line A², from point 36 to point 39.
43. Draw line B², from point 39 to point 12.
44. Draw line C², from point 37 to the point of intersection, of lines B and Z.
45. Draw line D², from point 12 to point 3.
46. Draw line E², from point 3 to point 9.
47. Find ⅛th of the waist circumference; locate point 40 on line D², ½ of that distance from point 12 toward point 3.
48. Find ⅛th of the difference between the chest circumference and the waist circumference. Locate point 41 on line E², the distance of ½ of that amount from point 9 toward point 3.
49. Locate point 42 on line E², the same distance from point 41 toward point 3, as the distance from point 12 to point 40.
50. Locate point 43 on line D², twice as far from point 40 toward point 3, as the distance between point 9 and point 41.
51. Locate point 44 on line D², equidistant from point 40 and point 43.
52. Locate point 45 on line D,² ½ of the distance between point 12 and point 40, from point 43, toward point 3.
53. Locate point 46 on line D², the same distance from point 45 toward point 3, as the distance between point 40 and point 43.
54. Locate point 47 on line D², equidistant from point 45 and point 46.
55. Locate point 48 on line D², the same distance from point 3 toward point 46, as the distance between point 46 and point 47.
56. Locate point 49 on line E², the same distance from point 3 toward point 42, as the distance between point 3 and point 48.

UTILITY AND DECORATIVE ART. 61

57. Locate point 50 on line E^2, ½ of the distance between point 9 and point 41 from point 42, toward point 49.

58. Locate point 51 on line E^2, the same distance from point 50 toward point 49, as the distance between point 46 and point 48.

59. Locate point 52 on line E^2, the same distance from point 51 toward point 49, as the distance between point 42 and point 50.

60. Locate point 53 on line C^2, ⅜ of an inch further from point 37 toward the point of intersection of lines B, Z, and C^2, than the distance between point 12 and point 44.

61. Locate point 54 on line C^2, the same distance from point 53 toward the point of intersection of lines B, Z, and C^2, as the distance between point 37 and point 53.

62. Draw line F^2, from point 20 to point 41.

63. Wrap one end of the tape-measure neatly and firmly around the pencil, near the sharpened end, and hold it firmly with the front two fingers and thumb of the right hand ; measure twice the waist-depth back from the pencil, on the tape-measure, either by the inches on the tape-measure or the square, at which point, hold it firmly between the forepart of the thumb and forefinger of the left-hand ; then placing and holding the point of the pencil firmly on point 19, stretch the tape-measure, and with the fore part and nail of the left-hand thumb or finger placed firmly on that length, press it upon the table in front of the draft, opposite point 12. Then, by repeated trials, find a point for the position of the left-hand, where, if the left-hand position is maintained, and the right-hand is moved and guided toward the waist, by the stretched tape-measure, the point of the pencil will exactly touch upon point 42, and again upon point 19, when moved back : then draw curved line G^2, from point 19 to point 42; as illustrated in the cut; and locate point 55, at the point of intersection of lines G^2 and Z.

64. Draw curved line H^2, from point 34 to point 55, by the same length of tape-measure, and principles and rule, as given under instruction 63d, above, only changing positions of both hands true to, and drawing from point 34 to point 55 ; and as illustrated in the cut.

65. Draw curved line I^2, by the same length of tape-measure, and principles and rule, as given under 63d, above; only changing positions of both hands true to, and drawing from point 55 to point 50 ; and as illustrated in the cut.

66. Draw curved line J^2, from point 33 to point 51, by the same length of tape-measure, and principles and rule, that you drew lines G^2, H^2, and I^2; only, changing positions true to, and drawing from point 33 to point 51; as illustrated in the cut ; and locate point 56, at the point of intersection of lines J^2 and Z.

67. Draw curved line K^2 from point 56 to point 52, by the same length of tape-measure, and principles and rule, as given for lines G^2, H^2, I^2, and J^2; only, changing positions true to, and drawing from point 56 to point 52.

68. Draw line L^2, from point 5 to point 49.
69. Draw line M^2, from point 5 to point 48.
70. Draw line N^2, from point 54 to point 47.
71. Draw curved line O^2, from point 54 to point 46 ; as illustrated in the cut.
72. Draw curved line P^2, from point 54 to point 45 ; as illustrated in the cut.
73. Draw line Q^2, from point 53 to point 44.
74. Draw curved line R^2, from point 53 to point 43 ; as illustrated in the cut.
75. Draw curved line S^2, from point 53 to point 40 ; as illustrated in the cut.
76. Locate point 57 in on line B, the distance of the hip depth from point 3, toward point 4.
77. Draw line T^2, at right angles with, and from line D, through point 57, to that distance in front of, and from the point of intersection of lines T^2 and E, which is ½ of the distance between point 38 and point 39 ; at which point, locate point 58.
78. Draw line U^2, from point 12 to point 58.
79. Draw line V^2, from point 58 to point 11.
80. Draw line W^2, from point 11 to point 4.

81. Draw line X^2, from point 4 to point 8.
82. Draw line Y^2, at right angles with line A, from point 44 to line W^2; and locate point 59 at the point of intersection of lines Y^2 and T^2.
83. Locate point 60 on line T^2, ½ as far from point 59 toward point 58, as the distance between point 40 and point 44.
84. Draw line Z^2, from point 40 through point 60, to line W^2.
85. Locate point 61 on line T^2, the same distance from point 59 toward point 57, as the distance between point 59 and point 60.
86. Draw line A^3, from point 43 through point 61, to line W^2.
87. Draw line B^3, at right angles with line A, from point 47 to line W^2; and locate point 62 at the point of intersection of lines B^3 and T^2.
88. Locate point 63 on line T^2, the same distance from point 62 toward point 61, as the distance between point 59 and point 61.
89. Draw line C^3, from point 45 through point 63, to line W^2.
90. Locate point 64 on line T^2, the same distance from point 62 toward point 57, as the distance between point 62 and point 63.
91. Draw line D^3, from point 46 through point 64, to line W^2.
92. Draw curved line E^3, from point 49 to point 57; as illustrated in the cut.
93. Locate point 65 on line T^2, twice as far from point 57 toward line D—as the distance between point 63 and point 64.
94. Draw curved line F^3, from point 48 to point 65; as illustrated in the cut.
95. Draw line G^3, at right angles with T^2, from point 65 to line W^2.
96. Find the difference between ½ of the chest circumference and ½ of the blade-width; subtract this difference from ½ of the hip circumference: Then locate point 66 on line T^2, one-fifth of this last difference from the point of intersection of lines D and T^2, toward point 65.
97. Draw line H^3, from point 41 to the point of intersection of lines T^2 and D.
98. Draw line I^3, from point 42 through point 66, to line X^2.
99. Locate point 67 on line T^2, at a point equidistant from point 66 and the point of intersection of lines T^2, H^3, and D.
100. Draw line J^3, from point 50 through point 67, to line X^2.
101. Locate point 68 on line T^2, twice the distance between point 66 and the point of intersection of lines T^2, H^3, and D, from point 67, toward point 57.
102. Draw curved line K^3, from point 51 to point 68; as illustrated in the cut.
103. Draw line L^3, at right angles with line T^2, from point 68 to line X^2.
104. Locate point 69 on line T^2, the same distance from point 57 toward point 67, as the distance between point 67 and point 68.
105. Draw curved line M^3, from point 52 through point 69, to line X^2; as illustrated in the cut.
106. Locate point 70 on line W, equidistant from point 22 and point 35.
107. Draw line N^3, from point 70 to point 42; and locate point 71 at the point of intersection of lines N^3 and Z.
108. Draw line O^3, from point 71 to point 50.
109. Locate point 72 on line O, equidistant from point 24 and point 27.
110. Indicate pointed line P^3, from point 72 to point 53; as illustrated in the cut.

Q. E. D.

N. B.—1. All possible kinds and styles of ladies' garments can quickly and accurately be traced upon the lining or material, by means of a tracing-wheel, from the above draft; which should be accurately drafted to the accurate measures of each customer; and upon good, firm paper, or special pattern material; and should be carefully preserved for that particular customer; until a sufficient change of form necessitates or demands a re-taking of the measures, and the making of a new draft.

2. Full explanations for the elimination of the various parts of the draft, by tracing its parts upon the material, will be given under "Tracing."

3. Instructions for drafting with only one dart, will be given under Plate V.

4. The indicated and pointed lines, along lines O and W, show the proper curves to be traced for basting and stitching; as do also, the indicated and pointed lines across the angles of the straight lines, on the waist, and elsewhere.

5. Indicated and pointed line P^3, from point 72 to point 53, is used for cutting vest-fronts, etc.; and for garments with seams to the shoulder, in front.

6. In cutting, all seams and extensions for finishing, must be allowed, every where.

7. The measures, the correct method of taking and proving them, and using them, and all the particulars and details of the above draft should be thoroughly mastered, before attempting to proceed further. It is the key to all the secrets of the world's greatest cutters.

8. If the chest-circumference does not exceed the waist-circumference by 6 inches, or if sickness or deformity has increased the waist-circumference, so as to be equal to, or even to exceed the chest-circumference; under each and all of the above conditions, carefully and accurately calculate the number of inches that would have to be added to the chest circumference to made it six inches more than the waist-circumference. Then add ½ of this necessary amount to ½ of the chest-circumference, and locate point 2 that distance from point 1. Then find ½ of the amount that you added to ½ of the chest-circumference, and add this to ½ of the blade-width; and locate point 3 that distance from point 2 toward point 1. Then draft, per instructions for general drafting in all other parts of the draft—to the actual measures, as taken on the form; remembering, however, that in this case, the parts of lines D and E between points 6 and 2, and 7 and 1, respectively, will not be at right angles with line C; but said lines will be at right angles with lines C and A, respectively, at the top and at bottom of the draft, as in general drafting.

It must also be remembered when coming to the "spacings" and "locations of points for darts," etc., on the waist, that the rules must be applied as given; and the same as though an actual difference of 6 inches existed between the chest-circumference and the waist-circumference. At first this may seem arbitrary, and somewhat difficult of comprehension to the beginner; but it will grow very simple to the experienced student. Furthermore, the circumstances which call for this special rule, are very rare; and the rule is only submitted, to thoroughly arm the student for every possible and conceivable emergency of the cutter.

THE INTERNATIONAL METHOD OF FRENCH AND PRUSSIAN

SCIENTIFIC TAILOR PRINCIPLES

FOR ALL KINDS AND STYLES OF

Compound Sleeve-Drafting.

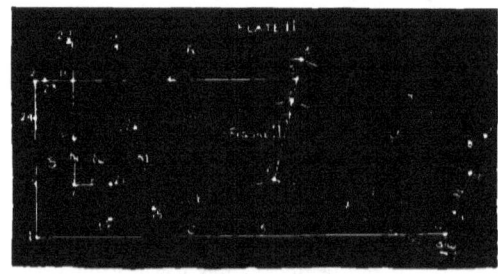

PLATE II.—FIGURE II.

INSTRUCTIONS.

1. Locate point 1 on the paper or material, 1 inch farther than the distance of the extreme length, from the bottom or right-hand edge, and 1½ inches back from the front edge.

2. Draw line A, at right angles with the front edge, from point 1 toward the opposite edge, to the distance of ½ of the arm's-eye circumference from point 1—at which point, locate point 2.

3. Draw line B, at right angles with line A, the distance of ¼ inch less than the elbow-depth from point 2, toward the right-hand edge—at which point locate point 3.

4. Find the difference between ½ of the extreme depth and the elbow-depth; subtract this difference from the extreme depth: Then draw line C, at right angles with line A, to the distance of this last difference, from point 1 toward the right-hand edge; at which point, locate point 4.

5. Locate point 5 on line C, the distance of ½ the extreme depth from point 1 toward point 4.

6. Draw line D from point 3 toward point 5, to the distance of ½ of the elbow-circumference from point 3; at which point locate point 6.

7. Draw line E, at right angles with line C, from point 4 toward the front edge, to the distance of ⅜ of the hand-circumference from point 4; at which point locate point 7.

8. Indicate pointed line F, from point 6 to point 7; as illustrated in the cut.

9. Locate point 8 on line F, ⅛ of an inch from point 7, toward point 6.

10. Indicate pointed line G, from point 8 to point 4—as illustrated in the cut.

UTILITY AND DECORATIVE ART.   65

11. Indicate pointed line H, at right angles with line F, from point 4 toward the opposite edge, to 5 times the distance between point 7 and point 4, from point 4; at which point, locate point 9.

12. Locate point 10 on line G, the same distance from point 4 toward point 9; as the distance between point 8 and point 4.

13. Indicate pointed line I, from point 6 to point 10; as illustrated in the cut.

14. Locate point 11 on line G, equidistant from point 9 and point 10.

15. Locate point 12 on line D, the same distance from point 3 toward point 6, as the distance between point 7 and point 4.

16. Indicate pointed line J, from point 12 to point 11; as illustrated in the cut.

17. Extend line D, toward the opposite edge from point 3, to the same distance from point 3, as the distance between point 12 and point 3; at which point locate point 13.

18. Indicate pointed line K, from point 13 to point 9; as illustrated in the cut.

19. Indicate pointed line L, from point 6 toward point 1, to the distance of the inside-height from point 6; at which point locate point 14; as illustrated in the cut.

20. Locate point 15 on line L, ¼ of the distance between point 14 and point 6, from point 14 toward point 6.

21. Draw line M, at right angles with line L, from point 15 toward the opposite edge, to the distance of ⅔ of the arm-circumference from point 15; at which point locate point 16.

22. Locate point 17 on line M, equidistant from point 15 and point 16.

23. Locate point 18 on line B, the distance of the outside height from point 3 toward point 2.

24. Draw line N, at right angles with line B, from point 18 toward line C, to the distance of ⅓ of the arm's-eye circumference from point 18; at which point, locate point 19.

25. Indicate pointed line O, from point 12 through point 17, to the point of intersection of lines O and N; at which point, locate point 20; as illustrated in the cut.

26. Draw line P, at right angles with line N, from point 19 toward line M, to the distance of ⅛ of the arm's-eye circumference from point 19; at which point, locate point 21.

27. Indicate pointed and curved line Q, from point 20 through point 21, to point 14; as illustrated in the cut.

28. Indicate pointed line R, from point 13 through point 16, to the distance of ¼ inch farther from point 13, than the distance between point 13 and the point of intersection of lines R and N, if line N is extended toward the opposite edge, from point 18—and as illustrated in the cut; at which point, locate point 22.

29. Locate point 23 on line B, ¼ of the distance between point 2 and point 18 from point 2, toward point 18.

30. Locate point 24 on line A, ⅛ of the arm's-eye circumference from point 2, toward point 1.

31. Indicate pointed and curved line S, from point 22 through points 23, 24, and 19, to point 14; as illustrated in the cut.

Q. E. D.

N. B.—1. From the above principles, and construction lines A, B, C, and D, every conceivable and possible style of sleeve can instantly be drafted, by the thoughtful student.

2. Perfect, and elegantly-fitting sleeves, are the ornament of any garment; and especially so in ladies' and children's garments. Too much attention, therefore, cannot be bestowed in thoroughly mastering the principles of scientific measurement and drafting, as herein above taught. To be an elegant and master sleeve-cutter, is a passport into credit, favor, and honor in any first-class establishment; and generally, is a true index to the actual worth and general ability of a cutter.

5

3. Here as in body-drafting, both the measures and principles should be so thoroughly mastered and familiarized, that the whole process becomes a part of memory, as the letters of the alphabet.

First, master the alphabet; then you may form words, then phrases, then sentences, and finally become the possessor and master of the language of the Queen of Fashion—the adored empress of the world.

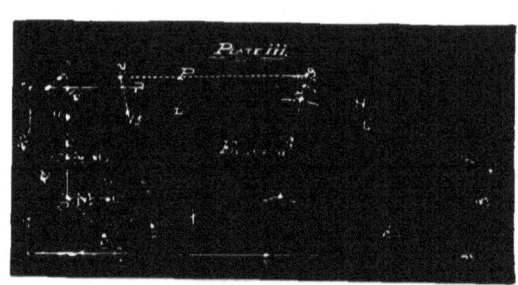

PLATE III.—FIGURE III.

INSTRUCTIONS.

1. Locate points 1, 2, 3, 4, 5, and 6, and draw lines A, B, C, and D, per instructions given under Figure II; and as you located these respective points and drew these respective lines, in Figure II.
2. Indicate pointed line E, from point 6 to point 4; as Illustrated in the cut.
3. Indicate pointed line F, at right angles with line E, from point 4 toward the opposite edge, to the distance of ½ of the hand-circumference from point 4, at which point, locate point 7; as illustrated in the cut.
4. Locate point 8 on line D, ½ inch from point 3, toward point 6.
5. Indicate pointed line G, from point 8 to point 7; as illustrated in the cut.
6. Extend line D from point 3, to the distance of ½ inch from point 3 toward the opposite edge—at which point, locate point 9.
7. Indicate pointed line H, from point 9 to point 7; as illustrated in the cut.
8. Indicate pointed line I, from point 6 toward point 1, to the distance of the inside height from point 6; at which point, locate point 10.
9. Locate point 11 on line I, ¼ of the distance between point 10 and point 6 from point 10, toward point 6.
10. Draw line J, at right angles with line I, from point 11 to the distance of ¾ of an inch more than ½ of the arm-circumference from point 11, toward the opposite edge—at which point, locate point 12.
11. Locate point 13 on line J, 1½ inches back from point 12 toward point 11.
12. Locate point 14 on line B, the distance of the outside-height from point 3, toward point 2.
13. Draw line K at right angles with line B, from point 14 toward line C, to the distance of ⅓ of the arm's-eye circumference from point 14; at which point, locate point 15.

UTILITY AND DECORATIVE ART. 67

14. Indicate pointed line L, from point 8 through point 13, to the point of intersection of lines K and L; at which point, locate point 16; as illustrated in the cut.

15. Locate point 17, on line K, equidistant from point 15 and point 16.

16. Draw line M, at right angles with line K, from point 15 toward line J, to the distance of ⅛ of the arm's-eye circumference from point 15; at which point, locate point 18.

17. Draw line N, at right angles with line K, from point 17 toward line J, to the distance of ½ of the distance between point 15 and point 18, from point 17; at which point, locate point 19.

18. Indicate pointed and curved line O, from point 16, through points 19 and 18, to point 10; as illustrated in the cut.

19. Indicate pointed line P, from point 9 through point 12, to the distance of ¼ inch farther from point 9, than the distance between point 9 and the point of intersection of lines P and K, if line K is extended; at which point locate point 20; as illustrated in the cut.

20. Locate point 21, on line B, equidistant from point 2 and point 14.

21. Locate point 22, on line A, ⅛ of the arm's-eye circumference from point 2, toward point 1.

22. Indicate pointed and curved line Q, from point 20, through points 21, 22, and 15, to point 10; as illustrated in the cut. Q. E. D.

N. B.—From the above examples of the production of directly opposite styles, by the same principles and construction lines A, B, C, and D, it will readily be seen that the possibilities in the degree of variation in the draft are commensurate with the possibilities in the degree of variation in styles; and therefore, equal to all emergencies of the cutter.

All of which applies with equal force to the system of principles and construction lines A, B, C, D, E, F, G, H, I, and K, in body-drafting, as illustrated and taught in Plate I, and under Figure I; from which, all kinds and styles of both ladies' and gentlemen's body-garments can instantly and accurately be drafted; by simply using the measures to those principles, and regulating the spacings and course of the seams, and the degree of closeness or ease of fit desired, or called for by the mandates of fashion. The entire work, in all its various departments, will be found to be based upon a system of principles, and not of rules merely; all of which have been attained by long and hard study, and a varied and extensive practice.

THE INTERNATIONAL METHOD OF FRANCO-PRUSSIAN
Scientific Tailor Principles
FOR DRAFTING ALL KINDS AND STYLES OF CHILDREN'S COSTUMES

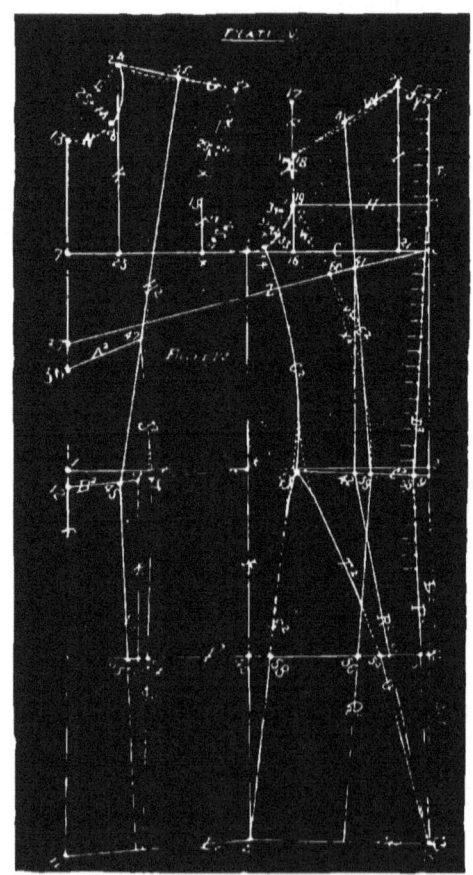

PLATE IV.—Figure IV.

MEASURES FOR PRACTICE.

		Inches.
1.	Neck-Circumference	9
2.	Skirt-Depth	12
3.	Blade-Width	12
4.	Chest-Circumference	24

UTILITY AND DECORATIVE ART. 69

	INCHES.
5. Waist-Depth	7
6. Back-Waist Depth	7¼
7. Back-Height	5
8. Front-Waist Depth	7½
9. Front-Height	3½
10. Arm's-Eye Circumference	12
11. Back-Shoulder Height	5¼
12. Front-Shoulder Height	6
13. Shoulder-Depth	4¾
14. Bust-Circumference	25
15. Waist-Circumference	20
16. Hip-Depth	5
17. Hip-Circumference	32
18. Sleeve: 1. Extreme Depth	13
2. Arm's-Eye Circumference	12
3. Elbow-Circumference	8
4. Elbow-Depth	8
5. Hand-Circumference	7
6. Inside-Height	4¾
7. Arm Circumference	9½
8. Outside-Height	6¾

CHILDREN'S GARMENT-DRAFTING.

PLATE IV.—FIGURE IV.

Instructions.

1. Locate points 1, 2, 3, 4, 5, 6, 7, 8, 9, 10, 11, 12, and 13, and draw lines A, B, C, D, and E, per instructions for locating these points and drawing these lines respectively, as given for drawing Figure I, under Plate I.
2. Locate point 14 on line C, the distance of ⅛ the arm's-eye circumference from point 5 toward point 7.
3. Draw line F, at right angles with line C, from point 14 toward the top, to the distance of ⅛ of the arm's-eye circumference from point 14; at which point, locate point 15.
4. Locate point 16 on line C, the distance of ⅛ of the arm's-eye circumference from point 5, toward point 6.
5. Draw line G, at right angles with line C, from point 16 toward the top, to the distance of ⅜ of the arm's-eye circumference from point 16; at which point, locate point 17.
6. Locate point 18 on line G, the distance of ⅓ of the arm's-eye circumference from point 16, toward point 17.
7. Locate point 19 on line G, equidistant from point 16 and point 18.
8. Draw lines H, I, J, K, L, M, N, O, P, Q, R, S, T, U, V, W, X, and Y, and locate points 20, 21, 22, 23, 24, 25, 26, 27, 28, 29, 30, 31, 32, 33, 34, and 35, per instructions for drawing these lines, and for locating these points, respectively, as given for drawing Figure I, under Plate I—only continuing curved line Y to point 33, from point 19; as illustrated in the cut.
9. Locate point 36 on line E, equidistant from point 7 and point 12.
10. Locate point 37 on line E, at a point which is the distance of ¼ of the distance between point 36 and point 7 from point 36, toward point 7.
11. Draw line Z, from point 37 to point 6.
12. Drawn line A^2, from point 36 to the point of intersection of lines B and Z.

13. Draw line B^2, from point 12 to point 3.
14. Draw line C^2, from point 3 to point 9.
15. Find the ⅛ of the difference between the chest-circumference and the waist-circumference: then locate point 38 on line C^2 that distance from point 9, toward point 3.
16. Draw line D^2, from point 20 to point 38.
17. Find ⅛ of the waist-circumference : then locate point 39 on line C^2, ½ of that distance from point 38 toward point 3.
18. Locate point 40 on line W, equidistant from point 22 and point 35.
19. Draw line E^2, from point 40 to point 39; and locate point 41, at the point of intersection of lines E^2 and Z.
20. Locate point 42 on line C^2, the same distance from point 39 toward point 3, as the distance between point 9 and point 38.
21. Draw line F^2, from point 41 to point 42.
22. Locate point 43 on line C^2, ½ inch further from point 42, toward point 3, than the distance between point 38 and point 39.
23. Locate point 44 on curved line X, equidistant from point 33 and point 5.
24. Draw curved line G^2 from point 44 to point 43, per instructions given under Plate I, for drawing lines G^2, H^2, I^2, J^2, and K^2 in Figure I—only changing the length of tape-measure true to the waist-depth given, and changing positions true to, and drawing from point 44 to point 43; as illustrated in the cut.
25. Locate point 45 on line B^2, the same distance from point 12 toward point 3, as the distance between point 42 and point 43.
26. Locate point 46 on line O, equidistant from point 24 and point 27.
27. Draw line H^2, from point 46 to point 45; and locate point 47, at the point of intersection of lines H^2 and A^2.
28. Locate point 48 on line B, the distance of the hip-depth from point 3, toward point 4.
29. Draw line I^2, at right angles with and from line E, through point 48, to line D; and locate point 49 at the point of intersection of lines I^2 and D.
30. Locate point 50 on line B^2, the distance of the ¼ of the difference between the chest-circumference and the waist-circumference from point 45, toward point 3.
31. Draw line J^2, from point 47 to point 50.
32. Locate point 51 on line B^2, equidistant from point 45 and point 50.
33. Draw line K^2, at right angles with line A, from point 51 to line I^2; and locate point 52, at the point of intersection of lines K^2 and I^2.
34. Draw line L^2, from point 11 to point 4.
35. Draw line M^2, from point 4 to point 8.
36. Locate point 53 on line I^2, ½ the distance between point 45 and point 51 from point 52, toward line E.
37. Draw line N^2, from point 45 through point 53, to line L^2.
38. Locate point 54 on line I^2, the same distance from point 52, toward point 48, as the distance between point 52 and point 53.
39. Draw line O^2, from point 50 through point 54, to line L^2.
40. Locate point 55 on line I^2, the same distance from point 49, toward point 48, as the distance between point 52 and point 54.
41. Draw line P^2, from point 38 through point 55, to line M^2.
42. Find ⅛ of the hip-circumference: then locate point 56 on line I^2, ½ of that distance from point 55, toward point 48.
43. Draw line Q^2, from point 39 through point 56, to line M^2.
44. Locate point 57 on line I^2, at a point equidistant from point 55 and point 56.
45. Draw line R^2, from point 42 through point 57, to line M^2.
46. Locate point 58 on line I^2, twice the distance between point 55 and point 56 from point 57, toward point 48.
47. Draw curved line S^2, from point 43 through point 58, to point 4; as illustrated in the cut.

48. Find the total sum of the distances between the point of intersection of lines E and I², and point 53; plus the distance between point 55 and point 56; plus the distance between point 57 and point 58: then subtract this sum of distances from ½ of the hip-circumference; locate point 59, on line I², the distance of that difference from point 54, toward point 49.

49. Draw curved line T², from point 43 to point 59; as illustrated in the cut.

50. Draw line U², from point 59 to point 8.

51. Indicate pointed and curved lines V², W², and X², per instructions given for drawing curved lines G², H², and I², in Figure I, under Plate I—only changing the length of tape-measure, true to the waist-depth given, and changing positions true to, and indicating from point 19 to point 39; and from point 34, to point 60; and from point 60, to point 42; as illustrated in the cut.

Q. E. D.

N. B.—1. All possible kinds and styles of children's dresses and costumes can easily and instantly be traced from the above draft, by the thoughtful student; by simply arranging the seams and spacings, to the desired style and degree of closeness or neatness of fit.

2. The short horizontal lines, intersecting with lines D, D², and P², the first of which (at the top) is marked Y², show the position for the buttonholes and buttons; and lines Z², A³, B³, and C³, show the extension for finishing, outside of the fit; when the dress is made to open at the back. The amount depends, upon the method of finish.

3. The pointed lines, along the shoulder-lines and across the angles, show the proper curves to be given, when tracing the various parts of the draft, separately, upon the lining or material.

4. Both dress and coat-sleeves for children, are drafted per instructions for drafting ladies' sleeves; and as given under Plate III, for drafting Figure III.

5. When the dress is opened at the back, line E is placed on the fold of the lining and the material; when open at the front, the buttonholes are cut from within to line E, and the buttons on the other side, are set upon line E; and the same extension is allowed in front for finish, which is indicated at the back; as illustrated in the cut.

THE INTERNATIONAL METHOD
OF
SCIENTIFIC TAILOR PRINCIPLES
FOR ALL KINDS AND STYLES OF

Boys' Coat, Vest, and Pantaloon-Drafting.

:o:

INSTRUCTIONS.

1. For boys' coats, vests, and overcoats, use the same method of measurement and drafting as given for drafting Figure IV, under plate IV; excepting the number and location of the seams, at the waist and hip lines; and also, the degree of closeness desired in the garment: which, are at your option, and according to will, taste, style, and fashion. The fit about the neck, shoulders, arm's-eye, and all general principles elsewhere, being the same as in Figure IV; only that it is left almost plain and nearly straight in form, from the construction line C, down to the full length; and is shaped at the back and front, at will; as given above.

2. The double-breast, and the collars, are made by the instructons for the same, under that special part of the work ; given elsewhere.

3. The sleeves are drafted per instructions for drafting Figure III, under plate III.

Boys' Pantaloons and Knee-Pants.

THE MEASURES.

Instructions for Measuring.

1. By means of pins, chalk, or a crayon-pencil, locate the following points upon the garment worn, at each side of the person; divesting the form of the coat and vest, before measureing, viz :
   a. At the height you wish the top of the pants ;
   b. At the most prominent bone, at the hips ;
   c. At the side of the leg, at a point even with, and on a horizontal line with the fullest point of the front knee-cap ;
   d. At the length desired below the knee, at the side.

2. Measure as follows, viz :
   a. The hip-depth ; from first point, to point second.
   b. The knee-depth ; from point second, to point third ;
   c. Outside-depth ; from point second to point fourth ;
   d. Inside-height ; from the point on the inside of the leg, opposite from, and on a horizontal line with point fourth, up along the inside of the leg, close up to the inside junction of the leg with the body.
   e. Waist-circumference ; around the form, on a horizontal line with point first.
   f. Hip-circumference; around the form, on a horizontal line even with point second.
   g. Thigh-circumference ; around the thigh, on a horizontal line even with the inside junction of the leg with the body.
   h. Leg-circumference ; around the leg, on a horizontal line with point fourth.
   i. The degree of closeness and length, is entirely at your option ; but, the measures should all be taken with great care and exceeding accuracy ; as everything depends upon the measures here, as elsewhere.

N. B.—In very close work, it is necessary also, to have the knee-circumference.

MEASURES FOR PRACTICE.

		Inches.
1.	Hip-Depth	6
2.	Knee-Depth	12
3.	Outside-Depth	22
4.	Inside-Height	19
5.	Waist-Circumference	20
6.	Hip-Circumference	23
7.	Thigh-Circumference	16
8.	Leg-Circumference	11

Boys' Pantaloon-Drafting.

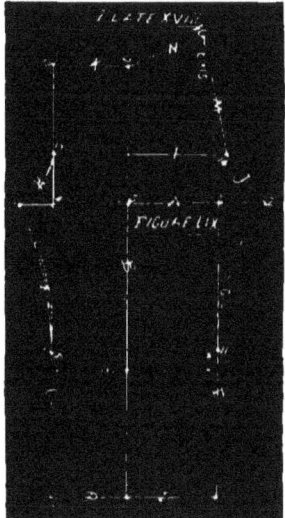

PLATE XVIII.—Figure LIX.

INSTRUCTIONS.

1. Add one inch to the inside-height; locate point 1, on the paper or material, that distance from the bottom, and one inch back from the front edge of the same.
2. Draw line A, at right angles with the front edge, from point 1 toward the opposite edge, to the distance of the thigh-circumference; at which point, locate point 2.
3. Locate point 3 on line A, the distance of $\frac{1}{6}$ of the thigh-circumference from point 2, toward point 1.
4. Locate point 4 on line A, the distance of $\frac{1}{8}$ of the thigh-circumference from point 1, toward point 3.
5. Locate point 5 on line A, at a point which is $\frac{1}{2}$ inch further from point 2 toward point 1, than $\frac{1}{2}$ of the thigh-circumference.
6. Draw line B, at right angles with line A, from point 5 toward the bottom, to the distance of the inside-height; at which point, locate point 6.
7. Extend line B, in the opposite direction, to the distance of the sum of the hip-depth, plus the outside-depth, from point 6; at which point, locate point 7.
8. Draw line C, at right angles with line A, from point 4 toward the bottom, to the distance of the inside-height from point 4; at which point, locate point 8.
9. Draw line D, at right angles with line C, and through point 6, to the distance of the leg-circumference, from point 8; at which point, locate point 9.
10. Draw line E, from point 3 to point 9.

11. Locate point 10 on line C, the distance of ⅔ of the knee-depth from point 4, towards point 8.

12. Draw curved line F, from point 1 to point 10; as illustrated in the cut.

13. Locate point 11 on line E, at a point which is ⅔ of the knee-depth from point 3, towards point 9.

14. Draw curved line G, from point 2 to point 11; as illustrated in the cut.

15. Extend line C, towards the top, to the distance of the sum of the hip-depth, plus the outside-depth, from point 8; at which point, locate point 12.

16. Locate point 13 on line C, at a point which is the distance of the hip-depth, from point 12, towards point 4.

17. Draw curved line H, from point 13 to point 1; as illustrated in the cut.

18. Draw line I, at right angles with line C, from point 13 toward the opposite edge, to the distance of ½ of the hip-circumference from point 13; at which point, locate point 14.

19. Draw curved line J, from point 14 to point 2; as illustrated in the cut.

20. Draw line K, from point 12 to point 7.

21. Extend line K, by indication, from point 7 toward the opposite side, to the distance of ½ of the waist-circumference from point 12; at which point, locate point 15; as illustrated in the cut.

22. Indicate pointed line L, at right angles with line K, from point 15 toward the top, to the distance of ⅓ of the hip-depth from point 15; at which point, locate point 16; as illustrated in the cut.

23. Draw line M, from point 16 to point 14; as illustrated in the cut.

24. Draw line N, from point 7 to point 16; which process, completes the draft for boys' pantaloons. The front of which, is formed and included within lines C, H, F, D, B, and K; and the back, within lines B, D, E, G, J, M, and N; as illustrated in the cut.

Q. E. D.

N. B.—1. The pointed lines O, P, and Q, show the knee-pants; and can be governed, as to their degree of closeness and length, by the will and choice of the cutter.

2. Line B, shows the seam at the side. The patterns must be traced down, or cut out, and all seams allowed everywhere; otherwise, one inch can be allowed between the back and front; and line B, can be repeated; and the allowances made in drafting, directly on the cloth.

3. Of course, we need hardly explain, yet lest some might possibly not comprehend the real import of the draft, we will merely add that it simply represents one side or part of the pantaloons, *i. e.*, the back and front of one leg; the parts necessarily having to be duplicated in the linings and goods. That there will necessarily have to be a duplicate of the front part, included within the lines C, H, F, D, B, and K, and also, of the back part, included within the lines B, D, E, G, J, M, and N; as illustrated in the cut.

4. The style and lengths, whether knee-pants are desired or otherwise long pants, are entirely at the will and option of the cutter. The same principles and the same draft being used, in each and all cases.

5. By using all the principles of measurement and general drafting, at the neck, shoulder, arm's-eye, sleeve, chest, bust, etc., and simply drafting straight down from the chest line to the desired length, and regulating the degree of ease or closeness of fit, as well as the seams at will, all kinds and styles of either double or single-breasted body-garments may be cut, for either boys or girls, from Figure IV, under Plate IV.

6. In the display of good taste, as to method of design, trimming, etc., children's garments call forth the highest degree of accomplishment, both in the artist, in finishing, and the scientist, in cutting.

UTILITY AND DECORATIVE ART. 75

Ladies' and Children's Under-Garments.

1. For the body part, from the close measures, per instructions for the general draft, as given in Figure I, for ladies, and Figure IV, for children; excepting the seams, and the degree of looseness; which are designed from the fit, afterwards, and at will; and the close fit is cut away at the neck and arm's-eye, to the desired depth of your design; on the same principle as given elsewhere in the work, for "low bodices" and "ball dresses."

2. For the lower part of the body, any designs for both ladies and children can instantly be cut from Figure LIX, on Plate XVIII.

3. There is not a design in any Fashion journal, but that can instantly be reproduced upon the paper or goods, from the actual measures; and according to the plates, as given above.

4. The artistic finish, of course, is entirely at will; and depends upon the costliness of the display desired in trimming, and the display of good taste in finish, etc. Comfort and utility, however, should be the chief considerations.

Gentlemen's and Boys' Shirt-Cutting.

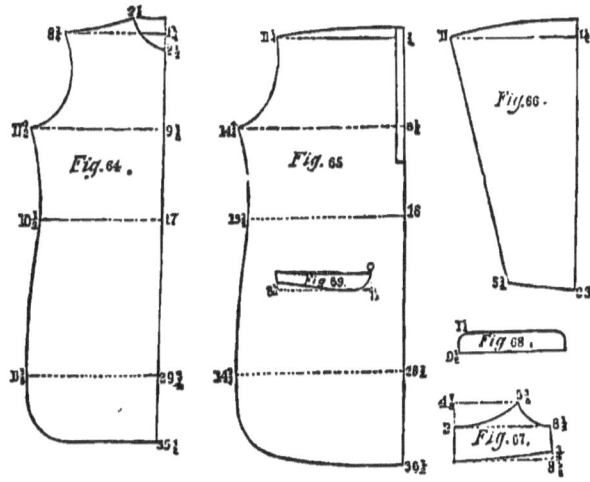

PLATE XXXIV.—FIGURES LXIV, LXV, LXVI, LXVII, LXVIII, LXIX.

EXPLANATIONS AND INSTRUCTIONS
OF
Plate XXXIV.—Figures 64, 65, 66, 67, 68 and 69.

The measures and methods of drafting are the same as for general body-drafting. The fit is afterwards cut to the desired shapes and degrees of ease, at will, or according to fashion; at each respective given point. We have, in the above Plate, of which Figures 64, 65, 66, 67, 68, and 69, represent respectively the front, back, sleeve, yoke, wrist-band, and neck-band, to a neck 15¼, sleeve 24½, chest-circumference 36, waist-circumference 33, etc., given, what would be the general method of measurement, or of cutting the close fit; all of which will be comprehended by the mere novice, at a glance; the scale being based upon the mathematical unit of measurement, or the mathematical inch. It is cut whole at the front, opening at the back. Place the points of yoke 5⅜ and 2, to shoulder at 2⅞ and 8¾ respectively; and also from ¾ and O, to O and 11¼ of the back shoulder; fulling and gathering the surplus of the same neatly thereon. The sleeve at O, joins to back at 11¼. The point of neck-band at O, joins to 8¼ on yoke. The rest will be comprehended at a glance.

N. B.—1. All possible kinds and styles of shirts can readily and accurately be drafted from the general construction lines of Plate I, by applying said principles; as given above.

2. It is best to draft an accurate pattern, and then lay this upon the muslin and trace or mark around the edges of the same; and allow all seams, extensions, finish, etc., outside of the actual fit.

THE
INTERNATIONAL METHOD

—OF—

ENGLISH AND AMERICAN

SCIENTIFIC TAILOR PRINCIPLES,

—FOR ALL—

KINDS AND STYLES

—OF—

BODY-DRAFTING,

BY TRANSFERRING THE MEASURES OR INDICES
OF THE MATHEMATICAL MOULD OR PER-
FECT FIT OF THE HUMAN FORM;

DIRECTLY UPON THE GOODS,

—WITH ANY—

SQUARE OR INCH-RULE

—AND—

TAPE-MEASURE ALONE.

THE FRONT.

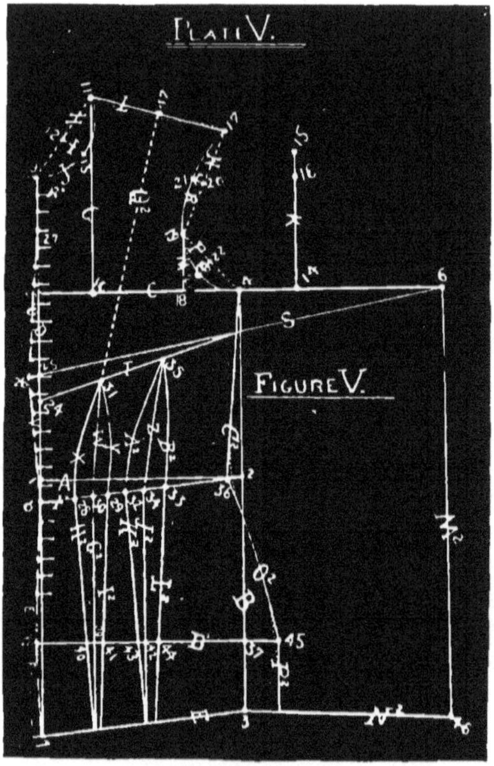

PLATE V.—FIGURE V.

INSTRUCTIONS.

1. Add 1½ inches to the skirt-depth. Locate point 1 on the goods, at a point which is that distance from the bottom, or right-hand edge, and 1½ inches back from the front edge, toward the opposite edge of the goods.

2. Find the difference between ½ of the chest-circumference and ½ of the blade-width. Draw line A, at right angles with the front edge, from point 1 toward the opposite edge, to the distance of that difference from point 1; at which point, locate point 2.

3. Draw line B, at right angles with line A, from point 2 toward the bottom, to the distance of the skirt-depth from point 2; at which point, locate point 3.

4. Extend line B, in the opposite direction, to the distance of the waist-depth from point 2; at which point, locate point 4.

5. Draw line C, at right angles with line B, from point 4 toward the front, to the distance between point 1 and point 2; at which point, locate point 5.

UTILITY AND DECORATIVE ART. 79

6. Extend line C, in the opposite direction, to the distance of ½ of the blade-width from point 4; at which point, locate point 6.

7. Find the sum of the front waist-depth, plus the skirt-depth. Draw line D, at right angles with line C, from point 5 toward the bottom, and through point 1, to the distance of that sum of depths; at which point, locate point 7.

8. Locate point 8 on line D, the distance of the front waist-depth from point 5, toward point 7.

9. Extend line D, toward the top, to the distance of the front-height from point 5; at which point, locate point 9.

10. Draw line E, from point 7 to point 3.

11. Draw line F, from point 8 to point 2.

12. Find ¼ of the neck-circumference; locate point 10 on line C, ½ inch less than that distance from point 5, toward point 4.

13. Draw line G, at right angles with line C, from point 10 toward the top, to the distance of the front shoulder-height from point 10; at which point, locate point 11.

14. Indicate pointed line H, from point 11 to point 9; as illustrated in the cut.

15. Locate point 12 on line H, at a point equidistant from point 11 and point 9.

16. Indicate pointed line I, at right angles with line H, from point 12 toward the opposite edge, to the distance of one-twelfth of the neck circumference from point 12; at which point, locate point 13; as illustrated in the cut.

17. Draw curved line J, from point 11 through point 13, to point 9; as illustrated in the cut.

18. Locate point 14 on line C, ⅙ of arm's-eye circumference from point 4, toward point 6.

19. Draw line K, at right angles with line C, from point 14 toward the top, to the distance of two-fifths of the arm's-eye circumference from point 14; at which point, locate point 15.

20. Locate point 16 on line K, at a point which is the distance of ⅓ of the arm's-eye circumference from point 14, toward point 15.

21. Draw line L, from point 11 toward point 15, to the distance of ¾ of an inch less than the shoulder-depth from point 11; at which point, locate point 17.

22. Locate point 18 on line C, at a point which is ⅙ of the arm's-eye circumference from point 4, toward point 5.

23. Draw M, at right angles with line C, from point 18 toward the top, to the distance of ⅙ of the arm's-eye circumference from point 18; at which point, locate point 19.

24. Indicate pointed line N, from point 17 to point 19; as illustrated in the cut.

25. Locate point 20 on line N, at a point equidistant from point 17 and point 19.

26. Indicate pointed line O, at right angles with line N, from point 20 toward line G, to the distance of ¼ of the distance between point 15 and point 16; at which point, locate point 21; as illustrated in the cut.

27. Indicate pointed line P, from point 19 to point 4; as illustrated in the cut.

28. Locate point 22 on the line P, at a point equidistant from point 19 and point 4.

29. Indicate pointed line Q, from point 22 toward point 18, to the distance of ½ of the distance between point 22 and point 18, from point 22; at which point, locate point 23; as illustrated in the cut.

30. Draw curved line R, from point 17 through points 21, 19, and 23, to point 4; as illustrated in the cut.

31. Locate point 24 on line D, at a point equidistant from point 5 and point 8.

32. Locate point 25 on line D, at a point which is ¼ of the distance between point 24 and point 5 from point 24, toward point 5,

33. Draw line S, from point 6 through point 25, to the distance of ½ inch more than ½ of the bust circumference from point 6; at which point, locate point 26.

34. Draw line T, from point 24 to the point of intersection of lines S and B.

35. Locate point 27 on line D, at a point equidistant from point 9 and point 5.

36. Draw line U, from point 27 to point 26.

37. Draw line V, from point 26 to point 8.

38. Find ⅛ of the waist-circumference; locate point 28 on line F, at a point which is ½ of that distance from point 8, toward point 2.
39. Find ⅛ of the difference between the chest-circumference and the waist-circumference. Locate point 29 on line F, at a point which is that distance from point 28, toward point 2.
40. Locate point 30 on line F, at a point which is equidistant from point 28 and point 29.
41. Locate point 31 on line T, at a point which is ⅜ of inch farther from point 24 toward line B, than the distance between point 8 and point 30.
42. Draw line W, from point 31 to point 30.
43. Draw curved line X, from point 31 to point 28; as illustrated in the cut.
44. Draw line Y, from point 31 to point 29, as illustrated in the cut.
45. Locate point 32 on line F, at a point which is ½ as far from point 29, toward point 2, as the distance between point 8 and point 28.
46. Locate point 33 on line F, at a point which is as far from point 32, toward point 2, as the distance between point 28 and point 29.
47. Locate point 34 on line F, at a point which is equidistant from point 32 and point 33.
48. Locate point 35 on line T, at a point which is as far from point 31, toward line B, as the distance between point 24 and point 31.
49. Draw line Z, from point 35 to point 34.
50. Draw curved line A^2, from point 35 to point 32; as illustrated in the cut.
51. Draw curved line B^2, from point 35 to point 33; as illustrated in the cut.
52. Locate point 36 on line F, at a point which is as far from point 2 toward point 33, as the distance between point 34 and point 33.
53. Draw line C^2, from point 4 to point 36.
54. Locate point 37 on line B, the distance of the hip-depth from point 2, toward point 3.
55. Draw line D^2, at right angles with line B, from point 37 toward the front edge, to a point in front of the point of intersection of lines D^2 and D, which is ½ as far from the point of intersection of lines D^2 and D, as the distance between point 25 and point 26; at which point, locate point 38.
56. Draw line E^2, from point 8 to point 38.
57. Draw line F^2, from point 38 to point 7.
58. Draw line G^2, at right angles with line A, from point 30 to line E; and locate point 39 at the point of intersection of lines G^2 and D^2.
59. Locate point 40 on line D^2, ½ as far from point 39 toward point 38 as the distance between point 28 and point 30.
60. Draw line H^2, from point 28 through point 40 to line E.
61. Locate point 41 on line D^2, the same distance from point 39 toward point 37, as the distance between point 39 and point 40.
62. Draw line I^2, from point 29 through point 41, to line E.
63. Draw line J^2, at right angles with line A, from point 34 to line E; and locate point 42 at the point of intersection of lines J^2 and D^2.
64. Locate point 43 on line D^2, the same distance from point 42 toward point 41, as the distance between point 39 and point 41.
65. Draw line K^2, from point 32 through point 43, to line E.
66. Locate point 44 on line D^2, the same distance from point 42 toward point 37, as the distance between point 42 and point 43.
67. Draw line L^2, from point 33 through point 44, to line E.
68. Extend line D^2, from point 37 toward the opposite edge, to a point as far from point 37 as twice the distance between point 43 and point 44; at which point, locate point 45.
69. Find the sum of the back-waist depth, plus the skirt-depth. Draw line M^2, at right angles with line C, from point 6 toward the bottom, to the distance of that sum of depths; at which point, locate point 46.
70. Draw line N^2, from point 3 to point 46.

71. Draw curved line O^2, from point 36, to point 45; as illustrated in the cut.
72. Draw line P^2, at right angles with line D^2, from point 45 to line N^2.
73. Locate point 47 on line L, equidistant from point 11 and point 17.
74. Indicate pointed line Q^2, from point 47 to point 31; as illustrated in the cut.
75. Locate the buttonhole lines, the first of which, at the top, is marked R^2; commencing at point 8, and spacing up and down from that point, at the desired distances. Then draw lines R^2, at right angles with the front edge of the goods, from lines D, U, V, E^2, and F^2, to the desired length of the buttonholes; as indicated in the cut.
Q. E. D.

N. B.—1. If only one dart is desired, locate the centre point on line F, equidistant from points 29 and 32; the centre point, on line T, at the top, equidistant from points 31 and 35; and enclose in the dart on line F, the distance between points 30 and 34. Now find the sum of the distances between points 28 and 30, and 34 and 33; subtract from this sum, the distances between points 29 and 32: Then take out ½ of this last difference in the front curve, on line F, from point 8 toward point 28; and ¼ of this same difference at the opposite side of the draft, from point 36 toward point 33; and the other ¼, or an equal amount, more than the general rule for drafting calls for, from the side front on the seam that joins lines C^2 and O^2; and complete the dart below line F, by the same, rules and principles, according to the ratio of the given spaces; as given for drafting for two darts.

2. For a half-fitting garment, allow ½ of the difference between the chest-circumference and the waist-circumference, for looseness on the waist lines; and draft accordingly, by the general rules, as given above. In the same manner and by the same principles, it is easy to draft a ⅔-fitting, or a ¾-fitting garment; either by making new drafts, or indicating the changes on the close-fitting draft, to the degree of closeness or ease of fit desired. This principle of calculating the degrees of fit from the perfect close fit, or the mathematical mould or contour, is the chief corner-stone in the temple of correct science, in the art of garment-cutting. The degree of tightness or looseness of a garment, is quite a different thing from the fit, which may be very bad in a very tight-fitting garment, and very good in a very loose-fitting garment, or *vice-versa*.

3. Every conceivable and possible style, in the location and course of seams or spacings, can quickly and easily be drafted, by applying the same principles in the same method of ratios, to the desired style. As in the science of the mathematics of surveying, the principles of the science of garment-cutting never change. The true science of garment-cutting, therefore, is founded upon the mathematical mould, contour, or dimensions of each particular form, for which the garment is intended; independent of seams, and the particular fashions of any age; whether that term expresses either an absurd incongruity, or the combined beauty of elegance and grace. Where mathematics is king, the cutter is a prince, or a princess; and fashion is cheerfully submissive, and is willingly "clothed in royal purple" by the same hand that embellishes architecture, and brings harmony and music out of discord. The spacings or particular location of the seams, as given in all the draftings of this work, are not intended as necessarily fixed, therefore, as they are located. But they were chosen because they represent scientific, mathematical, and symmetrical proportions, true to the measures given; and will from their very nature, therefore, always remain the standard method of spacing.

4. The pointed curved line at the shoulder, represents the proper method of indicating the curve for basting and stitching; the same is true of the indicated lines across the angles.

5. Pointed line Q^2, from point 47 to point 31, is used for cutting vest-fronts, etc.; and for garments with a seam to the shoulders in front.

6. The pointed lines in front of the draft, show the extensions for finish.

7. For diagonal and double-breasted garments, locate point 1, the width of the desired extra amount further back from the front, toward the opposite edge of the goods; and allow the desired amount and shape or style of the extension, in front of the draft. All of which, applies also with equal force to Plate I.

The Side-Front or "Under-Arm Gore."

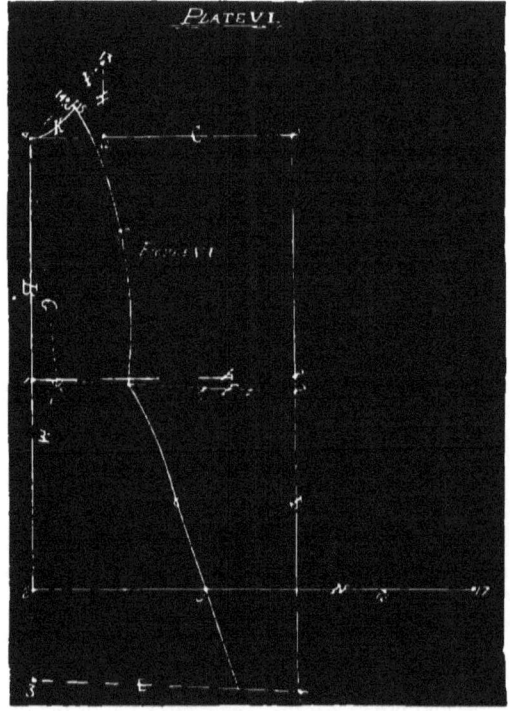

PLATE VI.—FIGURE VI.

INSTRUCTIONS.

1. Add 1½ inches to the skirt-depth; locate point 1 on the material, that distance from the bottom, or right hand edge, and 1½ inches back from the front edge, toward the opposite edge of the goods.

2. Draw line A, at right angles with the front edge, from point 1 toward the opposite edge, to the distance of ½ of the blade-width from point 1; at which point, locate point 2.

3. Draw line B, at right angles with line A, from point 1 toward the bottom, to the distance of the skirt-depth from point 1; at which point, locate point 3.

4. Extend line B, in the opposite direction from point 1, to the distance of the waist-depth from point 1; at which point, locate point 4.

UTILITY AND DECORATIVE ART. 83

5. Draw line C, at right angles with line B, from point 4 toward the opposite edge, to the distance of ½ of the blade-width from point 4 ; at which point, locate point 5.

6. Find the sum of the back-waist depth, plus the skirt-depth ; draw line D at right angles with line C, from point 5 toward the bottom, to the distance of the sum of these depths from point 5; at which point, locate point 6.

7. Draw line E, from point 3 to point 6.

8. Locate point 7 on line D, the distance of the back waist-depth from point 5, toward point 6.

9. Draw line F, from point 1 to point 7.

10. Find ⅛ of the difference between the chest-circumference and the waist-circumference ; locate point 8 on line F, that distance from point 7, toward point 1.

11. Find ⅛ of the waist-circumference ; Locate point 9 on line F, ½ of that distance from point 8, toward point 1.

12. Locate point 10 on line F, ½ of the distance between point 7 and point 8 from point 1, toward point 9.

13. Draw line G, from point 4 to point 10.

14. Locate point 11 on line F, at a point equidistant from point 9 and point 10.

15. Locate point 12 on line C, at a point which is one-sixth of the arm's-eye circumference from point 4, toward point 5.

16. Draw line H, at right angles with line C, from point 12 toward the top, to the distance of one-sixth of the arm's-eye circumference from point 12 ; at which point, locate point 13.

17. Indicate pointed line I, from point 13 to point 4 ; as illustrated in the cut.

18. Locate point 14 on line I, at a point that is equidistant from point 13 and point 4.

19. Indicate pointed line J, from point 14 toward point 12, to the distance of one-seventh of the distance between point 14 and point 12 from point 14 ; at which point, locate point 15; as illustrated in the cut.

20. Draw curved line K, from point 15 to point 4 ; as illustrated in the cut.

21. Wrap one end of the tape-measure neatly and firmly around the pencil, near the sharpened end, and hold it securely with the front two fingers and thumb of the right-hand ; measure twice the waist-depth back from the pencil, on the tape-measure, either by the inches on the tape-measure or the square, at which point, hold it securely between the forepart of the thumb and forefinger of the left-hand ; then, placing and holding the point of the pencil firmly on point 15, stretch the tape-measure, and with the forepart and nail of the left-hand thumb or finger placed firmly on that length, press it upon the table, in front of the draft, opposite point 1 : then by repeated trials, find a point for the position of the left-hand, where, if the left-hand position is maintained, and the right-hand moved and guided toward the waist line, by the stretched tape-measure, the point of the pencil will exactly touch upon point 15, when moved back ; then draw curved line L, from point 15 to point 11; as illustrated in the cut.

22. Locate point 16 on line B, the distance of the hip-depth from point 1, toward point 3.

23. Draw curved line M, from point 10 to point 16 ; as illustrated in the cut.

24. Find the difference between ½ of the chest-circumference and ½ of the blade-width. Subtract this difference from ½ of the hip-circumference; then draw line N, from point 16 toward the opposite edge, to the distance of this last difference from point 16 ; at which point, locate point 17.

25. Find one-eighth of the hip-circumference ; locate point 18 on line N, ½ of that distance from point 17, toward point 16.

26. Locate point 19 on line N, at a point equidistant from point 16 and point 18.

27. Draw curved line O, from point 11 through point 19, to line E.

Q. E. D.

N. B.—In very large forms, it is best to cut the garment with two under-arm gores. By a mere glance at the compound draft, it will readily be seen that by spacing the distance between points 31 and 33, on curved line X, into 3 equal parts instead of two, and then drawing a new line B, and changing the position of the original line B, and proceeding by the same principles for the two pieces, and the space they include, as you did for the same space, if but one gore was desired, that it will become an easy matter for the careful student to eliminate them, from the compound draft; and to draft each, independently from the other, directly upon the goods. But this is only advisable or proper, in very large forms.

2. The pointed or indicated lines across the angles, in the above cut, represent the proper curves to be used in basting and stitching.

THE BACK.

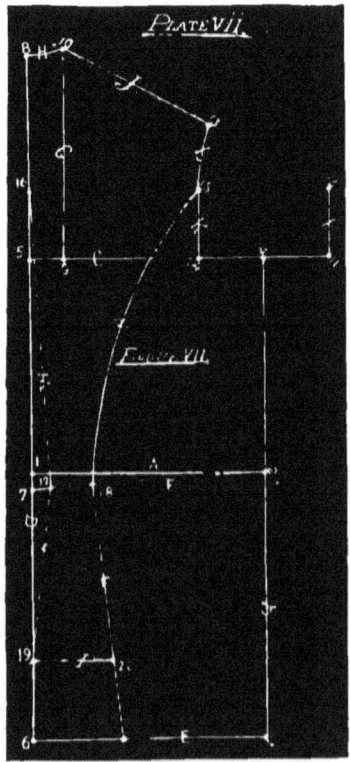

PLATE VII.—FIGURE VII.

INSTRUCTIONS.

1. Add 1½ inches to the skirt depth; locate point 1 on the material, that distance from the bottom, or right-hand edge, and 1½ inches from the front edge, toward the opposite edge.

UTILITY AND DECORATIVE ART. 85

2. Draw line A, at right angles with the front edge, from point 1 toward the opposite edge, to the distance of ½ of the blade-width from point 2; at which point, locate point 2.

3. Draw line B, at right angles with line A, from point 2 toward the bottom, to the distance of the skirt-depth from point 2; at which point, locate point 3.

4. Extend line B, in the opposite direction, to the distance of the waist-depth from point 2; at which point, locate point 4.

5. Draw line C, at right angles with line B, from point 4 toward the front edge, to the distance of ½ of the blade-width from point 4; at which point, locate point 5.

6. Find the sum of the back waist-depth, plus the skirt-depth; draw line D, at right angles with line C, from point 5 toward the bottom, to the distance of the sum of these depths from point 5; at which point, locate point 6.

7. Draw line E, from point 6 to point 3.

8. Locate point 7 on line D, the distance of the back waist-depth from point 5, toward point 6.

9. Draw line F, from point 7 to point 2.

10. Extend line D, from point 5 toward the top, to the distance of the back-height from point 5; at which point, locate point 8.

11. Find ⅛ of the neck-circumference; locate point 9 on line C, ¼ inch less than that distance from point 5, toward point 4.

12. Draw line G, at right angles with line C, from point 9 toward the top, to the distance of the back shoulder-height from point 9; at which point, locate point 10.

13. Extend line C, from point 4 toward the opposite edge, to the distance of ⅙ of the arm's-eye circumference from point 4; at which point, locate point 11.

14. Draw line I, at right angles with line C, from point 11 toward the top, to the distance of ⅙ of the arm's-eye circumference from point 11; at which point, locate point 12.

15. Draw line J, from point 10 toward point 12, to the distance of the shoulder-depth from point 10; at which point, locate point 13.

16. Locate point 14 on line C, the distance of one-sixth of the arm's-eye circumference from point 4, toward point 9.

17. Draw line K, at right angles with line C, from point 14 toward the top, to the distance of ⅙ of the arm's-eye circumference from point 14; at which point, locate point 15.

18. Draw curved line L, from point 13 to point 15; as illustrated in the cut.

19. Locate point 16 on line D, the distance of ⅙ of the arm's-eye circumference from point 5, toward point 8.

20. Find ⅛ of the difference between the chest-circumference and the waist-circumference; locate point 17 on line F, at a point which is ½ of that distance from point 7, toward point 2.

21. Find ⅛ of the waist-circumference; locate point 18 on line F, at a point which is ½ of that distance from point 17, toward point 2.

22. Draw curved line N, from point 15 to point 18, by the same length of tape-measure, and principles and rule, as given under Plate VI., for drawing line L, in Figure VI; only locating the fixed position for pressing and holding the tape-measure on the table, on the opposite side of the draft, opposite point 2; and arranging the positions of both hands, and the stretched tape-measure and point of pencil, true to, and drawing from point 15 to point 18; as illustrated in the cut.

23. Locate point 19 on line D, the distance of the hip-depth from point 1 toward point 6.

24. Draw line O, from point 17 to point 19.

25. Find ⅛ of the hip-circumference; draw line P at right angles with line D, from point 19 toward the opposite edge, to a point which is ½ of that distance from point 19; at which point, locate point 20.

26. Draw line Q, from point 18 through point 20, to line E.

Q. E. D.

N. B.—1. All seams everywhere, and extensions for finishing, must be allowed, outside of the lines representing the fit.

2. When extension for plaiting or drapery, is desired in the back, locate point 1 the desired extra distance farther, from the front edge or fold of the material, and draft per instructions as given above; and when drafted, indicate the necessary or desired extension, from line O, in the opposite direction.

3. The indicated or pointed line, along line J, or the shoulder, represents the proper curve to be used in basting, and stitching.

The Side-Back or "Side-Body."

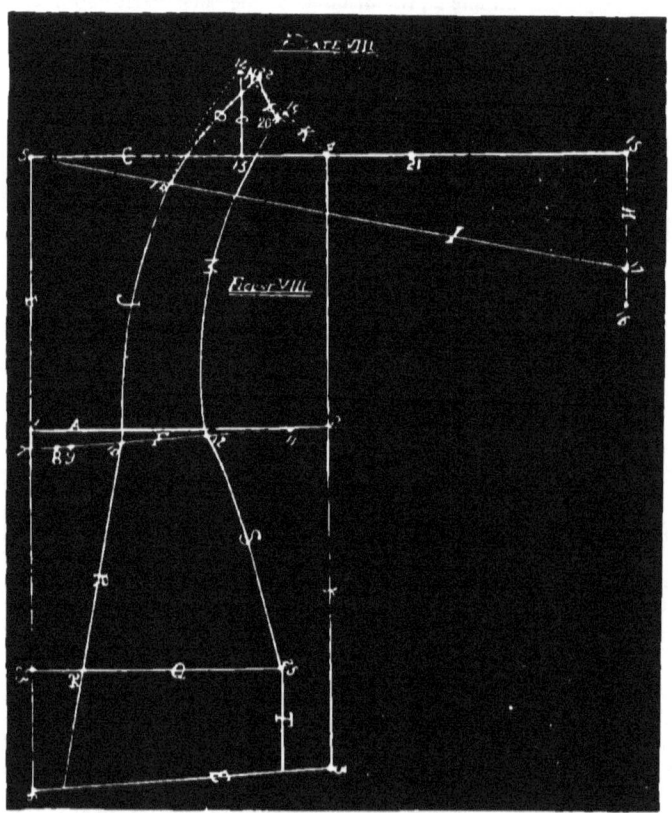

PLATE VIII.—FIGURE VIII.

INSTRUCTIONS.

1. Add 1½ inches to the skirt depth; locate point 1 on the material, that distance from the bottom or right-hand edge, and ¾ of an inch back from the front, toward the opposite edge.

UTILITY AND DECORATIVE ART.   87

2. Draw line A, at right angles with the front edge, from point 1 toward the opposite edge, to the distance of ½ of the blade-width from point 1 ; at which point, locate point 2.

3. Draw line B, at right angles with line A, from point 2 toward the bottom, to the distance of the skirt-depth from point 2 ; at which point, locate point 3.

4. Extend line B, in the opposite direction, to the distance of the waist-depth from point 2 ; at which point, locate point 4.

5. Draw line C, at right angles with line B, from point 4 toward the front, to the distance of ½ of the blade-width from point 4 ; at which point, locate point 5.

6. Find the sum of the back waist-depth, plus the skirt-depth ; draw line D, at right angles with line C, from point 5 through point 1, toward the bottom, to the distance of that sum of depths from point 5 ; at which point, locate point 6.

7. Draw line E, from point 6 to point 3.

8. Locate point 7 on line D, at a point which is the distance of the back-waist depth from point 5, toward point 6.

9. Draw line F, from point 7 to point 2.

10. Find ⅛ of the difference between the chest-circumference and the waist-circumference ; locate point 8 on line F, at a point which is ½ of that distance from point 7, toward point 2.

11. Locate point 9 on line F, ½ of the distance between point 7 and point 8 from point 8, toward point 2.

12. Find ⅛ of the waist-circumference ; locate point 10 on line F, ½ of that distance from point 9 toward point 2.

13. Locate point 11 on line F, the same distance from point 2 toward point 10, as the distance between point 7 and point 9.

14. Locate point 12 on line F, at a point which is equidistant from point 10 and point 11.

15. Locate point 13 on line C, at a point which is ⅙ of the arm's-eye circumference from point 4, toward point 5.

16. Draw line G, at right angles with line C, from point 15 toward the top, to the distance of ⅙ of the arm's-eye circumference from point 13 ; at which point, locate point 14.

17. Find the difference between ½ of the chest-circumference and ½ of the blade-width ; extend line C, from point 4 toward the opposite edge, to the distance of that difference from point 4 ; at which point, locate point 15.

18. Find ½ of the front waist-depth ; draw line H, at right angles with line C, from point 15 toward the bottom, to that distance from point 15 ; at which point, locate point 16.

19. Locate point 17 on line H, at a point which is ¼ of the distance between point 15 and point 16 from point 16, toward point 15.

20. Draw line I, from point 5 to point 17.

21. Indicate, and draw curved line J, from point 14 to point 10, by the same length of tape-measure, and principles and rule, that you drew curved lines L and N, under Plates VI and VII, in Figures VI and VII, respectively ; only changing the position for pressing the tape-measure on the table, to the opposite side, and opposite point 2 ; and arranging the position of the stretched tape-measure and the point of the pencil, true to point 14 and point 10 ; and indicating curved line J, from point 14 to line I : and drawing the same line, from the point of intersection of lines J and I, to point 10 ; as illustrated in the cut.

22. Locate point 18 at the point of intersection of lines J and I.

23. Indicate pointed line K, from point 14 to point 4 ; as illustrated in the cut.

24. Locate point 19 on line K, at a point equidistant from point 14, and point 4.

25. Indicate pointed line L, from point 19 toward point 13, to a point which is one-seventh of the distance between point 19 and point 13 from point 19 ; at which point, locate point 20; as illustrated in the cut.

26. Draw curved line M, from point 20 to point 12, by the same length of tape-measure, and principles and rule, that you drew curved line J; only arranging the

positions, and the stretched tape-measure and point of pencil true to, and drawing from point 20 to point 12 ; as illustrated in the cut.

27. Locate point 21, on line C, the distance of $\frac{1}{6}$ of the arm's-eye circumference from point 4, toward point 15.

28. Indicate pointed line N, from point 14 toward point 21, to a point twice as far from point 14 as the distance between point 19 and point 20 ; at which point, locate point 22; as illustrated in the cut.

29. Draw curved line O, from point 22 to point 18, by the same length of tape-measure, and principles and rule, that you drew curved lines J and M; only changing positions, and the stretched tape-measure and point of pencil true to, and drawing from point 22 to point 18 ; as illustrated in the cut.

30. Draw curved line P, from point 22 to point 20 ; as illustrated in the cut.

31. Locate point 23 on line D, the distance of the hip-depth from point 1, toward point 6.

32. Find $\frac{1}{8}$ of the hip-circumference. Draw line Q, at right angles with line D, from point 23 toward line B, to $\frac{1}{4}$ of that distance from point 23; at which point, locate point 24.

33. Draw line R, from point 10 through point 24, to line E.

34. Find the difference between $\frac{1}{2}$ of the chest-circumference and $\frac{1}{2}$ of the blade-width; subtract this difference from $\frac{1}{3}$ of the hip-circumference. Then extend line Q, from point 24 toward line B, the distance of two-fifths of this last difference from point 24; at which point, locate point 25.

35. Draw curved line S, from point 12 to point 25 ; as illustrated in the cut.

36. Draw line T, at right angles with line Q, from point 25 to line E.

Q. E. D.

N. B.—1. That part of curved line J, indicated from point 18 to point 14, is discarded for curved line O, from point 18 to point 22; and is only used in this draft, to illustrate the proper dart, taken out, to produce the correct and scientific fit around the curve of the body; and at the junction of the back, side-back, and the sleeve.

2. The pointed lines across the angles of the straight and curved lines, here as elsewhere, indicate the proper curves, in basting and stitching.

3. In all the drafts, allow the necessary seams and extensions for finishing, outside of the lines representing the fit.

THE SLEEVE.

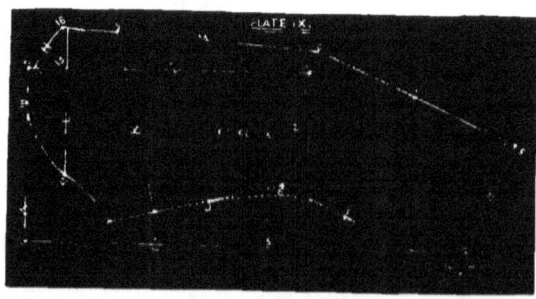

INSTRUCTIONS.

PLATE IX.—FIGURE IX.

1. Add 1 inch to the extreme depth; locate point 1, on the material, that distance from the bottom or right-hand edge, and 1½ inches back from the front, toward the opposite edge.
2. Draw line A, at right angles with the front edge, from point 1 toward the opposite edge, to the distance of ½ of the arm's-eye circumference from point 1; at which point, locate point 2.
3. Draw line B, at right angles with line A, from point 2 toward the right-hand edge, to the distance of ¼ inch less than the elbow-depth from point 2; at which point, locate point 3.
4. Find the difference between ½ of the extreme depth, and the elbow-depth; subtract this difference from the extreme depth: Then draw line C, at right angles with line A, from point 1 toward the right-hand edge, to the distance of this last difference from point 1; at which point, locate point 4.
5. Locate point 5 on line C, the distance of ½ of the extreme depth from point 1, toward point 4.
6. Draw line D, from point 3 toward point 5, to the distance of ½ of the elbow-circumference from point 3, toward point 5; at which point, locate point 6.
7. Indicate pointed line E, at right angles with line C, from point 4 toward the front edge, to the distance of ⅛ of the hand-circumference from point 4; at which point, locate point 7.
8. Draw line F, from point 6 toward point 7, to a point which is ⅛ of an inch less from point 6, than the distance between point 6 and point 7; at which point, locate point 8.
9. Draw line G, from point 8 to point 4.
10. Draw line H, at right angles with line F, from point 4 toward the right-hand or opposite edge, to a point which is 5 times the distance between point 7 and point 4 from point 4; at which point, locate point 9.
11. Extend line D, from point 3 toward the opposite edge, to the same distance from point 3 as the distance between point 7 and point 4; at which point, locate point 10.
12. Draw line I, from point 10 to point 9.
13. Draw line J, from point 6 toward point 1, to the distance of the inside-height from point 6; at which point, locate point 11.
14. Locate point 12 on line J, ¼ of the distance between point 11 and point 6 from point 11, toward point 6.
15. Draw line K, at right angles with line J, from point 12 toward the opposite edge, to the distance of ⅔ of the arm-circumference from point 12; at which point, locate point 13.
16. Locate point 14 on line B, the distance of the outside-height from point 3, toward point 2.
17. Draw line L, at right angles with line B, from point 14 toward the front edge, to the distance of ⅓ the arm's-eye circumference from point 14; at which point, locate point 15.
18. Draw line M, from point 10 through point 13, to the distance of ¼ inch further from point 10, than the distance from point 10 to the point of intersection of lines M and L, if line L is extended toward the opposite edge; at which point, locate point 16; as illustrated in the cut.

19. Locate point 17 on line B, ¼ of the distance between point 2 and point 14 from point 2, toward point 14.
20. Draw curved line N, from point 16 through points 17, 18, and 15, to point 11; as illustrated in the cut.

Q. E. D.

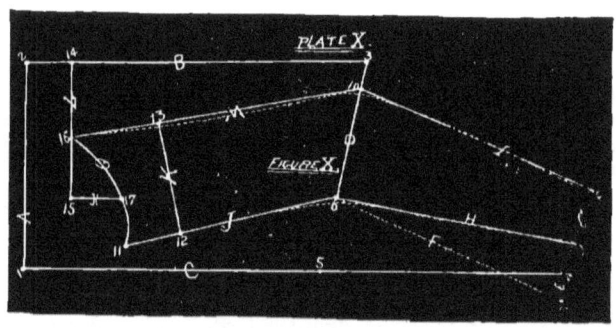

PLATE X.—FIGURE X.

INSTRUCTIONS.

N. B.—1. Locate points 1, 2, 3, 4, 5, and 6, and draw construction lines A, B, C, and D, by the same rules given for locating these respective points, and drawing these respective lines, under Plate IX, for drawing Figure IX.

2. Indicate pointed line E, at right angles with line C, from point 4 toward the front edge, to the distance of ⅛ of the hand-circumference from point 4; at which point, locate point 7; as illustrated in the cut.

3. Indicate pointed line F, from point 6 to point 7; as illustrated in the cut.

4. Draw line G, at right angles with line F, from point 4 toward the opposite edge, to the distance of 3 times the distance between point 7 and point 4 from point 4; at which point, locate point 8.

5. Locate point 9 on line G, the same distance from point 4 toward point 8, as the distance between point 7 and point 4.

6. Draw line H, from point 6 to point 9.

7. Locate point 10 on line D, the same distance from point 3 toward point 6, as the distance between point 4 and point 9.

8. Draw line I, from point 10 to point 8.

9. Draw line J, from point 6 toward point 1, to the distance of the inside-height from point 6; at which point, locate point 11.

10. Locate point 12 on line J, the distance of ¼ of the distance between point 11 and point 6 from point 11, toward point 6.

11. Draw line K, at right angles with line J, from point 12 toward the opposite edge, to the distance of ⅓ of the arm-circumference from point 12; at which point, locate point 13.

12. Locate point 14 on line B, at a point which is the distance of the outside-height from point 10, toward point 2.

13. Draw line L, at right angles with line B, from point 14 toward the front edge, to the distance of ⅓ of the arm's-eye circumference from point 14; at which point, locate point 15.

14. Draw line M, from point 10 through point 13, to line L; and locate point 16 at the point of intersection of lines M and L.

15. Draw line N, at right angles with line L, from point 15 toward the right-

UTILITY AND DECORATIVE ART.

hand, to the distance of ⅛ of the arm's-eye circumference from point 15; at which point, locate point 17.

16. Draw curved line O, from point 16 through point 17, to point 11; as illustrated in the cut.

Q. E. D.

N. B.—1. From the above illustrations and cuts, in Figures IX and X, representing the upper and under parts of the sleeve, it will be readily and easily seen, by studying them in connection with Plates II and III, and Figures II and III, under the Franco-Prussian method, how either Figure III, or any other possible style of sleeve can, in like manner, by means of the above principles and construction lines, be drafted directly and independently of each part, upon the material.

In the above principles and construction lines, as taught and illustrated, the thoughtful student will find the key that will unlock all the secrets of the greatest and most popular and successful sleeve-cutters of the world; which is true also, and applies with equal force to the principles of measurement, drafting, and construction lines, in the art of body-drafting; as enunciated and taught in the preceding part of this work. But to be simply a perfect cutter, is quite a different thing from being a perfect cutter and an artist! The very dullest pupil can thereby reach a high standard of perfection and success; but it is left for the artist, to excel.

2. The curved pointed lines along lines M, I, J, and W, as elsewhere, show the proper curves for basting and stitching.

3. All seams and extensions for finishing, must be allowed outside of the lines representing the fit.

4. Children's sleeves, are drafted according to Plate III, either in the compound or direct drafts.

The International Method of Franco-Prussian Scientific Tailor Principles, for all Kinds and Styles of
SKIRT-DRAFTING.

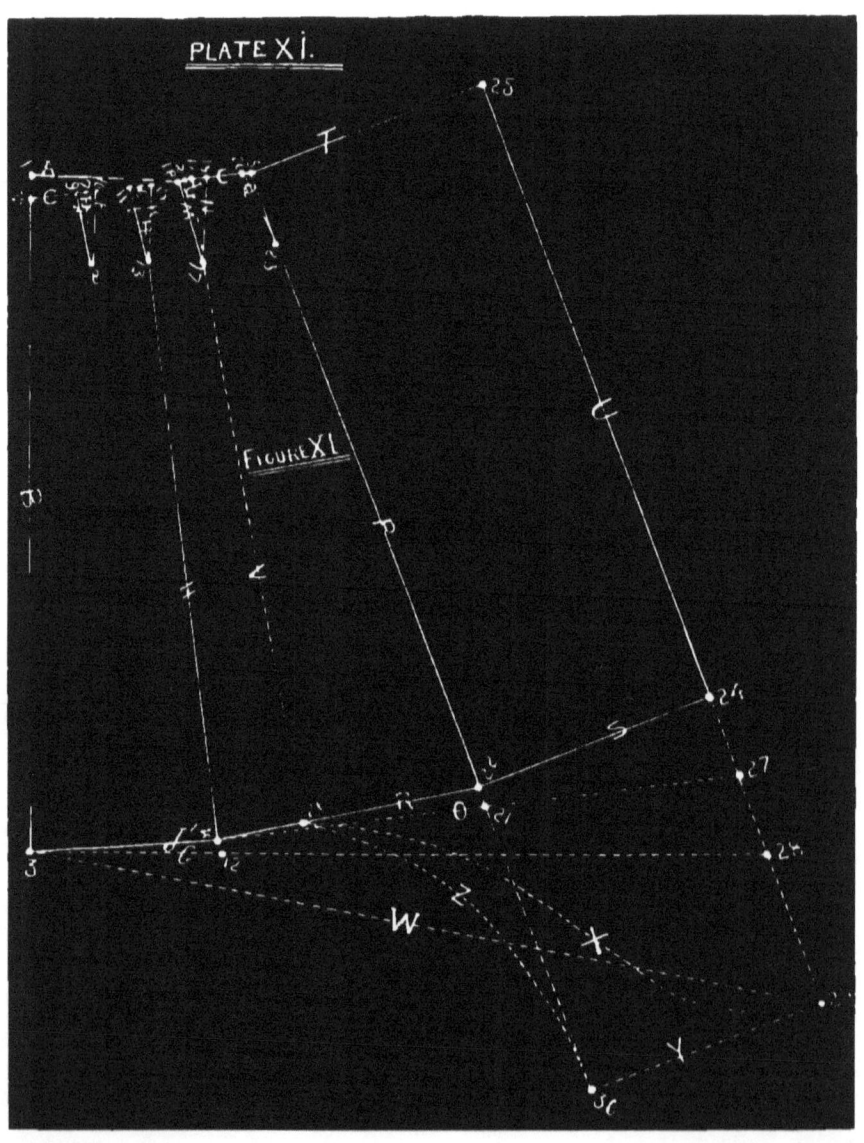

UTILITY AND DECORATIVE ART. 93

INSTRUCTIONS.

PLATE XI.—FIGURE XI.

1. Add 1½ inches to the skirt-depth; locate point 1, on the paper, that distance from the bottom or right-hand edge, and 1 inch back from the front edge, toward the opposite edge,
2. Find the difference between ½ of the chest-circumference, and ½ of the blade-width; draw line A, at right angles with the front edge, from point 1 toward the opposite edge, to the distance of that difference from point 1; at which point, locate point 2.
3. Find the difference between the front waist-depth and the waist-depth; add this difference to the skirt-depth; then draw line B at right angles with line A, from point 1 toward the bottom, to that distance from point 1; at which point, locate point 3.
4. Locate point 4 on line B, the distance of the skirt-depth from point 3, toward point 1.
5. Find ⅛ of the difference between the chest-circumference and the waist-circumference; add this amount of that difference to ⅛ of the waist-circumference: then draw line C, from point 4 through point 2, to the distance of that sum from point 4; at which point, locate point 5.
6. Locate point 6 on line C, equidistant from points 4 and 5.
7. Locate point 7 on line C, equidistant from points 4 and 6.
8. Indicate pointed line D, at right angles with line C, from point 7 toward the bottom, to the distance of ⅓ of the hip-depth from point 7; at which point, locate point 8; as illustrated in the cut.
9. Find ⅛ of the difference between the chest-circumference and the waist-circumference; locate point 9 on line C, ½ of that distance from point 7, toward point 4.
10. Draw curved line E, from point 9 to point 8; as illustrated in the cut.
11. Locate point 10 on line C, the same distance from point 7, toward point 6, as the distance between points 7 and 9.
12. Draw curved line F, from point 10 to point 8; as illustrated in the cut.
13. Locate point 11 on line C, the same distance from point 6, toward point 10, as the distance between points 7 and 10.
14. Measure and note the distance between points 4 and 6; then indicate pointed line G, at right angles with line B, from point 3 toward the opposite edge, to twice that distance from point 3; at which point, locate point 12.
15. Indicate the pointed part of line H, from point 6 toward point 12, to the distance of ⅓ of the hip-depth from point 6; at which point, locate point 13; extend line H, from point 13 toward point 12, to the distance of the skirt-depth from point 6; at which point, locate point 14.
16. Draw curved line I, from point 11 to point 13; as illustrated in the cut.
17. Draw line J, from point 3 to point 14.
18. Locate point 15 on line C, the same distance from point 6 toward point 5, as the distance between points 6 and 11.
19. Draw curved line K, from point 15 to point 13; as illustrated in the cut.
20. Locate point 16 on line C, equidistant from points 5 and 6.
21. Indicate pointed line L, at right angles with line C, from point 16 toward the bottom, to the distance of ⅓ of the hip-depth from point 16; at which point, locate point 17; as illustrated in the cut.
22. Locate point 18 on line C, the same distance from point 16 toward point 15, as the distance between points 6 and 15.

23. Draw curved line M, from point 18 to point 17; as illustrated in the cut.

24. Locate point 19 on line C, the same distance from point 16 toward point 5, as the distance between points 16 and 18.

25. Draw curved line N, from point 19 to point 17; as illustrated in the cut.

26. Locate point 20 on line C, ⅓ of the distance between points 18 and 19 from point 5, toward point 19.

27. Measure, and note the distance, between points 15 and 20; then indicate pointed line O, at right angles with line H, from point 14 toward the opposite edge, to three-times that distance from point 14; at which point, locate point 21; as illustrated in the cut.

28. Draw line P, from point 5 toward point 21, to the distance of the skirt-depth, from point 5; at which point, locate point 22.

29. Locate point 23 on line P, the distance of ⅓ of the hip-depth from point 5, toward point 22.

30. Draw curved line Q, from point 20 to point 23; as illustrated in the cut.

31. Draw line R, from point 14 to point 22.

32. Find the sum of the distances between points 3 and 14, plus the distance between points 14 and 22; subtract this sum from ½ of the desired width of the skirt at the bottom: then draw line S at right angles with line P, from point 22 toward the opposite edge, to the distance of that difference from point 22; at which point, locate point 24.

33. Draw line T at right angles with line P, from point 5 toward the opposite edge, to the same distance from point 5, as the distance between points 22 and 24; at which point, locate point 25.

34. Draw line U, from point 25 to point 24.

Q. E. D.

N. B.—1. This is a compound draft. The method of tracing or eliminating the various parts will be found under "TRACING." It is a superior and exceedingly accurate method, presenting principles foreign to any ever published; and is worthy of the student's most careful analysis and study. All possible kinds and styles of skirts can be easily and quickly drafted by the same principles, from the same construction lines; as given and illustrated above.

2 The pointed line V, from point 17 to point 26, and the extended and new pointed lines G, O, U, P. W, X, Y, and Z, below the draft of the plain walking-skirt, represent the plain "demi-promenade," "demi-train," and "full-trained" skirts; and are drafted, as follows:

A.—The Demi-Promenade:—(1). Locate point 26 on line R, ⅓ of the distance between points 14 and 22 from point 14, toward point 22.

(2). Indicate pointed line V, from point 17 to point 26; as illustrated in Figure XI.

(3). Extend pointed line O, from point 14 to the point of intersection of lines O and U—if line U is extended—as illustrated; at which point, locate point 27.

(4). Draw a line for the bottom of the skirt, from point 26 to point 27.

B.—The Demi-Train:—(1). Extend pointed line G from point 12 to the point of intersection of lines G and U, if line U is extended, as illustrated; at which point, locate point 28.

(2). Draw a line from point 26 to point 28, for the bottom of the skirt.

C.—The Full Train:—(1). Extend line U from point 28 toward the bottom edge, to the same distance from point 28 as the distance between points 24 and 28; at which point, locate point 29.

(2). Indicate pointed line W, from point 3 to point 29; as illustrated in the cut.

(3.) Then draw the shape desired; as represented by the indicated line X, from point 26 to point 29.

(4). ☞ For the square train, indicate pointed line Y, at right angles with line U, from point 29 toward the front, to the same distance from point 29, as the distance between points 24 and 22; or to the point of intersection of lines P and Y—if line P is extended; at which point, locate point 30; as illustrated in the cut.

(5). Then draw the desired shape from point 26 to point 30; as illustrated by pointed line Z, in the cut. Q. E. D.

P. S.—The amount of dart space between the "side gore" and the back is regulated by taste, or the degree of closeness of fit to the form, desired. Be careful in figured goods, that you cut it all to run the same way; and not to cut duplicate pieces, either of the skirt, waist, or sleeve; or in other words, two pieces for one and the same side. Fashion dictates the width of skirts, from 2 to 6 yards—at times having 3 full back breadths—one in front, and a broad gore at each side. At present, it is often only 2 yards wide, and close-fitting; with slashes or cuts in the bottom of the front breadth, 3½ inches deep, braided and nicely finished, and hidden by the trimmings.

The International Method

OF

Franco-Prussian

AND

English and American

SCIENTIFIC TAILOR PRINCIPLES,

FOR ALL KINDS AND STYLES OF

EITHER COMPOUND OR DIRECT

"PRINCESS"-DRAFTING.

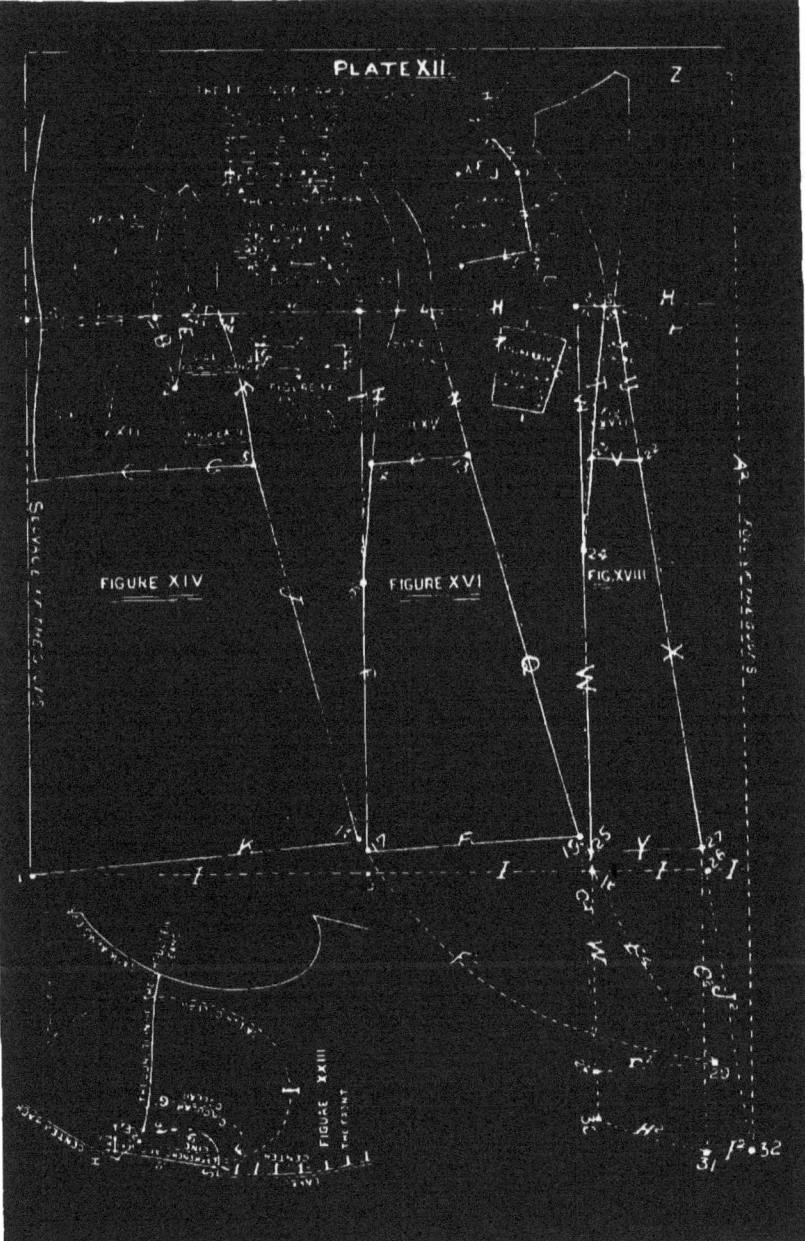

INSTRUCTIONS.

PLATE XII.—FIGURE XIV.

1. Figures XII, XIII, XV, and XVII, represent the traced and cut linings of the front, side-front, side-back, and back, respectively, upon which the construction lines A (of the original drafts), and the lines representing the fit, are traced; and which are drafted, traced and cut, per instructions for "body-drafting;" as given in the preceding part of this work.

2. Letter the construction line on Figure XII, A; the curved line near the opposite edge, B; the line representing the lower edge, C; as illustrated in the cut.

3. Letter the construction line on Figure XIII, D; the curved line near the front edge, E; the curved line near the opposite edge, F; the line representing the lower edge, G; as illustrated in the cut.

4. Locate point 1 at the point of intersection of lines A and B.

5. Locate point 2 on curved line B, the distance of the hip-depth from point 1, toward the bottom.

6. Locate point 3 at the point of intersection of lines E and D.

7. Locate point 4 at the point of intersection of line F, with the waist-line of Figure XIII; as illustrated in the cut.

8. Locate point 5 at the point of intersection of lines F and G.

9. Find the sum of the desired depth of the train, plus the desired depth of the hem; locate point 6 on the selvage or front edge of the goods, that distance from the bottom or right-hand edge; as illustrated in the cut.

10. Locate point 7 on the selvage or front edge of the goods, the distance of the skirt-depth from point 6, toward the top.

11. Draw line H, at right angles with the selvage or front edge, from point 7 toward the opposite edge, to the distance of $\frac{1}{4}$ of the skirt-width; at which point, locate point 8.

12. Place Figure XII, or the lining of the front, in such a position on the goods, that the fullest part of the curve in front, shall be even with the selvage or front edge of the goods, and line A exactly upon line H; as illustrated in the cut: then baste or fasten Figure XII securely, in that position, upon the goods.

13. Place Figure XIII, or the lining of the side-front upon the goods, on the opposite side of Figure XII, in such a position, that the arc of the arm's-eye circumference will be toward the front; line D exactly upon line H, and lines E and B, intersecting at point 2; as illustrated in the cut: then baste or fasten Figure XIII, or the lining of the side-front, securely in that position, upon the goods.

14. Indicate pointed line I, at right angles with the selvage or front edge of the goods, from point 6 toward the opposite edge, to the distance of $\frac{1}{4}$ of the desired skirt-width from point 6; at which point, locate point 9: as illustrated in the cut.

15. Draw line J, from point 5 toward point 9, to the distance of the desired skirt-depth from point 4; at which point locate point 10; as illustrated in the cut.

16. Draw line K from point 6 to point 10: which process completes the entire "front" of the "Promenade-Princess."

17. Letter the construction line on Figure XV, L; the curved line near the side joining with the front, M; the line near the opposite edge below line L, N; the line representing the edge at the bottom, O; as illustrated in the cut.

18. Locate point 11 at the point of intersection of lines L and M.

19. Locate point 12 at the point of intersection of line N with the waist-line of Figure XV.

UTILITY AND DECORATIVE ART. 99

20. Locate point 13 at the point of intersection of lines N and O.
21. Locate point 14 at the point of intersection of lines M and O.
22. Extend line H, from point 8 toward the opposite edge of the goods, to the distance of ⅙ of the desired skirt-width from point 8; at which point, locate point 15.
23. Draw line P, at right angles with line N, from point 8 toward the bottom, to the distance of ½ of the desired skirt-depth from point 8; at which point locate point 16; as illustrated in the cut.
24. Place Figure XV, or the lining of the side-back upon the goods, opposite line P, in such a position that the arc of the arm's-eye circumference will be toward the front, line L exactly upon line H, and lines P and M (if line M were extended toward the bottom) intersecting at point 16; as illustrated in the cut: then baste or fasten Figure XV securely in that position, upon the goods.
25. Extend line P, from point 16 toward point 9, to the distance of the desired skirt-depth from point 11; at which point, locate point 17; as illustrated in the cut.
26. Indicate the extension of pointed line I from point 9 toward the opposite edge, to the distance of ⅙ of the desired skirt-depth from point 9; at which point locate point 18; as illustrated in the cut.
27. Draw line Q from point 13 toward point 18, to the distance of the desired skirt-depth from point 12; at which point, locate point 19: as illustrated in the cut.
28. Draw line R from point 17 to point 19; which process, completes Figure XVI, or the side-back of the Promenade-Princess.
29. Letter the construction line on Figure XVII, or the lining of the back, S; the line near the edge joining the side-back, below line S, T; the line near the opposite edge, below line S, U; and the line representing the edge at the bottom, V; as illustrated in the cut.
30. Locate point 20 at the point of intersection of lines S and T.
31. Locate point 21 at the point of intersection of line U with the waist-line of Figure XVII.
32. Locate point 22 at the point of intersection of lines U and V.
33. Locate point 23 at the point of intersection of lines T and V.
34. Extend line H, from point 15 to the opposite edge, or fold of the goods; as illustrated in the cut.
35. Draw line W at right angles with line H, from point 15 toward the bottom, to the distance of ½ of the desired skirt-depth from point 15; at which point, locate point 24; as illustrated in the cut.
36. Place Figure XVII, or the lining of the back, upon the goods, opposite line W, in such a position, that the arc of the arm's-eye circumference will be toward the front, line S exactly upon line H, and lines W and T (if line T were extended toward the bottom) intersecting at point 24; as illustrated in the cut: then baste or fasten Figure XVII securely in that position, upon the goods.
37. Extend line W from point 24 toward point 18, to the distance of the skirt-depth from point 20; at which point, locate point 25: as illustrated in the cut.
38. Indicate the extension of pointed line I, from point 18 toward the opposite edge, to the distance of one-twelfth of the desired skirt-width from point 18; at which point, locate point 26; as illustrated in the cut.
39. Draw line X from point 22 toward point 26, to the distance of the desired skirt-depth from point 21; at which point, locate point 27; as illustrated in the cut.
40. Draw line Y, from point 25 to point 27; which process, completes Figure XVIII, or the entire "back" of the "Promenade-Princess."

Q. E. D.

N. B.—1. The above diagrams show both the compound and direct method, jointly illustrated; and can be easily eliminated and drafted either way, by a little thought and care, according to the principles of the above instructions.
2. Wrappers, morning wrappers, new markets, ulsters, demi-ulsters, water-proofs, etc., are all drafted and cut by the same principles as given above; only varying from the circumference measures of the "close fit," to the desired degree in "ease of fit," and to the demands of style, or the texture and weight of the material chosen.

3. The indicated lines Z and A² show the extension for plaits, gathers, or shirring, etc., in the back, from the fold of the goods, and according to the various tastes and styles of finishing; and which may be left the entire length from the neck, at line Z to the floor, or which may be cut at line B², below the waist, and only left in the garment, from that point to the bottom; to any degree of depth or fullness desired, or that the goods may allow.

4. The indicated lines C², D², E², F², W, G², H², I², and J², at the bottom of the draft, illustrate the train, which may be drafted either as an extension of the original parts entire, or cut separately; and arranged in finish, so as to be attached, or detached, at will.

5. Draft the train, as follows:

  a. Indicate pointed line C² at right angles with line I, from point 19 toward the bottom, and to the distance of four-fifths of the desired depth of the train from point 19; at which point locate point 28; as illustrated in the cut.

  b. Indicate line D² at right angles with line C², from point 28 toward the opposite edge, to the distance of ⅔ of the desired extra width of the train; at which point locate point 29; as illustrated in the cut.

  c. Indicate the slightly curved line E² from point 19 to point 29; as illustrated in the cut.

  d. Indicate curved line F² from point 17 to point 29; as illustrated in the cut.

  e. Indicate the extension of line W, from point 25 toward the bottom, to the same distance from point 25, as the distance (along the curved line E²) between point 19 and point 29; at which point locate point 30; as illustrated in the cut.

  f. Indicate pointed line G² at right angles with line I, from point 27 toward the bottom, to the distance of the desired depth of the train, from point 27; at which point locate point 31; and as illustrated in the cut.

  g. Indicate the slightly curved line H², from point 30 to point 31; and as illustrated in the cut.

  h. Indicate pointed line I² at right angles with line G², from point 31 toward the opposite edge, to the distance of ⅓ of the desired extra width of the train; at which point locate point 32; and as illustrated in the cut.

  i. Indicate pointed line J², from point 27 to point 32, and as illustrated in the cut; which process, completes the "Princess-Train."

Q. E. D.

P. S.—Fashion dictates both the style, shape and extent of the train; but whether it be large or small, square, circular, or oblong, the principles of drafting remain the same; at least, until the laws of mathematics change. Mathematics ever regulates the principles of the fit; and upon this province, art and fashion never can encroach; they have to do only with the design, shape and decorative finishings. The above mathematical principles are the leafless tree of winter, with its trunk and naked branches. Art, like the hand of Spring, covers them with leaves, and buds, and blossoms; each in its season, and according to its kind.

The International Method of Scientific Tailor Principles for all Kinds and Styles of
COLLAR-DRAFTING.
PLATE XII.—FIGURE XIX.
THE BERLIN, OR "CHOKER"- COLLAR.

INSTRUCTIONS.

1. Locate point 1 on the fold of the paper or material, 1 inch from the bottom or right-hand edge, toward the top.
2. Draw line A, at right angles with the fold, toward the front, to the distance of ½ of the neck-circumference from point 1; at which point, locate point 2: as illustrated in the cut.
3. Draw line B, exactly on the edge of the fold, or at right angles with line A, from point 1 toward the top, to the distance of one-twelfth of the distance between point 1 and point 2 from point 1; at which point, locate point 3.
4. Extend line B, from point 3 toward the top, to the distance of the desired height of the collar from point 3; at which point, locate point 4: as illustrated in the cut.
5. Draw line C, from point 3 to point 2.
6. Draw line D, at right angles with line A, from point 2 toward the top, to the distance of the desired height of the collar, from point 2; at which point, locate point 5.
7. Draw line E, from point 4 to point 5; which process, completes the draft of the "Berlin-Collar," as formed and indicated within lines B, C, D, and E; as illustrated in the cut.

Q. E. D.

PLATE XII.—FIGURE XX.
"THE ROLLING-CHOKER" COLLAR.

INSTRUCTIONS.

1. Locate point 1 on the fold of the paper or material, 1 inch from the bottom or right-hand edge, toward the top.
2. Draw line A, at right angles with the fold, from point 1 toward the front, to the distance of ½ of the neck-circumference, from point 1; at which point, locate point 2.

3. Find one-twelfth of the distance between point 1 and point 2. Draw line B, exactly on the fold, or at right angles with line A, from point 1 toward the top, to ½ of that distance from point 1; at which point, locate point 3: as illustrated in the cut.

4. Extend line B, from point 3 toward the top, to the distance of the desired width of the collar from point 3; at which point, locate point 4: as illustrated in the cut.

5. Locate point 5 on line A, equidistant from point 1 and point 2.

6. Draw line C, at right angles with line A, from point 5 toward the top, to the distance of the desired width of the collar from point 5; at which point, locate point 6.

7. Draw line D, at right angles with line A, from point 2 toward the top, to the same distance from point 2, as the distance between point 1 and point 3; at which point, locate point 7: as illustrated in the cut.

8. Extend line D, from point 7 toward the top, to the distance of the desired width of the collar from point 7; at which point, locate point 8: as illustrated in the cut.

9. Draw line E, from point 5 to point 3.
10. Draw line F, from point 7 to point 5.
11. Draw line G, from point 6 to point 4.
12. Draw line H, from point, 8 to point 6; which process, completes the draft of the "Rolling-Choker" Collar, as formed and indicated within lines B, D, E, F, G, and H; as illustrated in the cut.

Q. E. D.

PLATE XII.—FIGURE XXI.
"THE BYRON" COLLAR.

INSTRUCTIONS.

1. Locate point 1 on the fold of the paper or material, 1 inch from the bottom or right-hand edge, toward the top.

2. Draw line A, at right angles with the fold, from point 1 toward the front, to the distance of ½ of the neck-circumference from point 1; at which point, locate point 2.

3. Find one-twelfth of the distance between point 1 and point 2. Draw line B, exactly on the fold, or at right angles with line A, from point 1 toward the top, to ½ of that distance from point 1; at which point, locate point 3: as illustrated in the cut.

4. Extend line B, from point 3 toward the top, to the desired distance of the width of the collar, from point 3; at which point, locate point 4: as illustrated in the cut.

5. Locate point 5 on line A, equidistant from point 1 and point 2.

6. Draw line C, at right angles with line A, from point 5 toward the top, to the same distance from point 5, as the distance between point 1 and point 4; at which point, locate point 6.

7. Draw line D, at right angles with line A, from point 2 toward the top, to the same distance from point 2, as the distance between point 1 and point 3; at which point, locate point 7: as illustrated in the cut.

8. Extend line D, from point 7 toward the top, to the distance of ⅓ of the neck-circumference, from point 2; at which point, locate point 8: as illustrated in the cut.

9. Locate point 9 on line D, 3 times as far from point 7 toward point 8, as the distance between point 2 and point 7.

10. Draw line E, at right angles with line D, from point 8 toward the front, to the same distance from point 8, as the distance between point 2 and point 9; at which point, locate point 10: as illustrated in the cut.
11. Draw line F, from point 10 to point 9.
12. Draw line G, at right angles with line D, from point 9 toward the front, to the same distance from point 9, as the distance between point 2 and point 9; at which point, locate point 11.
13. Draw line H, at right angles with line D, from point 7 toward the front, to the same distance from point 7, as the distance between point 9 and point 11; at which point, locate point 12.
14. Draw line I, from point 11 to point 12; as illustrated in the cut.
15. Draw line J, from point 7 to point 5.
16. Draw line K, from point 5 to point 3.
17. Draw line L, from point 6 to point 4.
18. Draw line M, from point 10 to point 6; which process, completes the draft of "The Byron Collar," as formed and indicated within the lines B, F, G, H, I, J, K, L, and M; as illustrated in the cut.

Q. E. D.

N. B.—1. All of the above collars may be cut either square or round, in front at the top, and arranged in width or shape, according to fancy or the dictates of fashion.

2. A graceful collar is, to any garment, what the encircling leaf of green is to the delicately-tinted opening bud. It is one of those "little things", that either mars the entire garment, or becomes a graceful ornament, that sheds a beautiful and rich lustre of finish to the whole; and invariably reflects either honor or dishonor, upon the artist.

3. The above principles alone, are worth the entire cost of this book, to any cutter; whose ambition it is to excel, in this important department of grace and elegance, in garment-making. Study these examples and their solutions, as you would the fundamental rules of arithmetic, if you aimed to become a mathematician; and you will be abundantly repaid for all time and labor expended; in the matchless grace and perfection you will obtain, in result, in the completed garment.

PLATE XII—FIGURE XXII.

THE ENGLISH ROLLING-COLLAR.

INSTRUCTIONS.

1. Locate point 1 on the fold of the paper or material, 1 inch from the bottom or right-hand edge, toward the top.
2. Draw line A, at right angles with the fold, from point 1 toward the front, to the distance of the required depth, from the centre at the back of the neck, to the lapel in front; at which point, locate point 2.
3. Draw line B, exactly upon the fold, or at right angles with line A, from point 1 toward the top, to the distance of the desired width of the collar; at which point, locate point 3.
4. Locate point 4 on line A, equidistant from point 1 and point 2.
5. Draw line C, at right angles with line A, from point 2 toward the top, to the distance of $\frac{1}{6}$ of the distance between point 1 and point 2; at which point, locate point 5: as illustrated in the cut.
6. Locate point 6 on line B, $\frac{1}{2}$ of the distance between point 2 and point 5, from point 1, toward point 3.

7. Extend line C, from point 5 toward the top, to the distance between point 6 and point 3 from point 5; at which point, locate point 7 : as illustrated in the cut.
8. Draw line D, from point 7 to point 3.
9. Locate point 8 on line D, the same distance from point 7 toward point 3, as the distance between point 2 and point 5.
10. Draw curved line F, from point 5 to point 4 ; which process, completes the draft of "The English Rolling-Collar," as formed and indicated within the lines A, B, D, E, and F; and as illustrated in the cut.

Q. E. D.

N. B.—1. The above collar may be either round or otherwise, on the top at the front, and any depth and width desired, or required.
2. "The Close-buttoned Paletot-Collar" is exactly the same, excepting that it requires only ½ the distance between point 2 and point 5; as is given and illustrated in Figure XXII.

PLATE XII.—FIGURE XXIII.

THE FRENCH COAT-COLLAR.

INSTRUCTIONS.

1. Place the patterns or linings of the upper part of the back, shoulder to shoulder with the upper part of the front; as illustrated in the cut: then baste or fasten them securely in that position.
2. Locate point 1 on the curved front neck-line, at the point you wish the fold of the lapel; as illustrated in the cut.
3. Locate point 2 at the point of intersection of the curved lines of the front and back-neck, with the junction line of the shoulders; as illustrated in the cut.
4. Indicate the pointed centre line from point 1 through point 2, to the distance of ⅛ of the neck-circumference from point 2; at which point, locate point 3: as illustrated in the cut.
5. Draw line A, at right angles with the centre line, from point 3 toward the front, to the distance of the desired width of the collar, when completed; at which point, locate point 4; as illustrated in the cut.
6. Draw line B, from point 4, parallel with the centre line, to the desired distance toward the lapel; at which point, locate point 5: as illustrated in the cut.
7. Draw line C, from point 5 to point 1.
8. Locate point 6 on the curved neck-line of the front, equidistant from point 1 and point 2.
9. Draw curved line D, from point 1 to point 6, touching the neck-line within ¼ of an inch, at the point equidistant between point 1 and point 6; as illustrated in the cut.
10. Draw line E, parallel with line B, from point 6 toward the back, to the point of intersection of lines E and A, (if line A, is extended from point 3 toward the arms'-eye); as illustrated in the cut: which process, completes the draft of the French "coat-collar," as formed and indicated within the lines A, B, C, D, and E; and which is transferred from the diagram, either upon paper or the material, by the use of the tracing-wheel.

Q. E. D.

N. B.—1. The "Circular-Collar" is formed by tracing the entire half of the neck-circumference, for the fit; and designing its shape, at will: as illustrated in Figure XXIII, of Plate XII, by lines F, G, and H.

2. The "Sailor-Collar" is traced at the neck, precisely as the "Circular-Collar"; and is also designed at will, and according to the demands of fashion, in its shape; as on Plate XII, in Figure XXIII; and as indicated by the curved line I.

3. Both the circular and sailor collars, should be left ¼ inch easier in the fit, in the back, than the seam of the centre back; as illustrated in the cut.

Q. E. D.

PLATE XII.—FIGURE XXIV.

THE FRENCH TAILOR-CUFF.

INSTRUCTIONS.

1. Draft the sleeve to the measures, according to Plate II, or Plate III. Place either the pattern, or the basted linings, folded or basted properly, upon the goods; fasten the same firmly in the proper position with the thread of the material, or the lining of the cuff. Then outline your design from the Fashion-Plate, upon this lining or pattern of the exact fit of the sleeve; and by means of the tracing-wheel, transfer the same, upon the lining or material, upon which you desire it; as illustrated in the cut.

Q. E. D.

N. B.—All kinds and styles of cuffs are easily and quickly drafted, by the same principles.

PLATE XII.—FIGURE XXV.

THE NEW-MARKET HOOD.

INSTRUCTIONS.

1. Either using the linings or cut patterns of the upper part of the back and the front, place them shoulder to shoulder, with the centre-back seam toward the front; as illustrated in the cut, by the lines A, B, C, D, E, F, G, and H.

2. Locate point 1 at the point of intersection of lines A and E.

3. Locate point 2 on line E, the desired length of the hood in the back, from point 1, toward the bottom.

4. Draw line I, at right angles with line E, from point 2 toward the opposite edge, to the distance of ½ of the desired width of the hood, in the centre; at which point, locate point 3: as illustrated in the cut.

5. Draw line J, at right angles with line E, from point 1 toward the opposite edge, to ½ the desired width of the hood at the shoulders; at which point, locate point 4: as illustrated in the cut.

6. Draw line K, from point 4 toward point 3, to a point 1 inch less from point 4, than the distance between point 1 and point 2; at which point, locate point 5: as illustrated in the cut.

7. Draw line L, from point 2 to point 5.

8. Locate point 6 at the point of intersection of lines F and H.

9. Draw line M, from point 4 toward point 6, to the point of intersection of lines M and F; at which point, locate point 7; which process, completes the draft of "The New-Market Hood," as formed and indicated within the lines A, F, M, K, L and E.

Q. E. D.

N. B.—1. This is simply half of the hood, as shown in the design. Cut the hood, with line E on the lengthwise fold of the goods, to avoid a centre seam. Cut the lining, as the hood. Close the seams of each, separately. Arrange the lining over the outside; and seam all, except the neck edges. Then turn the hood rightside out; and form a backward plait, at each side of the neck. Bind the neck with the material; and fasten about the neck, with a hook and loop.

2. It is sometimes left to extend further front, and a dart of one inch taken out; and the outside edge brought over against line F, and shaped by it; after the dart is folded or cut. Any style can easily be cut from the draft.

THE INTERNATIONAL METHOD

OF

SCIENTIFIC TAILOR PRINCIPLES,

FOR ALL KINDS AND STYLES OF

POLONAISE-DRAFTING.

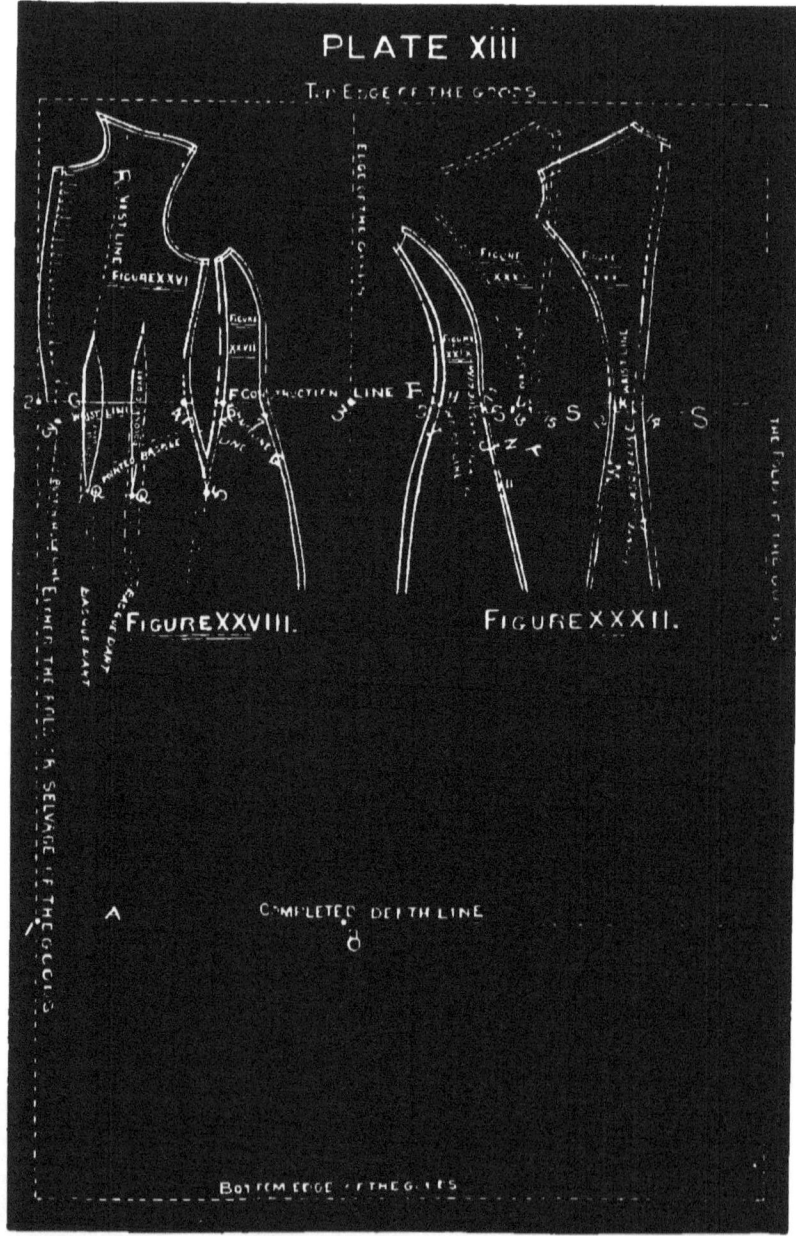

INSTRUCTIONS.

PLATE XIII.—FIGURE XXVIII.

1. Carefully observe the following, viz :
    a. Where the chosen design opens, back or front; and arrange the fold of the goods, to suit.
    b. How many seams have plaitings and loopings, and where located; and arrange according to the following instructions, for the extensions required; and to the proper depth fancied, or required.
    c. Accurately calculate, from the number and width of the plaitings or loopings in the front, if called for in the design, the amount of goods required for the same, plus the amount desired for hem and finishing ; locate point 1 at that distance from the bottom or right-hand edge, on the fold or selvage of the goods : then,
2. Indicate pointed line A at right angles with the front edge, from point 1 to the opposite edge.
3. Locate point 2 on the front edge, the desired or required depth of the completed front from point 1, toward the top.
4. Draw line B, or the construction-line, at right angles with the front edge, from point 2 to the opposite edge; at which point, locate point 3.
5. Letter the traced line running at right angles with and across the lining of the front or Figure XXVI, above the traced "waist-line," by the character C; as illustrated in the cut.
6. Letter the curved traced line, below the waist near the opposite edge of the same Figure, by the character D.
7. Letter the curved traced line, below the "waist-line," near the edge joining the front, of Figure XXVII, by the character E.
8. Letter the traced construction-line above the "waist-line" of the lining of the side-front, or Figure XXVII, by the character F.
9. Letter the curved traced line, near the opposite edge below the "waist-line" of the lining of the side-front or Figure XXVII, by the character G.
10. Locate point 3 at the point of intersection of the curved and traced "button-hole-line" near the front edge of the front, and the waist-line; as illustrated in the cut.
11. Locate point 4 at the point of intersection of line C and line D.
12. Locate point 5 on line D, the distance of the hip-depth from point 4, toward the bottom.
13. Locate point 6 on Figure XXVII at the point of intersection of lines E and F.
14. Locate point 7, at the point of intersection of the waist line and line G ; as illustrated in the cut.
15. Place Figure XXVI, or the lining of the front, on the goods, in the position in which, line C will be exactly upon the construction line B, with the fullest part of the curve of the front edge, even with the selvage or fold of the material; as illustrated in the cut: then baste or fasten Figure XXVI, securely on the goods, in that position.
16. Place Figure XXVII, or the lining of the side-front on the goods, opposite Figure XXVI, in the position in which line F will be exactly upon the construction line B, with the arc of the arm's-eye circumference toward the front, and lines D and E, intersecting at point 5 ; as illustrated in the cut : Then, baste or fasten Figure XXVII securely on the goods, in that position; which process, completes the draft of

the *front of the polonaise* upon the goods; and which, is indicated by the traced lines of the perfect fit—on Figures XXVI and XXVII ; the indications and outlines of which, form Figure XXVIII; when, continued to the desired length and slope from the lower edge of the linings—over the material.

17. Locate point 8 on the edge of the goods of Figure XXXII by the same rules and principles as given for locating point 1, on Figure XXVIII ; only applying said rules and principles to the back, instead of the front.

18. Extend the indication of line A, or the "completed depth-line," at right angles with the front edge of the goods, or Figure XXXII, from point 8 to the opposite edge ; as illustrated in the cut.

19. Re-locate point 3 on the edge of the goods of Figure XXXII, by the same rules and principles as given for locating point 2 on Figure XXXII ; only applying said rules and principles to the back instead of the front.

20. Extend line B, at right angles with the front edge of the goods, or Figure XXXII, from point 3 to the opposite edge or fold ; as illustrated in the cut.

21. Letter the traced line, above the "waist-line," at right angles with and across Figure XXIX, by the character H.

22. Letter the curved line below line H, of Figure XXIX, near the edge joining the front by the character I.

23. Letter the traced line below line H, near the opposite edge of Figure XXIX, by the character J.

24. Letter the traced line above the "waist-line," at right angles with and across Figure XXX, by the character K; and the same line on Figure XXXI, by the character L.

25. Letter the traced line below line K; near the front edge of Figure XXX, by the character M; and the same line on Figure XXXI, by the character N.

26. Letter the traced line below line K, near the opposite edge of Figure XXX, by the character O ; and the same line on Figure XXXI, by the character P.

27. Locate point 9 at the point of intersection of the "waist-line" and line I.

28. Locate point 10 at the point of intersection of the "waist-line" with line J; as illustrated in the cut.

29. Locate point 11 on line J, the distance of the hip-depth from the point of intersection of lines H and J, toward the bottom ; as illustrated in the cut.

30. Locate point 12 at the point of intersection of the "waist-line" and line M ; and locate point 13 at the point of intersection of the "waist-line" and line N.

31. Locate point 14 at the point of intersection of the "waist-line" and line O ; and locate point 15 at the point of intersection of the "waist-line" and line P; as illustrated in the cut.

32. Place Figure XXIX, or the lining of the side-back on the goods, in the position in which, line H will be exactly upon line B, the arc of the arm's-eye circumference toward the front, and line I intersecting the front edge of the goods at a point between point 3 and point 8, which will bring the desired or sufficient amount of fullness at the completed depth-line ; as illustrated in the cut : then, baste or fasten Figure XXIX, securely on the goods, in that position.

33. Place Figure XXX, or the lining of the back, upon the goods opposite Figure XXIX, in the position in which, line K will be exactly upon line B, the arc of the arm's-eye circumference toward the front, and lines J and M intersecting at a point sufficiently distant below line B, to bring the desired and required proportional amounts of fullness, for plaitings, loopings, or drapery—between Figures XXIX and XXX, and back of Figure XXX respectively ; if drapery is desired or called for between Figures XXIX and XXX : and place Figure XXX in the same position as regards line K (which now becomes line L,) and the arc of the arm's-eye circumference, with line M, (which now becomes line N), and line J intersecting at point 11, when no fullness or drapery is desired or called for, between Figure XXIX and Figure XXX, (which now becomes Figure XXXI); as illustrated in the cut. This process, completes the draft of the back of the polonaise upon the goods; and which, is indicated by the traced lines of the perfect fit on Figures XXIX, XXX, and

XXXI; and the indications and outlines of which form Figure XXXII, when continued to the desired length and slope, from the lower edge of the linings—over the material.
Q. E. D.

N. B.—1. Figure XXXI is simply Figure XXX in a different position; and its existence, and the lines and points upon it are only assumed, when the position of the back, or Figure XXX is abandoned, and the position represented by Figure XXXI is assumed.

2. The above is a thorough, exhaustive, and complete system of scientific principles; equal to and commensurate with every possible emergency of the cutter, in quickly and accurately drafting all kinds and styles of polonaises; and they deserve the student's most rigid and careful analysis.

3. The darts of the polonaise are generally closed at the distance of the hip-depth below line B, as illustrated in the cut; and the extra amount or surplus difference, between the amount taken up in the polonaise-dart and the basque-dart, is taken from the spring given to line D, of Figure XXVI, when drafting the linings; or the basque-dart may be used, which is closed at the hip-depth below line B, in the abrupt form as illustrated by the horizontal Q, in the cut; and the goods left, as plaiting or fullness, to be arranged as an adornment in certain kinds of drapery, at the front.

4. Line R is drawn from the top of the front dart, to a point on the shoulder-line which is ⅓ of the distance of the shoulder from the curved neck-line; and which, is used for cutting vest-fronts, tracing the front in two parts, as a compound draft; and allowing seams along line R, on both parts; and then proceeding with the draft of the polonaise, as before.

5. All kinds and styles of drapery can be arranged by simply leaving the entire outlines of the goods below the draft in the form of plain parallelograms, and folding it underneath, where not needed; often enhancing the appearance of the outside drapery, aside from serving the economical purpose of leaving the silk or velvet whole; for remaking, or remodeling the dress, at some future time. If the goods is limited, however, and the surplus is needed immediately; it is best for the beginner to fold tissue-paper, in large sheet form, to imitate the drapery as accurately as possible to the design, and cut it when thus plaited, and folded and looped; and then open it and giving the sharp points a rounded finish, cut the goods below the linings by that pattern, joining the lines at the top neatly and accurately, with the lines representing the perfect fit—on the waist linings.

6. Line S, indicated from line J in Figure XXIX, to the fold of the goods, represents the position or height of the plaiting or drapery, which is governed by the display of good taste, in its adaptation to the form for which it is intended; higher for long forms, and lower for short forms; and as fashion decides.

7. All actual complete depth-measurements should be from points 3, 4, 6, 7, 9, 10, 12, 13, 14, and 15, toward the bottom, when the draft is complete in all other respects; and the seams, arranged accordingly.

8. The folds and loopings should be marked, before the pattern is taken apart, according to the above instructions, and transferred from this pattern upon the material, when both are laid open and the pattern placed upon the material; both in polonaise, overskirt, and basque drapery; in order, to simplify the re-arranging of the same, for the beginner.

PLATE XIII.
BASQUES.
INSTRUCTIONS.

1. Cut the linings per instructions given, either in Figure I, or Figures V, VI, VII, and VIII; and the sleeves, per instructions given, in either of the desired sleeve Plates.

2. Form the draft on the goods, by the linings; in the same manner and by the same principles, as given under the same plate, for placing the linings for polonaises.

3. For plain round-basques, let all the seams, excepting the darts, intersect at the hip-depth below the construction line B; and allow the lines representing the seams, to run almost perpendicularly from the hip-line, to the depth desired. This principle underlies all close fitting-garments; over, and below the hips; whether it is a basque, sacque, redingote, newmarket, cloth tailor-suit, or ulster. Measure the desired distance for the depth of each seam, from points 3, 4, 6, 7, 9, 10, 12, 13, 14, and 15, toward the bottom; locating points at each seam, the desired depth from these respective points: and then draw lines from point to point, in horizontal directions, and trace according to these lines.

4. For the pointed-basque, proceed to locate the various depths at the various seams, per instructions for the round-basque. Then with the pencil or crayon, using the desired or fancied radius of tape-measure, on the principle you drew the curves in forming the draft for the linings in the back, form the curves at the bottom of the basque; as illustrated by the basque-lines, on Plate XIII.

5. The principle of using the vest-line, is the same as that given for the polonaise. This is also true of the plaitings, and drapery; and methods of reading them accurately, from books, for the use of designers; and from fashion plates. Although according to Bosworth, the basque is of Biscayan origin; the jaunty polonaise, on the authority of Guthrie, a gift from the Poles; and the queenly princess, the favorite of Cleopatra's many-tinted robes of royal purple, as well as the chosen form of the habiliments of Queens and Princesses—whose thrones, palaces, and forms, have long since crumbled to dust—they are all nevertheless, in science, the offspring of one common parent—draft; and when properly understood, are each and all, quickly and easily drafted from one and the same draft: and as given, either in Plate XII, or Plate XIII.

YOKES.

To eliminate the desired yoke from the draft or drafts, simply use the upper part of the draft for the close-fitting body, for the front and the back, to the depth desired; at which point, draw a line parallel with the construction line C, entirely across the draft; and trace and use the fit above said line, for said or desired yoke; into which, any desired amount of fullness of the goods or material can be plaited, shirred, or gathered.

LOW-BODICES AND BALL-DRESSES.

Complete the close fitting and high-necked draft, per instructions for the same; and after being thus completed, design the depth and shape you desire, in the completed bodice; and trace the exact fit everywhere, as indicated by the draft, up to said line, and as indicated; tracing the line indicating the depth and design, to which it is desired to have the dress completed; and allow all seams everywhere, as for the plain high-body.

UTILITY AND DECORATIVE ART.   113

TRACING.

1. Always prove your draft, by carefully measuring the parts which represent the goods or the fit on the chest, bust, waist, and hips; before tracing any parts of the draft, upon the goods or linings. If you have made a mistake, rectify the same in an equal division at the seams on said line or lines, as the case may be; so as to exactly agree with the measures called for in the measure-book, at said points.

2. Then placing the pressed linings double-ply upon the table, and making the desired allowance outside of the lines representing the fit, for extensions, seams, finish, etc., pin the draft in a true position with the linings, whether crosswise or lengthwise, upon the same.

3. With the tracing-wheel, trace the exact fit thoroughly, through both folds of the linings; first, the front, then the back, then the side-back, then the side-front; cutting each piece out as it is traced, before tracing the next; and allow the desired extensions for finish in front, and at the centre seam of the back (on the fold), and ¼ inch seams in the neck and arm's-eye; ½ inch seams everywhere else, excepting at the seam where the front and side-front join; where, it is best to allow a seam of 1 inch; in case of a possible desire to enlarge the garment, at some future time.

PLATE I—FIGURE I.

THE FRONT.

4. Trace the following lines in the following order, viz: A from point 1 to point 48; D^2 from point 12 to point 48; W^2, N; the indicated curved line along O; V^2, U^2, B^2, A^2; E from point 36 to point 13; Y^2, Q^2, Z^2, S^2, A^3, R^2, B^3, N^2, C^5, P^2, D^3, O^2, G^3, F^3, M^2, and X from point 5 through points 31, 15, and 29, to point 27.

THE BACK.

5. A, from the point of intersection of lines A and G^2, to the point of intersection of lines A and F^2; E^2 from point 42 to point 41; X^2 from the point of intersection of lines X^2 and I^3 to point 8; J, the indicated curved line along W, I^3, G^2, Y, from point 19 to point 35; H^3, F^2; and D, from point 20 to point 10.

THE SIDE-BACK.

6. A, from the point of intersection of lines A and J^2 to the point of intersection of lines A and I^2; E^2 from point 51 to point 50; X^2 from the point of intersection of lines X^2 and L^3, to the point of intersection of lines X^2 and J^3; X, from point 33 to point 34; L^3, K^3, J^2, J^3, I^2, and H^2.

THE SIDE-FRONT.

7. A, from the point of intersection of lines A and L^2 to the point of intersection of lines A and K^2; E^1 from point 49 to point 52; X^2 from point 4 to the point of intersection of lines X^2 and M^3; X from point 5 to point 33; B from point 4 to point 57; E^1 from point 57 to point 49; L^2, M^3, K^2; and J^2, from point 56 to point 33.

THE PRINCESS-BACK AND SIDE-BACK.

8. Trace the same as above, with the following exceptions, viz: where the back and side-back join, trace line N^3 for the back, and lines O^3 and line N^3 from point 71 to point 70 for the side-back.

8

THE VEST-FRONT.

9. Trace the same as given for the front, above, with the following exceptions, viz: Where the vest and the front join, trace Z^2, S^2, and P^3 for the vest; and A^3, R^2, and P^3, for the junction of the front with the vest.

PLATE II.—FIGURE II.

THE FRENCH-SLEEVE.

10. The top side—G, H, D, S, F, L, K and R.
The lower side—H from point 10 to point 11; D from point 6 to point 12; Q, I, L, J and O.

PLATE III.—FIGURE III.

11. The top side—F, D, Q, E, I, H and P.
The lower side—F, D from point 6 to point 8; O, E, I, G and L.

PLATE IV.—FIGURE IV.

THE CHILDREN'S DRAFT.

THE GABRIELLE.

12. *The Front.*—Place line E on the fold: A, from point 1 to the point of intersection of lines A and H^2; B^2 from point 12 to point 45; L^2 from point 11 to the point of intersection of lines L^2 and N^2; N, the curved line indicated along line O, from point 24 to line H^2; N^2, and H^2.

13. *The Side-Front.*—A, from the point of intersection of lines A and J^2 to the point of intersection of lines A and G^2; B^2 from point 50 to point 3; C^2 from point 3 to point 43; L^2 from the point of intersection of lines L^2 and O^2 to point 4; M^2, from point 4 to point 8; X, from point 44 through points 5, 31, 15 and 29 to point 27; O^2, J^2, H^2; the indicated curved line along O, from line H^2 to point 27; U^2, T^2, and G^2.

14. *The Back.*—A, from the point of intersection of lines A and E^2, to the point of intersection of lines A and D^2; C^2 from point 39 to point 38; M^2 from the point of intersection of lines M^2 and Q^2 to point 8; J, the curved line along line W from line E^2 to point 22; Q^2, E^2, P^2, D^2; and line D from point 20 to point 10.

15. *The Side-Back.*—A, from the point of intersection of lines A and G^2, to the point of intersection of lines A and F^2; C^2 from point 43 to point 42; M^2, from point 4 to point 8; X, from point 44 to point 19; Y from point 19 to point 35; the curved line indicated along line W from point 35 to line E^2; S^2, G^2, R^2, F^2, and E^2 from point 41 to point 40.

16. *The English-Back.*—Trace both parts the same, as given above, with the following exceptions, viz: where the back and side-back join, trace for the back, line V^2; for the side-back, X^2 and W^2, and curved line, from point 44 through point 33, to point 34.

17. *The French Back.*—This will be seen at a glance: simply using the entire back, tracing half as much more out at each side, at the waist, as the distance between points 39 and 42; and allowing for the difference in the extra amount of spring at each side, at the hip-line; as the draft indicates.

N. B.—Any style of garment, for a child, can instantly be traced from the above draft; in locating the desired seams, by a few points and lines upon the fit as it stands; and regulating the degree of fit and the style, at will.

CUTTING THE GOODS.

1. Trace the fit or mould of the form, upon the linings, first; then place the linings upon the goods, and cut the material according to the outlines of said linings; pinning each piece of the linings, according to the most economical arrangement, into position, before cutting any part. Be certain you have, in said position, allowed for the extra or required depths, below the waist-lines; also, that the position of the linings is true, with the goods. Locate the front and back, or the largest pieces, first; and the side-pieces—sleeves, collars, cuffs, etc., can, by a little thought and careful management, be cut out of the pieces remaining.

2. It is best for beginners, to place each part of the linings separately upon the goods, and not to cut the material close to the edge of the linings; simply, "blocking it out," pinning each part of the linings to its corresponding part of the material, and folding it and placing it by itself; this will obviate the danger of duplicating the same, or cutting double pieces for the same side of the body, or the sleeves; by having the pieces mixed with those not yet cut.

3. When a fabric is very narrow, cut the fronts on the "half-fold," cutting each front separately; and plain material, or that alike both up and down, as geometric patterns often are, you can add the necessary widths for gores of dresses, by joining the widths.

4. Tunics, tabliers, peplumes, double skirts, polonaises, or by whatever name the top or over-garment is termed, are all cut in a similar manner.

5. Double-width materials are folded at the centre, and the fronts of tabliers and tunics, placed along the fold. Cutting out mantles, sacques, cloaks, and dolman-sleeved wraps, are governed by the same principles as those given for other garments.

6. Jackets are usually cut as jacket-bodices, differing only in size and trimming.

BIAIS-CUTTING.

1. Biais-cutting presents some little difficulty, to the novice; yet, when the principle is solved, it becomes easy. It is most important that it should be thoroughly understood, and accurately taught; as it is much used in the connoisseurs', lady-tailors' and modistes' art.

2. Place the material, with the selvage to your front, upon the cutting-table; raise the corner and fold it to the top or upper selvage line, so as to place the straight cut end of the fabric along the selvage line; pin this carefully down, and crease the fabric where the natural fold occurs. This is the exact cross of the fabric; and this fold is cut through, leaving a half square piece (which is laid aside), and a true biais, or cross-cut line on the fabric. The width of the biais pieces, is next ascertained. If a six-inch biais is required, measure six inches at both selvadges, turn the material down, and pin into position firmly; then fold the other selvage down six inches, and crease the fabric as before, cutting on the creased line. Repeat this, as many times as the number of pieces desired, to equal the length called for. A six-inch flounce, requires a biais eight inches wide. Measure each piece exactly, and independently of the others; as using the first by which to cut the second, etc., is not accurate, and causes the biais to be untrue; and, consequently, not to hang accurate and true. Practice on paper until the art is mastered. Experienced cutters can take a shorter process, by square, chalk and pencil; which experience will teach the graduate, in the above.

BIAS-CUTTING ON TWILLED GOODS.

Great care is required in cutting bias on twilled fabrics. Place the material, right-side down, on the table; and the left-hand corner turned over, as for cutting ordinarily. This brings the right side uppermost, and the lines of the twill appear perpendicular. The same rule applies to cutting bias on crape.

CUTTING TRAIN-SKIRTS.

The promenade presents no difficulty whatever, after understanding the draft ; but the trains, might possibly puzzle the novice. Three breadths are required in the back, in very long trains; and the extra length must be divided between them according to the draft; and as illustrated. The centre breadth should not be made very much longer than the sides, and must be rounded in the centre; as illustrated, and not cut to a point.

If there is not material enough for this, it may be made in the following manner, viz: one whole breadth in the centre, and a gored breadth on each side of it; this must however be the full width of the material at the bottom, and slope-off to a width of five inches at the top. In cutting the gores, where there is an up-and-down pattern, they must, of course, be cut one beyond the other; and the parts left, can be used for sleeves, side-pieces, etc. This rule applies to velvet, velveteen, and all fabrics having a raised pile; which, require to be made with the pile brushed in one direction; which, according to common sense, best reason, and usage, is downward. To those who insist to reverse it, and cut it with the pile in the opposite direction, we most cheerfully grant that privilege; and, will not dispute their right to do so. It is simply a matter of taste and economy.

In cutting gores, outline each before cutting, by the fit of the draft upon the linings. The lengths of trains are, entirely, a matter of taste. This also applies to the gores. A demi-train may have six gores; one full gore in front and back, and four side-gores; and may be, from 5 to 20 inches longer than the front. Perhaps the most standard length for full trains is, from 1 to 2 yards longer than the front, and gored; as illustrated in the cut. The train is the same as the demi-train; only, longer and wider. The length at the back, and the width at the bottom, in a well-balanced train, should be symmetrical, and equal. The true designer and cutter will study the form, the combinations of color, the nature and texture of the material, and the place and purpose for which the garment is intended ; and by a careful study and application of the principles of the draft, as given in the illustration on the cut, and as taught in the instructions, it will be an easy matter for the dullest pupil to attain success; while it is left for the natural designer, to excel. Here, as in other departments, the true cutter is an artist ; the customer is the canvas; the materials are the colors ; the square, tape-measure, pencil, and scissors, are his brushes; and it is his province, to so mingle the shades and colors of form, material, purpose, and design, as to reproduce and reflect in his or her production, the highest possible human achievement, in imitation of the beautiful in Nature.

DRAPED-SLEEVES.

Draft the close-fitting sleeve, per instructions for drafting the sleeve. Design the desired depth you wish the sleeve; at which point, give the inside a slight curve, away from the arm ; and then design at will, by a curved extension, the depth of drapery at the opposite side, as well as its shape. It is exceedingly simple, if the above principle is honored carefully; the drapery adjusting itself to the draft of the perfect fit, along the upper part of the arm and shoulder.

BASTING.

For basting the linings separately, for the purpose of "trying-on," we have neither place nor use, in the system. It is not only a useless, but a positively injurious prejudice and process ; as it utterly unfits the linings for further use; changing, destroying, and utterly annihilating, the original mathematical mould or fit of the form, impressed upon them by the tracing-wheel, from the draft; which impression is, for the purpose of accurately basting, stitching, making, and finishing the material— upon said linings. If out of either prejudice or fear, in the customer or the cutter,

UTILITY AND DECORATIVE ART. 117

the garment is "tried-on," it should certainly not be done in such a manner as to destroy the very object of this whole work ; and so as to annihilate the impression of the fit, for the attainment of which, so much time, care, and labor has been expended. But in that case, it should be done after thorough and careful basting, and partial stitching of the goods or material upon said linings—immediately before completion. While this, even, is positively not necessary on any form, where the system has been accurately followed in all its details, we will certainly find no fault with any one, who out of fear or prejudice, either in the customer or cutter, will insist upon so-doing. But to those who insist upon first fitting the linings, and then use the same linings for said garment, we promise nothing more than an approximation to perfection; as it is a radical fault, and an injurious and deceptive mistake. "Fitting" garments, at most, when on the form, is only another mode of expression of "mis -fitting;" by pulling the garment out of its correct and truly perfect shape, and destroying its symmetry. Hence, many garments after being worn a short time, will show these defects, in skewing and crooked seams, and wry attitudes ; disfiguring, instead of adorning the person. But for the beginner, to practice and learn cutting, it is well to cut linings for mere practice; and baste them, and study and learn from their results ; and for this purpose, we give the following rules by which to baste linings, viz :

1. Place the various parts together, so that the construction and waist-lines, traced across the parts at the waist, will be exactly even.
2. Baste up-and-down, in each direction from said lines, beginning at the waist; and run a single stitch at a time, locating the stitches ¼ inch apart.
3. Commence with the back and side-forms, holding the back toward you; baste it a little full, and stretch the rounded edge of the side-forms.
4. Hold the back toward you, in basting the shoulder seams ; stretching the front shoulder seam, until it is even with that of the back.
5. Baste the darts very carefully, upon the trace-marks; holding them and basting by the same principle as given above, for basting the curved seams at the back.
6. All the pieces should first be very carefully and accurately pinned into position, at intervals along the seams; each way, from the waist.
7. In basting the sleeve, place the construction line across the same at the elbow, exactly even, which should bring it even at the front; then baste per instructions above : placing the top and under sleeve even at each end ; bringing any fullness allowed, equidistant each way, from the construction line across the sleeve, at the elbow.

BASTING THE MATERIAL.

In correctly and accurately basting the material upon the linings, and then said parts into one and the whole, lies the grand secret of garment-making ; if said linings are correctly cut, as taught in this work.

1. Always cut the linings one-half inch longer than the material. Place the material on the table, with the wrong-side up; and with the wrinkles neatly smoothed-out. Begin with the front. Carefully place the lining upon the goods; with a long basting-needle and basting-thread, black on white, or white on other goods, begin in the centre, at the top of the front; baste on a direct line, down to the top of the front dart, basting through the center of the dart, to the bottom ; taking one stitch at a time, one-half inch apart ; and, holding the lining (by an upward movement of the thimble of the left-hand, at the taking of each stitch), one-half inch full; equidistant each way, from above and below the waist. This is done at all points and sides of the garment, around the entire waist-line. It should at least extend, or be distributed, over a space of three inches above, and three inches below the waist. It is the sovereign remedy for inelegance of fit at the waist ; and for the presence of wrinkles in the material, on the outside ; as the linings in adjusting themselves to the form, stretch the material on the outside, and hold it in this stretched position.

2. Baste the second dart, in the same manner; beginning at the top of the dart, and basting through the centre; with the like fullness of the lining, to the bottom.

3. Begin at the lower opposite corner of the front, and baste all around the edge of the entire front; observing the law of the half-inch-fullness of the lining, above and below, and near the waist. Baste one inch from the edge.

4. Neatly trim the edges, and overcast them, around the entire piece; not drawing the stitches tightly.

5. Begin at the waist, one inch from the seam at the centre of the back, in basting the material upon the back. Baste in like manner, as for the front, from the waist toward the top; around, and within one inch of the edge, down to the waist; from thence, through the centre of the piece, to the bottom, observing the same law, for the necessary fullness at the waist, as in basting the front. Trim the edges, and overcast the same.

6. Begin at the top of the side-pieces, one inch from the edge, and at the centre; and baste down through the centre to the bottom; observing the same law for fullness at the waist; and, neatly fastening the basting thread, where you begin and quit, on all the pieces. Trim the edges neatly, as for the front and back; and overcast them neatly, all around on each side.

7. Begin at the darts, to baste the seams. Commence at the top of the dart, and begin carefully; after neatly pinning the seams into the best possible position at intervals, take one stitch at a time, one-fourth of an inch apart; stretching the seams of the darts on the waist, so as to join smoothly; and baste exactly on the trace-marks.

8. Stretch the seam at the junction of the side-front, above, on, and below the waist, where the fullness in the linings was allowed; so that they will join smoothly. Lay the pieces so the construction and waist-lines will exactly be even, and pin them, and baste up-and-down from the waist, each way, very accurately.

9. In basting the curved seams on the side pieces, the curved side should be basted easy, but not full; from a point $2\tfrac{3}{5}$ ways up from the waist, to within $1\tfrac{1}{2}$ inches of the top. Baste these seams with the back next to you; baste the right side first, beginning at the top; being sure, however, that it is first thoroughly and carefully pinned, at intervals; so as to be even at the waist and construction lines. In basting the left side, pin it as you did the right; removing the pins, as you baste.

10. Stretch the front shoulder seam thoroughly; until it is even at the neck and arm's-eye, with the seam of the back.

11. Stretch the lower part of the arm's-eye, from point 5 to point 15 in the draft, thoroughly; until it turns over at the edge. This prevents wrinkles in front of the arm. It is best to run a basting thread around the remainder of the arm's-eye, as well as around the neck; to keep those parts from stretching. All the various parts should be carefully handled; so as not to stretch the seams.

12. Baste the sleeve, into the arm's-eye of the body, with the fullest or highest part of the top of the sleeve, exactly on the top or highest part of the shoulder; and baste it full, from point 27 to point 29; as seen, in the draft. Before the sleeve is basted into position, it should, both lining and material, be overcast neatly; having the stitches $\tfrac{1}{3}$ of an inch apart, gathering the top with a running or overcast stitch. The linings should be of silk, or some soft material. The back seam, of the sleeve, should be turned toward the under part, and fastened to the lining. In heavy goods, the seams should be pressed open. The finishing is taught, elsewhere. In conclusion, we cannot but remark, that too much care cannot possibly be bestowed upon basting, in all its details. Accurate measurements are the roots; accurate drafting is the trunk; and accurate tracing and basting, are the branches and blossoms, of the tree of success: which, will yield the golden fruit of a rich reward, in the completed garment. And here, we part with the student; trusting we shall renew and extend our acquaintance, on the threshold of Part II.

PART SECOND.

ALL BRANCHES OF

LADIES' AND CHILDREN'S

Cloak and Wrap-Cutting,

AND

BASTING.

PREFATORY TO PART SECOND.

Everywhere in Nature, there is manifest "the eternal fitness of things." The right thing, in the right place; and, at the right time. And in the law of development, as taught and interpreted in her sanctuary, is reflected the correct method of receiving and imparting knowledge, on any subject. The year has its seasons; daylight its dawn ; and the oak that has withstood the storms of centuries, was once enclosed within the shell of an acorn ! And in like manner, from the lofty peaks to which the mind of Milton climbed, there is a ladder whose base rests upon the first character of the alphabet; every round of which he ascended, step by step ! All of which teaches us that we must first learn our letters, and patiently learn to spell the words, before we can read. Elements must first be mastered; from elements we must learn to form principles ; from principles the formation of characters, and the deductions and expressions, and the applications of truth. As we leave our narrow compass of vision in the valley, and ascend up the mountain's side, the outstretching plain lying along its base, grows larger and larger ; and as our journey is continued, our horizon enlarges, and the outlines of objects grow more distinct and clear ; and in every way, we begin to see things as we never saw them before ! And as we now stop here for a moment, to drink from the cold sparkling waters, bursting forth from beneath the overhanging rocks; and as we contemplate and enjoy, for a few moments of rest, the beautiful scene stretched out before us; like children, on their vacation ramble, we naturally desire to see more, and farther ; and we now therefore, invite the student to follow us, and "go up higher !"

In going to the designing and cloak rooms, we dismiss nothing of principles of measurement or drafting; but simply learn how to utilize and adapt, and apply said principles, in a larger and more extended field. How, that plain waist patterns are the buds of basques ; and how, that sacques are the blossoms of coats; and how, that trains, ulsters and long wraps, are the ripened fruit of all. How, that from the same and common parent-staff are born the twin sisters—the cape, and the circular; differing only in stature. And how, that Pelisses, and Dolman-sleeved circulars, wraps, etc., cling to the draft of the dolman, as the vines of the forest cling to the sturdy oaks. Armed with all the former principles of knowledge and practice, and having so successfully overcome all difficulties in "the close fit," surely it will be an easy matter for the student to simply regulate the degree of looseness desired, outside of said fit; along with the extra depth of waist at the back—to add grace of shape—in sacques and coats ; and to design, at will, their grace and comeliness.

COPYING.

To copy, is to imitate what we see in the productions of others.

INSTRUCTIONS.

1. Measure the entire size or height of the figure, on the plate; from this, fix a "scale," upon which to base your measurements of the depths; in the measurement of trains, draperies, points, shirrings, plaitings, angles of slope, vests, lappels, etc.

2. Example: For instance, the figure on the plate is 4 inches in height, from the bottom of the skirt to the neck; the depth of a point of drapery, at the side, is shown to be at a depth below the waist, at a point which is $\frac{1}{4}$ of the entire figure, or 1 inch from the bottom of the skirt. Now find the sum of the front height, plus the front-waist depth, plus the skirt depth of your customer; take $\frac{1}{4}$ of this distance: then drape so that the point hangs exactly that distance from the bottom of the skirt.

3. Proceed exactly by the same principle, when the widths are desired to be imitated, in vests, lappels, plaits, etc.; basing the ratio of proportion sought, upon the measure of the form for which it is intended. By a little thought and care, the most accurate copy can thus be read and reproduced, to the most complicated plate; where two sides are shown. Then attach this design, or peculiar style thus copied, to the draft of the perfect fit, at the waist and hip-lines.

4. To be a correct copyist, one should be accurate. If your memory is not good, use a pencil and note book; and accurately note:
   a. The number of seams.
   b. Their style.
   c. The finish and style at the neck.
   d. The style of sleeve, and its finish.
   e. The body of the garment; whether plain, plaited, scalloped, draped, etc.
   f. The skirt: its trimming, drapery, etc.
   g. The tailor-finishings, etc.
   h. Last, but not least, the combinations of materials, color, etc.

5. For beginners, it is positively necessary to practice in colored tissue-papers, until correct principles and elements of combination of colors, and in cut, finish, etc., are perfectly mastered; then, you can trust yourself on silks and velvets; when you have the measures, and the given design. All of which, is an easy and simple lesson; requiring only studiousness, industry, accuracy, and patience; without which, in anything in this life, there is no excellence.

DESIGNING.

DESIGN, comes from the Latin *De signo; de*, out, and *signo*, to mark: Its literal meaning, therefore, is to mark out, or point out by tokens; to designate; to form an idea, purpose, plan, or conceive the same; and then designate and describe the same. To establish or form for some end, would be an appropriate and correct definition of its meaning, in the art and science of garment-making. It means more than merely *copying*. There are plenty of copyists, but few designers. The designer may be aided in his conceptions of things, by studying the work of others; but, from the very nature of things, it would be impossible to rise higher than the fountain, from which this spring has its source. The true idea of designing means more than this; and the true designer will drink from a higher fountain: and will study beauty, as it is displayed and interpreted in nature.

Everywhere, in her sanctuary, we are taught in eloquent voices the grand and lofty conceptions and designs of her Author. From the muttering sounds of Heaven's thunder, the earthquake, or the mighty ocean's roar, down to the sweet low cadence of the whisperings of the soft breeze among the branches of the pine forest, on the calm and quiet summer evening; from the brilliancy of the angry flash of the lightning against the inky heavens of midnight, the burning sun, and the stars of heaven, down to the most delicate shades and tints of pink and white on the lily's fair face; from the infinite greatness of the conception and design of systems of suns, with all their heavenly train of stars and attendant planets, shoreless oceans of water and light, down to the the simplest pebble along the sea-shore; there is shown an eternal fitness and design of things, that defies the eloquence of all human language to express. Each bird has a plumage that defies imitation, in its true and magnificent conceptions, as to fitness, propriety, and beauty; and each shrub, and leaf and flower, is a miniature world of design, appropriateness, and matchless beauty that baffles even description—much less imitation! At best and most, we can but approximate in our feeble attempts at designing the works of the Great Artist and Teacher of all; whose halls of study prepared for man, compose the Universe; the walls of every department of which, are hung full of pictures, possessing a grace and beauty of design, that is matchless; and which is infinite, to the highest possible human conceptions. But while we can not equal design, as displayed in Nature, we can closely approximate the beauty of fitness and design of things we see around and about us, in Nature. And here alone, the lessons that teach correct designing, can ever be truly learned. A simple leaf gives more instruction on the "grace of curves," than one could learn from all the greatest Masters in Paris, Berlin, London, New York, or Philadelphia. And the distribution of the branches of the mighty oak of the forest, or the tiny shrub, teach a more eloquent lesson as to correct designing and the true propriety of symmetry, belonging to the various sized and shaped forms, than a life study would bring from the greatest productions from human hands; while, God has written upon sky and ocean, and the earth, the rock, the minerals, the animals, the plumage of the birds, and upon leaf and flower, the only true dictionary the world will ever find; as to the law and correct design, in combination, harmony, shading and blending of colors! Study these, rather than fashion journals. Correct taste, and true art is always in fashion. And there is a higher queen than that of fashion; to whose chariot wheels fashion is ten thousand times chained, as a slave! And it is the universal decision, by all the world's best and most learned critics, that many times those are most adorned, who according to fashion, are least adorned. Nature suggests to the designer, the correct science of symmetries and graceful designs, as to height, length, breadth and depth; as well as by the seasons' prevailing foliage, the colors, with their tints and shades, that are most appropriate to each season. She writes the true fashion journal for each season, in the colors she strews all over the earth. And it is the work of the artist to gather these tokens of favor, as they fall from her hand, and arrange them according to the best symmetry and grace, as to form; and according to correct and the most effective taste, for each peculiar customer, for each special occasion, and for each special purpose. New and correct combinations of color are read, in the wondrously mingled azure of the morning and evening sky; and from the royal and princely robes in leaf and flower; which, lend such rich lustre to nature's denizens, of both forest, field, and garden. While new cascades, in drapery, will spring into conception and existence, from the beautifully instructive cascade; as studied perchance, on the banks of the lonely mountain stream—when out on your summer's vacation.

Designing the arrangement of the bridal vail and orange blossoms, is one thing; the placing of the emblems and tokens of sorrow, is quite another. Study the fashion journal of man, but that of nature more. Cultivate true taste, and correct conceptions, and appropriateness, by communion with her teachings; in forest, field, and garden. Strive to rise above a mere slavery of form, in fashion; and be observing and reflective; compare and think; be studious, industrious, patient, and hopeful, with an ambition to excel that you may de good—by refining, educating, and ennobling humanity through your profession; and, success will crown your efforts.

The International Method of Foreign and American Scientific
Tailor-Principles, for all Kinds and Styles of

CAPE AND CIRCULAR-DRAFTING.

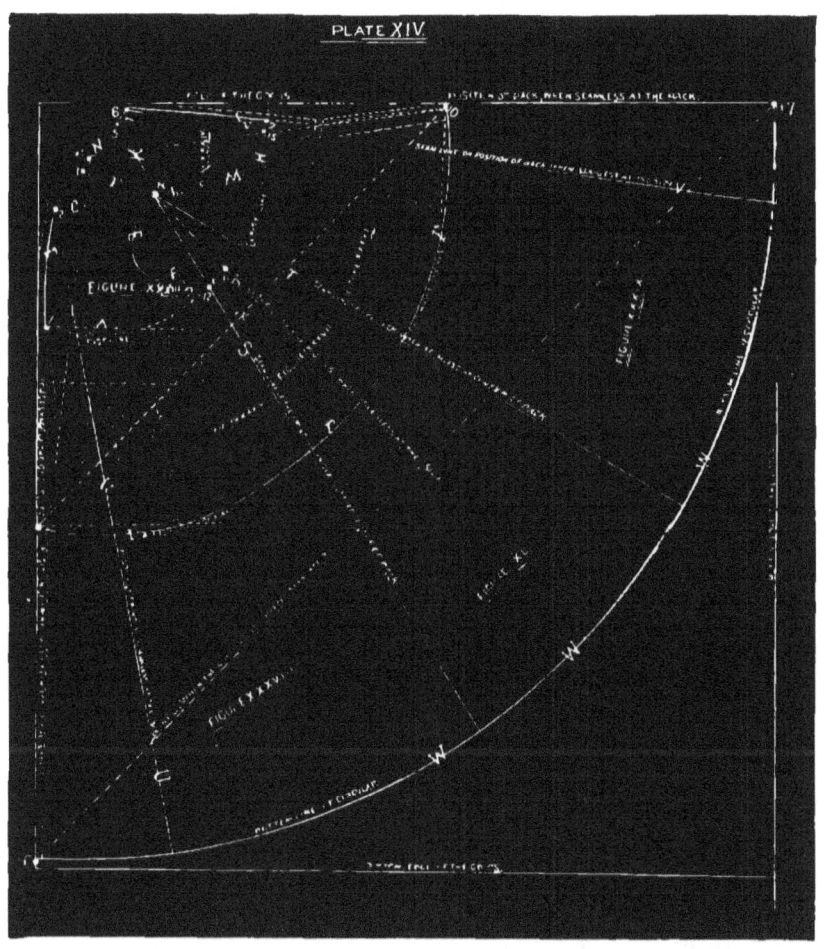

ELEMENTS AND PRINCIPLES OF CONSTRUCTION.

CAPES.

ELEMENTARY PARTS OF THE DRAFT.

PLATE XIV.—FIGURES XXIII AND XXIV.

INSTRUCTIONS.

1. Either trace accurate patterns of the upper parts of the front and the back, from the original "waist-draft," of the particular form for which the cape is intended, basing their lower edges on a line, parallel with, and equidistant from the construction lines A and C; or use the entire outlines, or complete front and back; either, or both of which methods are illustrated in the cut; by Figure XXXIII, or the pattern or linings of the front, as indicated and included within lines A, B, C, D, E, F, and G; and also Figure XXXIV, or the pattern, or linings of the back; as indicated and included within lines H, I, J, K, L, and M.

INSTRUCTIONS.
PLATE XIV.—FIGURES XXXV, XXXVI AND XXXVII.

1. Place Figure XXXIII upon the goods, opposite the selvage, sufficiently far from the right-hand edge, or bottom of the goods, to allow for the required depth, plus the hem; place XXXIV on the fold of the goods, opposite the bottom of the front, sufficiently far from the bottom at the fold, to allow for the required depth—plus the hem: Then arrange Figures XXXIII and XXXIV in the position in which, the arc of the neck-circumference of the back will be toward the front of the draft, the arc of the neck-circumference of the front toward the top, line A at right angles with the front edge, line H at right angles with the fold, line I on the fold, and line B the required distance from the selvage, for finishing; with lines D, E, K, and L intersecting at the same point: then, baste or fasten Figures XXXIII and XXXIV securely on the goods, in that position; as illustrated in the cut.

2. Locate point 1 at the point of intersection of lines A and B.
3. Locate point 2 at the point of intersection of lines B and C.
4. Locate point 3 at the point of intersection of lines C and D.
5. Locate point 4 at the point of intersection of lines D, E, K, and L.
6. Locate point 5 at the point of intersection of lines K and J.
7. Locate point 6 at the point of intersection of lines I and J.
8. Locate point 7 at the point of intersection of lines H and I.
9. Indicate line N, from point 6 to point 2.
10. Locate point 8 on line N, at a point equidistant from points 6 and 2.
11. With the end of the tape-measure wrapped around, and near the sharpened end of the crayon or pencil, and the distance, on a straight line between points 1 and 8, on the tape-measure as a radius, and point 8 as a centre, draw curved line O, or the "bust line," from point 1, through point 7, to the fold; as illustrated in the cut.
12. Having taken the "depth-measure" (on the form for which the cape or circular is intended), from the centre at the side of the neck, to the length or depth desired, locate point 9 on the selvage, that distance (in a direct line) from point 8; as illustrated in the cut.
13. Locate point 10 on the fold of the goods, the distance of the depth-measure (in a direct line) from point 8; as illustrated in the cut.

14. With the end of the tape-measure wrapped around, and near the sharpened end of the crayon or pencil, the distance of the depth-measure on the tape-measure as a radius, and point 8 as a centre, draw curved line P, or the "bottom line of the cape," from point 9 to point 10; as illustrated in the cut.

15. Indicate line Q, from point 9 to point 10; as illustrated in the cut.

16. Indicate line R, at right angles with line Q, from point 4 to line Q; and locate point 11, at the point of intersection of lines O and R; as indicated in the cut.

17. Having taken the "bust-measure" (on the form for which the cape or circular is intended) around the form over the fullest part of the bust and arms, to the desired degree of closeness, find ½ of said "bust-measure"; subtract this amount from the distance between point 1 and the point of intersection of line O and the fold, along curved line O: Locate point 12 on line O, the distance of ½ of this difference, from point 11 toward point 1.

18. Locate point 13 on line O, the same distance from point 11 toward point 7; as the distance between points 11 and 12.

19. Draw line S, from point 4 through point 12, to line P.

20. Draw line T, from point 4 through point 13, to line P; which process, completes the cape, seamless at the centre of the back: Figure XXXV, or the front of which, is indicated and included within lines P, B, C, D, and S; and Figure XXXVI, or the back of which, is indicated and included within lines P, the "fold of the goods" at the top, J, K, and T; as illustrated in the cut.
Q. E. D.

☞ For the cape, seamless at the side, abandon the "fold of the goods," "the selvage of the goods," and lines S and T, and proceed as follows, viz:

21. Locate point 14 on line O, the same distance from point 1 toward point 12, as the distance between points 11 and 12.

22. Without changing the position of Figure XXXIII, at the point touching upon point 4, move the lower part of the lining of the front back toward line R; to the position in which, line B will pass over point 14; and mark the line indicating that position, by the character U; as illustrated in the cut.

23. Locate point 15 on line O, the same distance from the point of intersection of line O with the "fold of the goods," toward point 13, as the distance between points 11 and 13.

24. Without changing the position of Figure XXXIV, at the point touching upon point 4, move the lower part of the lining of the back toward line R, to the position in which, line I will pass over point 15; and mark the line indicating that position, by the character V; as illustrated in the cut. Which process, completes the draft of the cape, seamless at the side; as formed and indicated within lines P, U, C, D, K, J, and V—the back and front, or Figure XXXVII.
Q. E. D.

N. B.—1. The "dolman-cape," or cape which is seamless at the back or side, having fullness on top of the shoulder, and a seam immediately in front of the arm, is easily cut from the same draft as given for the cape, in either form above; by applying the same principles of taking out the dart, and throwing the seam, as a mere extension of line E (where it begins to curve) down the front, to the bottom of the cape; and allowing the desired extra fulness, along the upper part of line E. A little thought, and practical experiment on paper, will easily solve the problem for the careful student. Or this cape may be cut from the upper part of the draft of the "dolman;" which, is given elsewhere.

CIRCULARS.

PLATE XIV.—FIGURES XXXVIII AND XXXIX.

INSTRUCTIONS.

1. Cut and locate Figures XXXIII and XXXIV exactly the same, and by the same rules and principles as given for the same, in drafting capes; and draw the construction lines, and locate all points, the same.

2. Having taken the "depth" and "bust"-measures, per instructions for drafting capes, locate point 16 on the selvage, the distance of the depth-measure, in a direct line from point 8, toward the bottom edge of the goods.

3. Locate point 17 on the fold of the goods, the distance of the depth-measure, in a direct line from point 8, toward the bottom edge of the goods.

4. With the end of the tape-measure wrapped firmly around, and near the sharpened end of the crayon or pencil, the distance of the depth-measure back from the pencil on the tape-measure, as a radius, and point 8 as a centre; draw curved line W, from point 16 to point 17.

5. Extend line S, from line P to line W.

6. Extend line T, from line P to line W; which process, completes Figure XXXVIII, or the front of the circular, that is seamless at the centre of the back; and as formed and indicated within lines W, extended line B, C, D, and line S; and also Figure XXXIX, or the back of the same circular; as formed and indicated within lines W, "the fold of the goods," J, K, and line T; as illustrated in the cut.

Q. E. D.

For the cape, seamless at the side, abandon the "selvage" and "fold" of the goods, and lines S and T; and proceed as follows, viz:

1. Without changing the position of Figure XXXIII, at the point touching upon point 4, move the lower part of the lining, or Figure XXXIII, back toward line R; to the position in which, line B will pass over point 14; and extend the continuation of line B to line W, lettering it by the character U; as illustrated in the cut.

2. Without changing the position of Figure XXXIV, or the lining of the back, at the point touching upon point 4, move the lower part of the same toward line R; to the position in which line I will pass over point 15; and extend the continuation of line I to line W, lettering it by the character V; as illustrated in the cut. Which process, completes the draft of the circular that is seamless at the side; as formed and indicated within lines W, U, C, D, K, J, and line V; or in other words, the front and back, or Figure XL; as indicated in the cut.

Q. E. D.

N. B.—1. The principle underlying the cutting of capes and circulars, is one and the same; differing as they do, in their depth only.

2. The above principles are exhaustive, and are commensurate with every possible emergency of the cutter; as touching the variations of form, style and material; and they deserve the student's most careful study and analysis.

3. In case the cape or circular is desired to fit very closely to the form, below the "bust-line," the desired amount of surplus goods may be taken out, equally toward the back and front; by simply changing the angle of lines S, T, U, and V, to the desired degree of slope; from the points of intersection of said lines, with the "bust-line." Every possible shade of style, as well as every degree of fit, is at the choice of the cutter; when armed with the above measures, principles, and draft.

The International Method of Foreign and American Scientific Tailor-Principles, for all Kinds and Styles of Sacque, Coat, Paletot, and Ulster-Drafting.

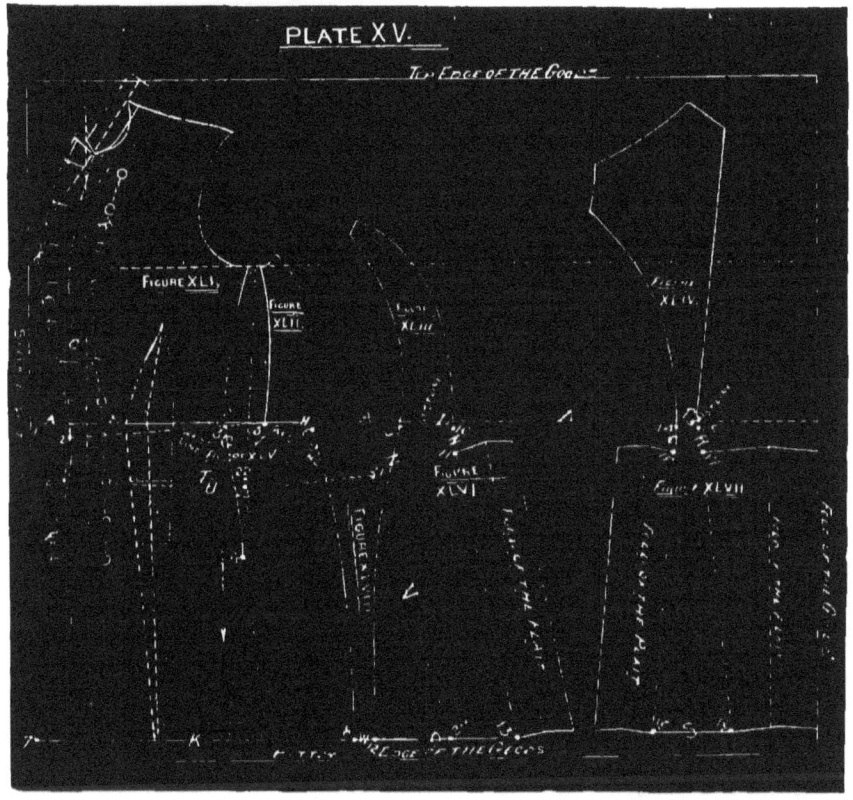

PLATE XV.—FIGURES XLI, XLII, XLIII AND XLIV.

EXPLANATORY INSTRUCTIONS.

1. Figures XLI, XLII, XLIII, and XLIV, are the linings of the front, side-front, side-back, and the back, respectively, of the dress-coat; having 1 dart, and double-box-plaiting, at the back. In attempting to draft either of the above standard style, or any plate desired, notice particularly, the following, viz :

a. The number of darts, and style of seams at the back; their number, and their location.

b. The extensions; where, and how many; and to what extent.

c. The length, and style of the skirt.

d. Whether double or single-breasted.

Then, provide for each particular demand in the design or plate, chosen, according to the following rules and principles; which, cover every possible emergency, whether it be a sacque, paletot, coat, or ulster; all having one and the same draft, and differing only, in length or depth; and in the degree of closeness of the fit, and styles of finish, etc.

2. Use the original close-fitting body and sleeve-measures, as taken per instructions for measuring the form, for dress-linings, and dress-sleeve-linings; only, using those measures, in their increased form; and according to the following table of ratios of increase, viz.:

THE CLOSE MEASURES FOR ALL KINDS OF GOODS.		Inches.	THE INCREASED, OR COAT-MEASURES. Inches. Medium-weight Goods.	Inches. Heavy-weight Goods.	Inches. Very Heavy-weight Goods.
1.	Neck-Circumference,	12	12½	13	13½
2.	Skirt-Depth,	10	10¼	10½	10¾
3.	Blade-Width,	18	18½	19	19½
4.	Chest-Circumference,	36	37	38	39
5.	Waist-Depth,	8	8	8	8
6.	Back-Waist Depth,	8½	8½	8½	8½
7.	Back-Height,	7½	7½	7½	7½
8.	Front-Waist Depth,	9	9	9	9
9.	Front-Height,	5	5	5	5
10.	Arm's-Eye Circumf.,	15	15½	16	16½
11.	Back Shoulder-Height,	7¾	7¾	7¾	7¾
12.	Front Shoulder-Height,	8½	8½	8½	8½
13.	Shoulder-Depth,	6¼	6⅜	6½	6¾
14.	Bust-Circumference,	37	38	39	40
15.	Waist-Circumference,	24	25	26	27
16.	Hip-Depth,	7	7	7	7
17.	Hip-Circumference,	48	49	50	51
18 Sleeve:	1. Extreme-Depth,	22	22¼	22½	22¾
	2. Arm's-Eye Circumf.,	15	15½	16	16½
	3. Elbow-Circumf.,	10	10½	11	11½
	4. Elbow-Depth,	13	13⅛	13¼	13⅜
	5. Hand-Circumference,	8	8½	9	9½
	6. Inside-Height,	8	8⅛	8¼	8⅜
	7. Arm-Circumference,	12	12½	13	13½
	8. Outside-Height,	11	11⅛	11¼	11⅜

N. B.—1. The height and depth-measures remain unchanged, excepting the shoulder and skirt-depths; the ratio elsewhere, being based on eighths, fourths, halves, and the standard unit of measurement, or the inch.

2. Draft Figure XLI, XLII, XLIII, and XLIV, or the plain patterns of the fit, or the linings, to the original close-fitting measures, and as increased in the above ratio; either by the compound, or by the direct method; the first of which, is illustrated and explained under Plate I, and the second, under Plates V, VI, VII, VIII.

3. Locate point 1 in both or either of the methods, the required distance further from the front edge, or edges, for double-breasted garments, diagonals, or extensions, than is given in the instructions for plain or single-breasted body-drafting.

UTILITY AND DECORATIVE ART.   129

4. Where two darts are called for, in the design, draft to the increased measure, per instructions for two darts, in body-drafting. Where only one dart is desired, draft per instructions given for drafting Figure V, under Plate V, with one dart; carefully and accurately locating the points for two darts first; and then eliminating from these given and fixed points, the position and amount to be enclosed, if but one dart is desired; and also, the amount, in that case, to come out as extra curve, in front at the waist; as well as at the side-front; where it joins the front, on the waist. All of which, can easily be done directly and at once, without first locating the points for two darts, after a little thought and experience, in the student.

5. Diagonals, here as in the dress, are simply surplus extensions, provided for in front of the draft, when locating point 1; and the same are entirely subject to fashion, and will, both as to depth, height, and design; being outside, and entirely independent of the fit; and which, is also true of the double-breast extension: The buttons being set as far inside of the curved line of the fit, at the front, as the front terminus of the buttonhole is in front, of the same curved line; the buttons and buttonholes being spaced exactly opposite each other, on these respective lines; so as to be symmetrical and true to each other, when completed and buttoned; and as all is illustrated in the cut.

PLATE XV.—FIGURES XLV, XLVI, AND XLVII.

Instructions for Forming the Draft upon the Goods.

Figures XLV, XLVI, and XLVII.

1. Having drafted, traced, and cut, either the patterns or linings, per instructions immediately preceding, and as given above, proceed as follows, viz: Find the difference between the front waist-depth, and the waist-depth; add this difference to the desired amount for hem; add this sum to the skirt-depth: then, locate point 1 on the selvage of the goods, the distance of this last sum from the bottom or right hand edge of the goods, toward the top; as illustrated in the cut.

2. Indicate line A, at right angles with the selvage of the goods, from point 1 to the fold or opposite edge; as illustrated in the cut.

3. Letter the following lines of Figure XLI, as follows, viz: the traced "construction line," at right angles with and across Figure XLI, above the waist line, B; the curved line near the front edge, representing the fit, C; the curved line in front of line C, near the edge of the extension for double-breast, representing the front terminus of the button-holes, D; the curved line representing the front edge of the double-breast, E; the curved line near the front, inside of line C, representing the line or location of points for the buttons, F; the curved line on the opposite edge, below line B, G; as illustrated in the cut.

4. Letter the following lines of Figure XLII, as follows, viz: The traced "construction line," at right angles with, and across Figure XLII, above the waist line, H; the curved line at the edge below line H, joining with the front or Figure XLI, I; the curved line on the opposite edge, below line H, J; as illustrated in the cut.

5. Locate the following points, at the following given points, on Figure XLI, viz: Point 2, at intersection of lines E and the "waist line;" point 3, at intersection of lines G and the "waist line;" point 4, on line G, the distance of the hip-depth from point 3, toward the bottom; as illustrated in the cut.

6. Locate the following points at the following given points on Figure XLII, viz: point 5 at point intersection of lines I and the "waist line;" point 6, at intersection of lines J and the "waist line;" as illustrated in the cut.

7. Place Figure XLI, or the lining of the front, upon the goods opposite the selvage, in the position in which, line B will be exactly upon line A; the arc of the neck-circumference toward the top and front; with line E, intersecting with the selvage

of the goods at a point, which is the distance of the skirt-depth (or the sum of the skirt-depth plus the hem, if a hem is desired), from the point of intersection of lines B and E, toward the bottom : Then locate point 7, at the point of intersection of line E and the selvage ; as illustrated in the cut. Then baste or fasten Figure XLI, securely on the goods, in that position,

8. Place Figure XLII, or the pattern or lining of the side-front, on the goods, opposite the front ; in the position in which, line H will be exactly upon line A, the arc of the arm's-eye circumference toward the top and front; with lines G and I intersecting at point 4: Then, baste or fasten Figure XLII, securely on the goods, in that position ; as illustrated in the cut.

9. Extend curved line J, in the direction given it by the draft, from point 6 toward the bottom; to the same distance from point 6, as the distance between the point of intersection of lines B and C, and point 7 ; at which point, locate point 8 : as illustrated in the cut.

10. Draw line K, from point 7 to point 8 ; which process, completes Figure XLV, or the entire front: as formed and indicated within line E, K, J, I, and G, below line A ; and by the outlines of Figures XLI and XLII, above line A ; and as illustrated in the cut.

11. Letter the following lines of Figure XLIII, or the pattern or lining of the said-back, by the following characters, viz: the traced line at right angles with and across Figure XLIII, above the "waist-line," L; the curved line below line L, at the edge joining with line J of the front, M; the line on the opposite side, below line L, N; and as illustrated in the cut.

12. Locate the following points, at the following given points on Figure XLIII, viz: point 9, at intersection of line M and the "waist-line;" point 10, at intersection of line N and the "waist-line;" point 11 on line N, the distance of the desired depth for the extension for plaiting, or finish, from point 10, toward the bottom; and as the same is illustrated in the cut.

13. Place Figure XLIII on the goods, opposite Figure XLV; in the position in which, line L will be exactly upon line A, the arc of the arm's-eye circumference toward the top and front; with line M (if extended at its given angle to the bottom edge) intersecting the bottom edge, at a point 1 inch farther toward the fold than the point of intersection of line J with the bottom edge—if line J were extended to the edge: then baste or fasten Figure XLIII securely on the goods, in that position; as illustrated in the cut.

14. Extend line M, at its given angle, toward the bottom edge, to the same distance from point 9, as the distance between points 6 and 8; at which point, locate point 12: and as the same is illustrated in the cut.

15. Extend line N, by indication at its given angle of slope toward the bottom, to the same distance, from point 10, as the distance between points 9 and 12; at which point, locate point 13; and as the same is illustrated in the cut.

16. Draw line O, from point 12 to point 13; and design the desired amount of extension, by a line back of and parallel with line N; and the same depth as the distance between points 11 and 13: which process, completes Figure XLVI, or the side-back; as formed and indicated within lines O, the lines bounding the extension for the plait, line M and line N, below line L; and as shown by the outlines of Figure XLIII, above line L; and as the same is illustrated in the cut.

17. Letter the following lines of Figure XLIV by the following characters, viz: the traced line, at right angles with and across Figure XLIV, above the "waist-line," P; the line at the edge joining with line N, of the side-back, below line P, Q; the line at the opposite edge, below line P, R.

18. Locate the following points, at the following given points on Figure XLIV, viz: point 14, at the intersection of line Q with the "waist-line;" point 15, at intersection of line R with the "waist-line;" point 16, on line Q, the distance of the desired depth of the extension for plaiting, from point 14, toward the bottom; point 17 on line R, the distance of the desired depth of the extension for plaiting, from point 15, toward the bottom; and as the same is illustrated in the cut.

19. Place Figure XLIV, or the pattern or lining of the back, on the goods, at the desired or required distance from the fold toward the front, in the position in which line P will be exactly on line A, the arc of the neck-circumference toward the top, and the arc of the arm's-eye circumference toward the front-edge, from the fold: then, baste or fasten Figure XLIV securely on the goods, in that position; and as the same is illustrated in the cut.

20. Extend line Q, by indication, at its given angle toward the bottom; to the same distance from point 14, as the distance between points 10 and 13; at which point, locate point 18; as the same is illustrated in the cut.

21. Extend line R, by indication, at its given angle toward the bottom; to the same distance from point 15, as the distance between points 14 and 18; at which point, locate point 19; and as the same is illustrated in the cut.

22. Draw line S, from point 18 to point 19; extending it to the fold, and toward the front, to the desired width of the extension; and designing the required extensions, as in Figure XLVI: and as the same are shaped and illustrated in the cut; which process, completes Figure XLVIII, or the centre-back; as the same is formed and indicated by lines S, the lines bounding the plaits or extensions, and Q and R below line P; and as formed and indicated by the outlines of Figure XLIV, above line P; and as the same is illustrated in the cut.

Q. E. D.

N. B.—1. If less goods is desired at the bottom, lines J, M, N, Q and R can be extended toward the bottom, from the distance of the hip-depth below line A, at right angles with line A; or a measure can be taken, and the exact desired amount drafted in the garment or costume, at that particular place; and to suit the fancy or taste of each and all.

2. Figure XLVIII, included within lines E, U, V, O, W and K, is "the English-coat or ulster-skirt"—the front, side-front and side-back being cut along line T; and the skirt in one entire part; joining with line Q of the centre-back, from the front, or line E. The distance between points 20 and 21 must be the same as the actual amount of cloth in the draft on line T, from line E to line Q; exclusive of darts and spaces, between those parts of the costume. The same principle governs the space or distance between points 7 and 24, on lines K, W, and O; as pertaining to the draft, on that line. Points 20 and 21 are located any desired distance below the "waist-line." Line T is a direct line, from point 20 to point 21. Point 22 is at the point of intersection of line "B" of the original body-draft, with line T.

Find ⅛ of the difference between the waist-circumference and the hip-circumference: point 23 is ¼ of that distance from point 22, on line "B," of the original body-draft. Line U, is a curved line, from point 20 through point 23, to point 21. The coat or ulster is cut on line T; and the skirt is cut by line U; and the latter should be slightly full, and carefully marked, and properly adjusted, in every particular; before going to the stitcher and presser.

3. Sacques, are only loose short coats; paletots, longer loose coats; and ulsters, simply long coats; each and all of which differ nothing in principle; only in depth, and the degree of closeness or ease of fit; which, principle of degrees, is the same as that given under Figure V, in Plate V; and which, deserves the student's most careful study and thorough analysis, in all its varied bearings; and which, when properly understood and mastered, will enable him or her, to easily find the "hidden end;" and unravel, what to many seems, a most vexatiously "tangled-skein."

N. B.—Riding-habits are cut exactly on the same principle as long coats—cutting the skirt at will, to any design or fashion plate.

FOREIGN AND AMERICAN

Scientific Tailor-Principles,

FOR

DRAFTING THE MATHEMATICAL MOULD OR PERFECT FIT OF THE HUMAN FORM, FROM THE INDICES OR MEASURES GIVEN, TRUE TO ALL KINDS AND STYLES OF

MANTEAUX, PELISSES, MANTELETS, DOLMANS, AND

DOLMAN-SLEEVED-WRAPS, ETC.

UTILITY AND DECORATIVE ART. 133

PLATE XVI.

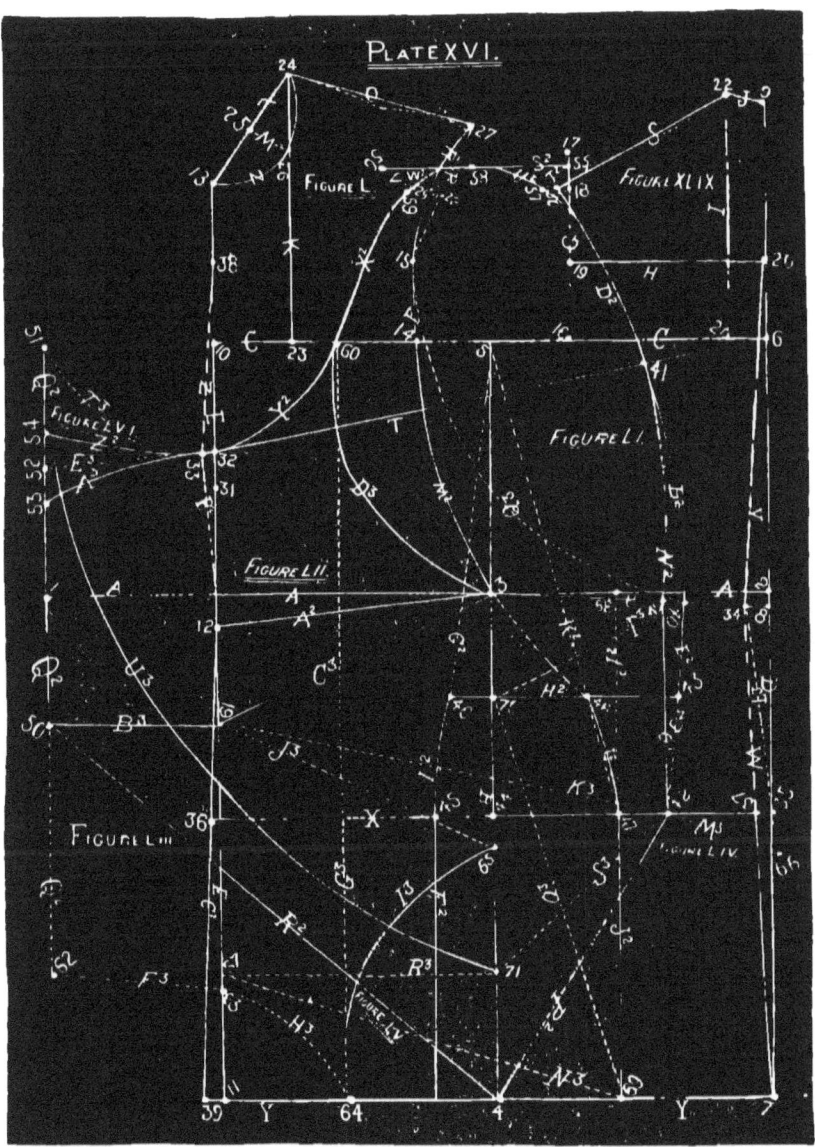

The International Dolman-Plate.

Dolman-Sleeved Wrap-Cutting.
PLATE XVI.—FIGURES XLIX, L, LI, LII.

EXPLANATION.

1. The above Figures, respectively, represent the back, front, top-sleeve, and under, or bottom-sleeve, of the Standard English-Dress-Dolman.
2. To simplify the process, use the exact measures given for practice, in close-fitting body-drafting; remembering however, that when drafting to any particular form, the close-fitting measures must first be taken, and the draft founded on the same measures, increased in the ratio given, for coat-cutting; and as illustrated under Plate XV.

INSTRUCTIONS.

1. Locate point 1 on the material upon which the draft is made, the distance of the desired skirt-depth, plus the hem (if a hem is desired), from the bottom or right-hand edge, and 1½ inches from the front, toward the opposite edge.
2. Add 1½ inches to the extreme arm-depth-measure: Draw line A, at right angles with the front edge, from point 1 toward the opposite edge, to that distance from point 1; at which point, locate point 2.
3. Locate points 3, 4, 5, and 6, and draw line B, and line C between points 5 and 6, per instructions given for locating the same points and drawing the same lines respectively, in Figure I, of Plate I.
4. Find the sum of the waist-depth, plus the skirt-depth: Draw line D, at right angles with line C, from point 6 toward the bottom, to that distance from point 6; at which point, locate point 7.
5. Locate points 8, 9 and 10, and extend lines D and C; per instructions given under Plate I, for locating the same points, and extending the same lines respectively, in Figure I.
6. Draw line E at right angles with line C, from point 10 toward the buttom, to the same distance from point 10 as the distance between points 6 and 7; at which point, locate point 11.
7. Locate points 12, 13, 14, 15, 16, 17, 18, 19, 20, 21, 22, 23, 24, 25, 26, 27, 28, 29, and extend and draw, and indicate lines E. F. G. H. I. J. K. L. M. N. O. P. and Q; per instructions given under Plate I, for locating the same points, and extending, drawing, and indicating the same lines respectively, in Figure I.
8. Draw curved line R, from point 27 through point 29, to point 15; and as the same is illustrated in the cut.
9. Draw line S from point 22, through point 18, to the distance of the shoulder-depth from point 22; at which point, locate point 30.
10. Locate points 31 and 32; per instructions given under Plate I, for locating the same points in Figure I.
11. Draw line T from point 6 through point 32, to the distance of ¼ inch more, or farther, from point 6 than ½ of the bust-circumference; at which point, locate point 33.
12. Draw line U, from point 3 to point 8.
13. Find ⅛ of the difference between the chest-circumference and the waist-circumference; locate point 34 on line U, ½ of that distance from point 8, toward point 3.
14. Draw line V, from point 20 to point 34.
15. Draw line W, from point 34 to point 7.
16. Locate point 35 on line D, the distance of the hip-depth from point 2, toward point 7.

UTILITY AND DECORATIVE ART. 135

17. Draw line X at right angles with line D, from point 35 toward the front, to that distance in front of the point of intersection of lines X and E, which is equal to ½ of the distance between points 32 and 33; at which point, locate point 36.

18. Locate point 37, at the point of intersection of lines X and W.

19. Draw line Y, from point 7 to point 11.

20. Locate point 38 on line E, at a point equidistant from points 10 and 13.

21. Draw line Z, from point 38 to point 33.

22. Draw line A^2, from point 12 to point 3.

23. Draw line B^2, from point 33 to point 12.

24. Draw line C^2, from point 12 through point 36, to the point of intersection of lines C^2 and Y, if line Y—is extended sufficiently far, toward the front, from point 11; and as the same is illustrated in the cut; at which point, locate point 39.

25. Find ⅙ of the waist-circumference; locate point 40 on line U, ½ of that distance from point 8, toward point 3.

26. With a radius of three-times the waist-depth from the point of the pencil, on the tape-measure, draw curved line D^2, from point 30 to point 40; per instructions and principles given under Plate I, for drawing curved lines G^2, H^2, I^2, and J^2, in Figure I: and then, locate point 41 at the point of intersection of lines D^2 and T.

27. Find ⅙ of the hip-circumference: Locate point 42 on line X, ½ of that distance from point 37, toward point 36.

28. Draw line E^2, from point 40 toward point 42, to the desired distance or depth of the extension, below the waist; at which point, locate point 43.

29. Locate point 44, at the point of intersection of lines B and X.

30. Find the difference between ½ of the hip-circumference and ½ of the waist-circumference: Locate point 45 on line X, ⅓ of that distance from point 44, toward point 36.

31. Draw line F^2, at right angles with line X, from point 45 to line Y.

32. Indicate line G^2, from point 5 to point 45; as illustrated in the cut.

33. Draw line H^2, at right angles with line D, from point 43 toward the front, to the point of intersection of lines G^2, and H^2; at which point, locate point 46.

34. Draw curved line I^2, from point 46 to point 45; as illustrated in the cut.

35. Locate point 47 on line X, at a point twice as far from point 44 toward point 42, as the distance between points 44 and 45.

36. Draw line J^2, at right angles with line X, from point 47 to line Y.

37. Indicate line K^2, from point 5 to point 47; as illustrated in the cut: and then locate point 48, at the point of intersection of lines K^2 and H^2.

38. With the same length of tape-measure, as a radius, draw curved line M^2, from point 15 to point 48; per instructions given, for drawing line D^2.

NOTE.—It is not positively necessary for line M^2, to pass over point 3; as illustrated in the cut. Draw it by the rule and principles given, above; whether passing through point 3, or not.

39. Locate point 49 on line U, the same distance from point 40, toward point 3, as the distance between points 8 and 34.

40. With the same radius of tape-measure, draw curved line N^2, from point 41 to point 49; per instructions and principles given, for drawing line D^2.

41. Draw line O^2, from point 49 to point 42.

42. Draw line P^2, from point 42 to point 4.

43. Draw line Q^2, at right angles with line A, from point 1 toward the bottom, to the distance of ½ of the waist-depth from point 1; at which point, locate point 50; as illustrated in the cut.

44. Draw line R^2, from point 50 to point 4.

45. Extend line Q^2, from point 1 toward the top, to the distance of the waist-depth from point 1; at which point, locate point 51.

46. Locate point 52 on line Q^2, equidistant from points 1 and 51.

47. Locate point 53 on line Q^2, at a point from point 52 toward point 1, which is ¼ of the distance between points 1 and 52.

48. Locate point 54 on line Q^2, the same distance from point 52 toward point 51, as the distance between points 52 and 53.

49. Locate point 55 on line G, at a point equidistant from points 17 and 18.

50. Draw line S^2, at right angles with line G, from point 55 toward the front, to the same distance from point 55, as the distance between points 16 and 17; at which point, locate point 56.

51. Indicate pointed line T^2, from point 55 toward point 14, to the same distance from point 55, as the distance between points 17 and 18; at which point, locate point 57; as illustrated in the cut.

52. Locate point 58 on line S^2, at a point equidistant from points 55 and 56.

53. Draw curved line U^2, from point 58 to the point of intersection of lines D^2 and H; as illustrated in the cut.

54. Indicate pointed line V^2, from point 56 toward point 16, to the same distance from point 56 as the distance between points 55 and 57; at which point, locate point 59: and, as the same is illustrated in the cut.

55. Draw curved line W^2, from point 59 to point 58; as illustrated in the cut.

56. Locate point 60 on line C, the distance of ¼ of the arm's-eye-circumference from point 5, toward point 10.

57. Draw curved line X^2, from point 59 to point 60; as illustrated in the cut.

58. Draw curved line Y^2, from point 60 through point 32, to point 33; and as the same is illustrated in the cut.

59. Draw curved line Z^2, from point 54 to point 33; as illustrated in the cut.

60. Draw line A^3, from point 53 to point 33, as illustrated in the cut.

61. Draw line B^3, at right angles with line Q^2, from point 50 to line E; and locate point 61, at the intersection of lines B^3 and E.

62. Draw line C^3, from point 61 to point 3.

63. Draw curved line D^3, from point 60 to point 3, with a radius of tape-measure, equal to the waist-depth; and per instructions and principles given for drawing lines D^2 and N^2, in the same cut; and as the same is illustrated in said cut; which process, completes the draft of "The Standard-Dolman." Figure XLIX, or the back, of which, is formed and indicated within lines D, V, W, Y, F^2, I^2, H^2, E^2, D^2, S, and J; Figure L, or the front, of which, is formed and included within lines E, Z, B^2, C^2, Y, J^2, L^2, M^2, R, O, and N; Figure LI, or the outside-sleeve, of which, is formed and indicated within lines P^2, O^2, N^2, D^2, U^2, W^2, X^2, Y^2, Z^2, Q^2, and R^2; Figure LII, or the under-sleeve of which, is formed and indicated within lines Q^2, B^3, C^3, D^3, Y^2, and A^3.

Q. E. D.

The Standard-Romanoff, or "Princess-Louise-Sleeve."

PLATE XVI.—FIGURE LIII.

INSTRUCTIONS.

1. Indicate pointed line E^3, from point 52 to point 32; as illustrated in the cut.

2. Indicate the extension of line Q^2, by points, from point 50, toward the bottom, to the same distance from point 50; as the distance between points 50 and 52; at which point, locate point 62: as illustrated in the cut.

3. Locate point 63 on line E, the same distance from point 61, toward the bottom, as the distance between points 61 and 32.

4. Indicate line F^3, from point 62 to point 63.

5. Indicate line G^3, at right angles with line C, from point 60, toward the bottom, to the distance of three-times the waist-depth from point 60; at which point, locate point 64; as illuttrated in the cut.

6. Indicate curved line H^3, the reverse of Y^2, from point 63 to point 64; as illustrated in the cut.
7. Locate point 65 on line B, the distance of the waist-depth from point 3, toward point 4.
8. Indicate line J^3 from point 61 to point 65: as illustrated in the cut.
9. Indicate line K^3 from point 61 to point 42; which process, completes Figure LIII, or the "Princess-Louise-Sleeve"—as formed and indicated within lines Q^2, F^3, H^3, I^3, J^3, K^3, O^2, N^2, D^2, U^2, W^2, X^2, Y^2, and E^3: the lower part which, when folded from underneath, to the upper part, will be exactly symmetrical, in all its parts—and as a whole.

Q. E. D.

The Standard French-Mantilla.

PLATE XVI.—FIGURES LIV, LV AND LVI.

INSTRUCTIONS.

1. Indicate the extension of line E^2, from point 43 to point 42; as illustrated in the cut.
2. Indicate line L^3, from point 34 through point 35, to the distance of the waist-depth from point 34; at which point, locate point 66; as illustrated in the cut.
3. Indicate line M^3, from point 42 to point 66, which process, completes Figure LIV, or the back; as formed and included within lines L^3, V, D, J, S, D^2, E^2, and M^3; as illustrated in the cut.
4. Locate point 67 on line E, 1½ times the waist-depth from the point of intersection of lines A and E, toward point 11.
5. Indicate the extension of line J^2, from point 47 to line A; and locate point 68 at the point of intersection of lines A and J^2; and as the same is illustrated in the cut.
6. Locate point 69 on line J^2, the distance of twice the waist-depth from point 68, toward the bottom; as illustrated in the cut.
7. Indicate line N^3, from point 67 to point 69.
8. Locate point 70, at the point of intersection of lines B and H^2.
9. Indicate line O^3, from point 69 to point 70.
10. Indicate line P^3, from point 70 to point 40.
11. With twice the waist-depth on the tape-measure, as a radius, indicate curved line Q^3, from point 40 to point 15, per instructions for drawing lines D^2, the same is indicated in the cut; which process, completes Figure LV, or the front, as formed and included within lines N^3, O^3, P^3, Q^3, R, O, N, and E; as illustrated in the cut.
12. Indicate line R^3, at right angles with line E, from point 67, to line B; and locate point 71, at the point of intersection of lines R^3 and B.
13. Indicate line S^3, from point 71 to point 42.
14. With the radius on the tape-measure, equal to the waist-depth, indicate the curved line T^3, from point 32 to point 51; per instructions, and principles given, for drawing curved lines D^2, M^2, N^2, etc., on the same plate; and as illustrated in the cut.
15. With the radius of three-times the waist-depth on the tape-measure, indicate curved line U^3, from point 51 to point 71; per instructions and principles given,

for drawing the curved lines, elsewhere, in the same draft: which process, completes Figure LVI, or the sleeve; as formed and included within lines S^3, O^2, N^2, D^2, U^2, W^2, X^2, Y^2, T^3, and U^3.

Q. E. D.

N. B.—1. The "Berlin-Romanoff-Pelisse" is simply an extension, in length, of the "English-Dolman"; having a double box-plait in the centre-seam of the back, and one at the side; and which is also graced with the "Princess-Louise-Sleeve."

2. "Pelisses," "Visites," "Dolman-Circulars," "Dolman-Sleeved-Cloaks," "Hortensian-Wraps," etc., are all drafted by the same principles; differing only, in the degree of closeness of fit, in the depth of the skirt and sleeve; and also, in the arrangement and draping of the sleeve, at the back and front.

3. If the back and sleeve are desired as a whole, or in but one piece, locate point 34 twice its former distance from point 8, toward point 3; draw lines V and W, to their new position; and abandon lines D^2, N^2, E^2, and O^2, tracing the back and sleeve as one piece; and as the same is taught, outlined and given in the preceding instructions and illustrations.

SACQUE, COAT, PALETOT, AND DOLMAN-SLEEVED WRAP-BASTING.

1. Baste per instructions for close-fitting garments; and "press-out" very carefully, all fullness allowed upon the linings everywhere.

2. The above draft and principles are not founded upon any imaginary or fanciful anatomical proportions of the human form, but upon actual measurement; and upon mathematically and practically demonstrated truths. And although, differing radically in its method of construction from anything ever offered by the profession, it embraces all the great and truly scientific principles of the world's greatest and most noted cutters and drapers; and that, too, in an exceedingly simplified and exhaustive form; all of which has cost the author no small amount of hard study, and expense of time and money; and which the earnest and industrious student will find an orchard rich with the perfume of blooming principles; that, by experience, will ripen into the "golden harvest" of a rich and cherished success. Here we bid adieu to the student, until we shall join company again at the entrance gate of a different field; and enter a garden in which bloom other flowers—or within the province of part third.

ALL BRANCHES OF

SEAMING, STITCHING, TAILOR-FINISHINGS, TAILOR - BUTTON-
HOLES, TRIMMING, DRAPING, STAMPING, EMBROIDERY,
CROCHETING, KNITTING, WORSTED-WORK, FANCY
AND ARTISTIC NEEDLE-WORK, LACE-WORK,
BEAD-WORK, MILLINERY, HAIR-WORK,
FEATHER-WORK, WAX FLOWERS,
AND WAX-WORK, SHELL-
FLOWERS AND SHELL-
WORK, TRANSFER-
WORK,

TRANSPARENCIES, LEAF-WORK, LEATHER-WORK, FANCY-
DYEING, WATER-COLOR PAINTING, CHINA AND
POTTERY PAINTING, GRECIAN OIL PAINT-
ING, LUSTRE PAINTING, LINCRUSTA
PAINTING, KENSINGTON
PAINTING,

AND

UTILITO-DECORATIVE ARTS.

PREFATORY TO PART THIRD.

In studying the development of mind, from the period of the thraldom and chains of the ignorance and fetichism of a barbarous, uncivilized, and uncultivated people, to the present high degree of civilization, refinement, and Christianity, we see a long, rugged, and toilsome way; and an intricate path through the various systems of symbols and abstractions of truth, which were set as stars in man's darkened heavens, by the Great Teacher of all, in order that He might lead man, and guide him in all that is noble, beautiful, and good. But in all cases, as in the above, the symbol is but a feeble token of the real. The sculptor sees in his completed shape of granite or marble but a poor reflector, and a very deficient representation of what he aimed at, when placing the chisel; the artist contemplates with disappointment the degree of deficiency, and the feebleness with which his completed painting, tokens or reflects his real conceptions. And yet, sometimes, the very deficiency of these instruments serve as most important indices, to the thoughtful and appreciative observer; and only lend richness of lustre, to the high standard of the author's designs; for which the given means, were inadequate to express. In the former parts of the work, we have had to deal with "mathematical figures" alone; and by a system of "forms" and "symbols," have attempted to teach by principle and suggestion, rather than by rules, outside of what they plainly express. The accurate and thoughtful student will recognize exactness of order, as well as that of correct science. Each step in the advancement of the work, upon careful investigation, will be found to be the legitimate fruit of a new graft, inserted into the original tree. In the former parts of the work, we have aimed to lead the student through the halls of the science of mathematics, touching our province; and we now, invite the reader to join us again; and go with us on an excursion into another province, where we shall receive some important lessons, from and under the ministrations of art.

INTRODUCTORY TO PART THIRD.

The principles elucidated and submitted in this department of the work, as in all other departments of the same, are not of doubtful import. They are not simply "theories," and things that are not practical, whether said theories be scientific or otherwise. Its matter will be found to be exceedingly practical and useful, as well as beautiful. It is the result of long and hard study, extensive practice, and careful and thoughtful observation. And to those, whose ambition is to excel in "the true, the beautiful, and the good," we commit and dedicate a book which has cost no little expense in time, labor, money, and thought; and that, too, in the midst of many complicated business matters and deep anxieties, coupled with delicate health, while thus imperfectly transferring our thoughts upon its lettered pages; in order to gratify the oft expressed wishes of our patrons and friends; and we cannot but feel, now that the work is completed, that it will, at least, be appreciated by these.

In the principles and suggestions given in the present part of the work from the ministering hand of art, it is sincerely believed there is supplied a long-felt-want, hitherto almost entirely overlooked. That it might profitably have been extended, and made still more exhaustive, we are aware; but this would have required a volume within and of itself, alone; in order that it might have been thoroughly exhaustive in all departments. But it is sufficiently comprehensive to be thoroughly practical; and, after all, that is the want of the present age. We have therefore sufficiently treated all subjects, for all practical purposes; and have, outside of this, taught by principle and suggestion, rather than by rule. If its varied pages will guide the novice, and enable the more experienced and skilful to grasp the conceptions of new ideas, and thereby the elements and principles of a higher, deeper, and broader knowledge, and a richer appreciation of the same, we have accomplished our aim. And it is the cherished hope of the author, whether the student or reader is experienced or inexperienced, young or old, educated or uneducated, rich or poor, that somewhere in its diversified fields, groves, and gardens, stretching over its province; or otherwise along the streams and lakes, and the beach of the waters that wash upon its shores; there may be found by each and all, some acceptable buds, and blossoms, and flowers; or otherwise some cherished shells, that will be made to adorn their earthly home; and the fragrance or lustre of which, will gladden and strengthen the hearts of those who gather there.

THE STITCHES.

1. The principal stitches used in dress-making are:

 a. The Running-Stitch.
 b. The Back-Stitch.
 c. The Overcast-Stitch.
 d. The French or Invisible-Stitch.
 e. The Chain-Stitch.
 f. The Button-Hole-Stitch; and
 g. The Embroidery-Stitch.

2. All of these are indicated by their name, as to their process; excepting the "French," "Button-Hole," and embroidery stitches; which will be explained in their applications, elsewhere.

3. Simple as is the running-stitch, great care must be exercised in taking the exact number of threads up, at each entry of the needle. Especially so, in running together silk breadths for skirts; and also grenadines, and other similar fabrics. It is used for seams of skirts, and putting on trimmings; and also in connection with felling and stitching; and for sleeve-seams, and French-seams. Do not draw the thread too tightly, so as to avoid drawing up the material. On heavy material, an occasional back-stitch is necessary, upon dress-skirts.

4. "Tacking" or "basting", is running upon an exaggerated scale; and is the process of introducing stitches from one to three inches apart.

5. Stitching for body-seams is exactly the same as that used in plain work.

6. In stitching the fronts and the side pieces to the back, the same number of threads must be taken up each time on the needle; in order to produce the pearl-like effect and appearance, which is so remarkable in the work of the accurate and popular "gentlemen-dressmakers" of Paris, Berlin, and in our own cities.

7. For stitching bodice-seams, the stitches may be less carefully executed; but certainly no careless work is allowable, in any kind of work.

8. Seams of bodices are usually both stitched and overcast; otherwise they are bound neatly, and afterward very accurately and carefully pressed.

9. The overcast-stitch is merely a sewing-stitch; which is taken from left to right, however, instead of from right to left; and which is also taken deeper into the fabric or goods, than the common stitches.

10. The grand secrets in all the stitches are the following, viz:

 a. Accuracy as to the direct, or the curved line.
 b. Closeness, and uniformity of their length.
 c. The degree of tightness with which they are drawn.
 d. The security, with which the work is begun and closed.
 e. The manner in which the work is held and handled, while operating.
 f. The manner in which it is pressed, when completed.

The novice should begin with the plainest and most simple stitch, and master every element and principle thoroughly, before experimenting on the second; and then again, the elements and principles of the second should be practically mastered, before attempting the third; and the student should continue thus, until all the stitches are thoroughly mastered; both theoretically and practically. Anything that is worth doing at all, is worth doing in the very best and most accurate manner possible ; and this is especially true of the art of stitchery. The stitchery of the garment is second only, to the design and cut of the same ; and it invariably stamps the grade of the garment. No amount of fine stitchery, or fine finish, will make an illy cut, and an ill-fitting garment, either beautiful or first-class ; but many elegantly designed and artistically cut garments—and that of the best and most costly materials, too,— have been utterly ruined, by imperfect and wretched stitchery. It is the stitchery and fine finish, more than anything else often, that stamps a dress or garment as "Tailor-made!" To a careful observation of all the details, both in cut and finish of the work, is due whatever ascendancy in reputation the "man-dressmaker" has over that of woman. There is no reason why the touch that is superior to man's, on the piano, and in nearly all the fine arts, should not outvie his, when it falls upon the needle. Artistic and elegant buttonholes and stitchery are to garments, what leaves and blossoms are to an orchard ; or what opening rosebuds are, to the green mantle of the rosebush ; and it is in the power and the province of our fair readers to win the prize, by most richly scattering these blossoms ; and by most deftly and artistically placing and arranging these rose-buds.

THE SEAMS.

1. The principal seams used, are the following, viz:
   a. The "Plain," or common-seam.
   b. The "Flat-Seam."
   c. The "French-Seam."
   d. The "Invisible Cloth-Seam."
   e. The "Double-Seam."
   f. The "Corded-Seam ;" and
   g. The "Stitched Hem-Seam."

2. Of the "Plain-Seam," we need not speak; as all will readily understand what is meant, and how it is done.

3. The "Flat-Seam," is used on cloth. It is made as follows, viz: Back-stitch the edges together, on the wrong side, ¼ inch from the edge. Press open the seam, on the wrong side, very heavily; and then hem the edges down, without catching the needle through, to the right side.

4. The "French-Seam," is required for very thin materials, lace-fabrics, etc.; and where lace and net are worn, over colored silks; and it is also used for transparent materials, where the inside is to have an equal finish, with that of the outside. Run the seam as closely as possible, to the edge, and on the right-side of the fabric; turn the seam over, and stitch on the wrong-side, just below the turned-in portion. In cloth and very thick materials, it is accomplished by being turned over, and then stitched down. For sewing-machine-work, it should be first "tacked." "The stitched-hem" is very pretty for muslin dresses, and light summer fabrics. For children, its beauty is enhanced by working the hem, with a beautiful silk, contrasting in color; giving the finish the effect of a "Russia-braid." Tarletan "ball-dress-flounces," stitched with white, or with colored silks, look admirably; and they are thus neatly trimmed, at a trifling expense. The Wilcox and Gibbs machine-stitch, can be used to an advantage here; being aided and improved also, by the sizes of the hemmers.

5. The "Invisible Cloth-Seam," is used for piecing; and is, in case of accurate cutting and careful needle-work, invisible. It is made as follows, viz: Hold the pieces flat, close together, and sew on the wrong side, with very fine "overseam-stitches;" only taking up, one-half of each side of the cloth. The thread must be drawn firmly, with each stitch; in which case, the right-side will only show a delicate line, to mark the seam. It is very simple, useful, and admirably beautiful.

6. The "Double-Seam," is made as follows, viz: Lay the upper edge, ⅓ of an inch from the edge of the under pieces; and place it in such a manner, that both right-sides lie together; then stitch a seam, close to the edge of the upper material; after which, lay it over and stitch it down beyond the first seam, where only two thicknesses of cloth come together. It will be found, easy of comprehension and execution; and is very useful, where strength is required.

7. The "Corded-Seam:" This seam, lies perfectly flat on the wrong-side, and similates a heavy cord, on the right-side. It is made as follows, viz: Leaving the under piece straight, fold the upper inward, and stitch it down to the under part, the width of a small cord, from the edge. This stitching, must be perfectly even; or the effect will be spoiled. It is both useful and decorative.

8. The "Stitched-Hem," is made as follows, viz: Lay the material over half an inch, on the wrong-side; and then stitch it from the right-side. As the raw edge of heavy goods does not ravel easily, it is not necessary to make the second turning; which, would be very thick and clumsy. The strip which forms the binding, is laid on the outside of the material; and is then, run down as far from the edge as the

binding should be wide. When this is done, then fold it over the edge, and on the under side; now sew it down, so that the stitches will be invisible. For making this hem, take a few threads of the material on the needle, and then, run it through the strip. Make the next stitch, the same. This is called "the invisible-hem;" and is very useful and pretty.

9. The "French Half-Hem:" Another mode of hemming, is used by French modistes; which, is called the "Half-Hem." It is used for holding dress-linings, in their proper position; the stitches are taken farther apart, and the needle inserted slantingly, or at an angle, so as to take up the least piece, at a time; and so as not to show on the right-side. This is easy enough, on thick fabrics; such as cloth, rep, and poplin; but it becomes very difficult, on thick silk, as it is likely to show the stitches; which, must be much closer, and at exactly even distances from each other, in this seam.

10. "The False-Hem;" is added to the edges of dresses, and is also used in many other ways. It is made as follows, viz : Stitch the hem to the lining. The lining and false hem are pieces of material run along the edge of the skirt, on the right-side; and which are turned up, and over, on the wrong-side; and then, lightly hemmed down, by "slip-stitches." Various fabrics are used for this hem. When economy is no object, the dress-material is used ; or otherwise silk, or sarcenet, matching in color; and which is lined with muslin—which, should be cut the selvage-way—in order to secure long pieces. Small plaits, are then laid, where requisite; but, the more careful student, will cut the lining the exact shape of the dress.

11. "Pocket-slit-Seam," and binding : In all heavy materials, for Vests, Jackets, New-markets, and Ulsters, or Coats, etc., cut the slit the desired length; and then make small slits, the width of a narrow seam, in each end, the opposite way. Bind the edges; then, sew to the upper edge a piece of the material, three inches deep; now bringing the pieces through, to the wrong side, stitch it the width of a cord, above the edge. On the right side of the under edge, run a bias strip; then lay it over on the wrong-side, and stitch it down the width of three-sixteenths of an inch from the edge neatly, and stitch a pocket to said pieces; and then fasten down the end of the cut, over the top-piece. This is most neatly performed, by the use of the "buttonhole-stitch." The stitches should be very fine, and neatly constructed; as well as regularly made, as to the degree of their length. If it is desired to give the pocket the "corded-finish," it is done thus, viz : First, sew in a cord, having the small strips added, in the same manner as has already been described. In this case however, the stitches should not pass through, on the outside. You can finish the ends in a similar manner; by the use of the "buttonhole-stitch." It should be made with the finest and best of silk.

GAUGING.

1. Gauging, is often termed "gathering." It is an ingenious method of economizing in time and labor, in the process of garnishing, over the other more laborious and slow methods; and at the same time, to obtain in many instances, a more beautiful result. At present, plaiting has almost supplanted gathering, entirely. Formerly, all dresses were gathered. And but a few centuries since, it was almost universally used ; and that, too, in the garments and costumes of both sexes. It is very often used now, at the "black-plait." And in a special manner, is it still used, in the "Abbess-plait."

It is also extensively used for the various styles of "drawn-sleeves;" also, in children's costumes. It is used for all "gathered-flounces."

2. To present a fine appearance, gathering must be done very accurately, and evenly. In each row, the same threads must be taken up, as was done in the one that preceeded it. The regularity of the stitches, and their correct alternations, stamp the work with that peculiar gracefulness of finish, seen in the productions of the masters, in this art. Grace of finish, should be wedded to strength and durability.

3. **The Gathered-Bouillion.**—The "bouillion," when gathered, is the reverse of the above. In this case, you gather the top first, on the same principle

as for general gathering; and when you have performed that part of the work accurately, next gather the lower edge, by arranging the stitches so as to take up all the intervals, or intervening spaces, left in the first gathering. It is in this little "nicety of arrangement" that, alone, lies the beauty of the "bouillion." After having done all the above "gathering," sew the gathers neatly and firmly, into a band. It is then "tucked down," under rouleaux; or otherwise, it is nicely adjusted underneath some other form of superposed-trimming.

4. **Gathered-Skirts.**—A "gathered-skirt," is simply a skirt having portions of the same gathered, by having the gauging longer; and, in addition to sewing the work in, "fold by fold," one or two rows of "slip-stitches" are taken; at a point, one inch below the lowest. This is done to keep the gathered folds, in their exact and correctly located position. In "gauging," use the very best and strongest of waxed-cotton, if you have not the silk-cord. Of course, the latter is by far, preferrable, to any kind or quality of cotton. The "gathered-flounce" is carefully and firmly, and in a neat manner, sewn into a band. All of which, by the mere novice, will be found to be exceedingly simple; and, a very easy process. The way to do a thing, is to feel that what has been and is done, you can do; and, then go determinedly, and patiently to work, and simply do it.

5. **Double-Gauging.**—This, is simply a repetition of the above principles and rules, for "single-gathering;" only, that it is repeated, at the proper and desired distance apart. It may be repeated as often as desired; and in this way, we have "double," "treble," and "quadruple-gathering," etc.

6. **Attaching, or Securing the Gathers.**—If, the goods gathered, is very wide, as it sometimes is at the back of children's dresses and costumes, proceed as follows, viz.:—First, gather the material very evenly and carefully, upon a good strong silk cord; now draw it evenly up, to within the given or desired space; then, placing a pin firmly in the goods or material, wind the thread or cord "under and over" this pin; so as to secure the work as tightly as may be required, during the process of attaching the gathers, or sewing them into the band. Secure each fold separately, and independently from the others, in a manner that will insure strength; as well as, neatness. Of course, the fullness is managed according to the quantity of material used, and the given space, upon which the gathers are placed. The band, represents said space. At the distance of 1¾ inches from the top, the gathers are usually "caught down." This should be done very neatly, by the use of the "slipped-stitch." This will insure security and firmness, to the work. You should always run a strong thread, through all the edges. It only requires a thorough and systematic method of accuracy, care, and patience; along with one or two experiments, for the mere novice, to grasp the entire process of "gathering;" and to become perfectly master of every detail, of this beautiful art of garniture—in garment-making.

CORDING.

1. Another beautiful and helpful method of finish, in garment-making, is the process of "cording." It, likewise, serves both for ornament, and for security and protection. Perhaps, even greater nicety of finish and care are demanded in this than, in "gauging." At least, equally as much. But we can find no place anywhere, for slovenliness or carelessness of work, in our entire province If "cording," or "piping," are worth doing at all, they are worth doing in the most artistic and best manner possible.

2. The kinds of "cording" are cassified as, follows, viz.:—
   a. "Single-Cording;"
   b. "Double-Cording;"
   c. "Treble-Cording;" and
   d. "Quadruple-Cording," etc.

3. The basis upon which the above classification rests, hinges upon the principle of the number of its repetitions, only. The same rules are used for "quadruple-cording," as for "single-cording." The former, is only a repetition of the latter, four-times!

4. The materials must be cut, on the true and exact bias ; having the proper lengths, and widths. For "single-cording" it should, of course, be narrower than for "double" or "quadruple-cording." In the work of "single-cording," proceed as follows, viz. :
   a. Cut the materials, the proper lengths and widths, on the exact bias ;
   b. Place the cord, in the centre of the bias, piece ;
   c. Fold the material over it nicely, and secure it by basting ; either with thread and needle, or otherwise with pins ;
   d. Then "run," it in its proper place ;
   e. Then turn the edges under, and " hem it down."
5. For making "double-cording," we usually fold both edges over the cord ; and when having brought and "run" everything into its proper place, we put the two cords together; and "run" or stitch them, to the material.
6. The trained and experienced hand, arranges and makes the "cording," as it is put on the work ; but this is not a very easy matter to do, accurately and correctly; and often, work is spoiled ; and both time and material are lost, thereby. It should not therefore, be encouraged in the work-room.
7. "Corded-Bodices," are sometimes fashionable. It is usually put on in "double-cording," around the edges. The material is usually, the same. But fashion, fancy, and good taste, must ever regulate such things. We are simply now, teaching *principles*. This, pertains not only to "cording," but to all subjects, in this work.

BINDING.

1. A handsome method of finish, and of garnishing, is termed "binding." It reflects both grace and protection ; and it is a very useful process. It is much used in all ways, and at all times. It is used both as "trimming," and as a secure " finish," at the edges of garments.
2. It is sometimes "double-hemmed;" *i. e.*, on both sides of the braid. And again, it is also "run" or stitched on evenly, on one side first ; and is then "hemmed," on the other side. And still another or third method is, as follows, viz.: fold the braid "in-half;" and then, "run" or stitch it on the material, by taking the stitches through, on the other side. This plan of finish is very nice, on very thin materials. It is often used in binding tarlatan-flounces, with satin ribbon, etc.
3. Finish the bias-flounce, at the edge, by a narrow hem of bias-silk. Be careful to cut both fabrics, exactly on the bias. Then place the silk with the right-side underneath, upon the right side of the flounce. Then stitch it on closely, but lightly ; at $\frac{1}{8}$ of an inch from the edge. It is then turned over, and is "felled" on the wrong-side, over the portion that was "turned in ;" the stitches being invisible, on the right-side.
4. Another method, with a similar effect is, as follows, viz. : Cut the bias-flounce, to a somewhat greater degree of length, than is actually required. Then neatly turn it down, on the right-side of the fabric ; at a point, which is $\frac{1}{8}$ of an inch more than is required for the "rolled-bind ;" now, stitch closely at a point, which is $\frac{1}{8}$ of an inch from the edge, and draw it rather closely ; and then, finish it off very securely. Now fold back the piece that was turned down, and fell it down," on the wrong-side. This, will result in an effect, almost like the first process ; but, it serves the purpose of variety, in processes ; and for which, many have a choice. Everything depends upon the accuracy and the nicety, in the execution of all these " little details," in the appearance, and the real worth, of the completed work. Better make a less number of garments thoroughly, and artistically, and get a first-class price, than to make four-times as many; and do fourth-rate work (which by the way is generally the hardest kind of work), and get only a fourth-rate price! The profession of "artistic-garment-making," is crowded "at the bottom ;" but there is plenty of room, at the top." And, there is an infinitely better opportunity for connoisseurs in this profession, in our country at the present time, than in many other professions ; where, both capital and mental training are extensively required ; and, where the

task is much more difficult, the honors no greater, and the remuneration, far less! The busy looms of the present age, have filled the earth with their choicest productions; the public is purchasing them, and is seeking able and accomplished artists to cut, drape, and complete the same; in the very best, and most useful, and most appropriate form. Aim high; strive hard, for excellency; and success will crown your efforts.

WHALE-BONES AND "CASINGS."

As the system entirely obviates the necessity of whale-bones or "stays," we have no use for them; but, as others might have a prejudice for them, we submit the following rules, in brief, viz:

a. They should reach a depth of $\frac{1}{3}$ of the hip-depth below the waist, and $\frac{1}{2}$ of the waist-depth above.

b. Cut the bones this length, nicely sloping and rounding the ends and edges.

c. The seams should be over-cast; and small casings, the size of the bones, should then be inserted.

d. Stitch the casings on the seam, and put the bones in; and then, over-cast them.

e. The casings should be cut on the bias; and should then be easily sewn on by hand; or, fancy ribbon may be used.

f. They are generally used in the two front gores, and at the side-waist-seams; but, it is at the option of the maker.

g. Fasten the bones by stitching through a hole, in the end of the bone.

WADDING.

"Shall we use wadding, or discard it?" This question must, of necessity, be left to the artist and to the customer. If the customer does not object to "wadded dresses," and you can enhance the appearance of the form and the fit, it is your privilege to do so; and perhaps necessity, in the particular case, would demand it. Certainly, many forms cannot possibly be elegantly fitted, and be made to reflect an elegant appearance without it. In such cases, it is a help to the artist to construct the proper degree of symmetry, which is lacking in the customer. To fit some forms very close and neat, would (without wadding) be in bad taste; while in all general cases, we have no use for it in our system. It was a grand invention to aid the deficient cutter! But when the garment was worn a few times, the effect was lost; and the customer disappointed! But where it is necessary, in order to give the figure or form a better appearance, we offer the following hints, viz:

It is generally placed back of, and over, and in front of the shoulder and arm. It should be so arranged, that the edges will taper down gradually to the mere thickness of a sheet of letter-paper. It should then be securely fastened, in its proper position; and, should be neatly and properly covered.

THE NECK AND SLEEVES.

Bind the neck with a narrow bias-band;—the sleeves, with a bias-facing, also. The band of the neck should be $\frac{1}{2}$ inches wide, so as to allow for turning. Stitch one side, on the outline-thread of the neck. Cut the "turning-in" quite within $\frac{1}{4}$ inch of the stitches; and then turn the bias over, and "hem it down" neatly, on the other side. Face the sleeves, in the same way; but turn down the outside, a little distance in from the edge, forming a fold; and in such a manner, that the facing does not show. Facings for coat-sleeves are from 3 to 4 inches deep; and they should be cut from the 'wrist-end" of the sleeve-pattern; and also, out of a piece of bias-material.

Nothing more enhances, or otherwise detracts from the beauty and elegance of a garment, than the finish given to the same, in the sleeve, and about the neck. The style and weight of the material, along with the occasion and purpose of the garment, should aid us in the choice of the design; but whatever the design, the finish must invariably be of the highest and best grade and quality of work. Never trust these "finishing-touches" to inexperienced hands, when a fine garment is to be completed. "Sleeve" and "neck-finishings" are to the entire garment, what the paint, or stain and varnish is to furniture.

"OPEN"-SLEEVES.

This sleeve is often cut in one entire piece, and folded on the bias; and the fold is turned toward the outside-length. It is joined by a seam, like that in the common sleeve. It can be left open, either at the inside or at the outside; and it can also be made of any length. If it is made narrow, it is sometimes draped with the other sleeve, at the back of the costume. Or, it may be of less length, but of extreme width; thus producing all kinds and styles of costumes; whether Oriental, mediaeval, Greek, or Roman; or otherwise those of the Cossack-form, used at balls in modern times. Among these, we find the quaint and somewhat fanciful names of "The Judge-Sleeve," "The Page-Sleeve," "The Religious-Sleeve," "The Leg-of-Mutton-Sleeve," etc.! And when short, and finished with lace, they are called the "Engageantes;" but this last, is rather an under-sleeve. These extreme forms are, then again, frequently gathered into a band; this band is the size of the arm's-eye-circumference, at the top; and that of the wrist, at the bottom. They may be varied in form:—first, by not being put into a wrist-band; but, by making up the gathering under a parement, or revers. They can also be gathered, or bouillioné, the whole way down; or they may be gathered at the top only. They are then cut very narrow at the edge, where they fit closely to the wrist. The trimming should correspond with the style of the dress. A fine method, for finishing sleeves of all kinds is as follows, viz: Place the right-sides of the material together, and the right-side of the lining together; stitch up, and turn them right-side-out. Both seams and stitches are then out of sight. Much time is also, saved in this way; but care must be exercised in turning it; or the material will, in some cases, be permanently creased. This principle of finish can sometimes be used to advantage, in other parts of the garment. The parement, or the cuff, is an important factor in enhancing the beauty of sleeves. It can be really added, to the sleeve; or, it may be merely simulated by the trimming. "Cording" sleeves, at the arm's-eye, is sometimes fashionable. We simply arm the student with principles, in the art of finishing; which can be applied at such times, and in such ways, as fancy or fashion may dictate. We simply furnish the method; the design is governed by fashion, and the artist.

WEDDING-TROUSSEAUX, AND LOW-BODICES FOR BALL-DRESSES.

The method of cutting "low bodices" is given elsewhere, in the department pertaining to cutting. The depth to which the bodice is cut, or the lowness of the corsage, depends entirely upon the figure, fashion, and correct taste; and upon the correct education and true refinement, and the modesty of the one ordering it. If left too high, it disfigures, rather than enhances, the beauty of its appearance; while if cut too low, it is both vulgar and inappropriate, as it is immodest and offensive to public sentiment and decency; and, cannot be too highly censured. The cut of this line, is the key of beauty in attaining elegance, in the completed garment. The upper part is turned down, outside of the dress; and is creased down, at the line where it is to be corded. The common material is, fine lawn; which is soft, and yet firm enough to bear the strain of lacing. "Ball-dresses," in the main, are made in either one of the following styles, viz:

 a. Pointed-Back and Front.
 b. With Round-Waists.
 c. With Basques of various shapes, cut "in one," with the Bodice.
 d. With Front-Basques, and Pointed-Back.
 The whalebones are cut in blunt points, so as not to cut the dress ; and are then pierced, encased, and secured, as in ordinary work. Perhaps the favorite is, the "Pointed-Basque." The edges are generally corded ; as are also, the sleeves :—either double or single. The top is then corded to match. "Eyelet-holes," are worked on the inside of the back whalebones ; usually in white silk, and in the close overcast-stitch ; which, is preferable to the buttonhole-stitch. Dresses are laced, from the top to the edge ; therefore, the silk lace is put in at the top left-hand eyelet-hole. "Round-waisted-bodices" are cut off at the waist-line, and are then nicely corded ; and the skirt is then sewn firmly, underneath the cord. A sash is always worn with these simple, but pretty, corsages. Basques are hemmed or corded, and according to the nature of the garniture. Colored silks look pretty, at the edges. "Basqued-bodices," are laced up at the back. By a union of the two styles, a bodice with front-basques and pointed-back, is easily arranged ; and in some figures, is very much more becoming and graceful, than the other styles given. Low-bodices are trimmed with folds of silk, tulle, or some rich lace ; and arranged in various forms, and termed, "Berthae ;" which is made on a shape cut from pliable net, or tulle ; and on a tarlatan foundation, taken double ; but, the experienced draper will fold the Berthae upon a paper pattern, running the folds in place, and trimming with lace ; which arrangement produces a fine effect. A "Bouillionné-Berthae" demands a foundation of stiff net, or tarlatan. The bouillonnés are usually finished by lace ; or otherwise with a frilling of the material. The sleeves are always trimmed to match the Berthae ; and, often consist of puffs of tulle. A frill of crepe-lisse, or "Frou-Frou tucker," is run inside of the cording of the neck. Whatever is chosen as a "neck-tucker," must find its repetition on the sleeves. Ball-dresses are worn over "slips" of silk ; or otherwise consist of rich silk, or faille, trimmed with tulle, net, blonde, crepe-de-chine, and other diaphanous fabrics. "Ball-dress-skirts," are made with long trains ; and the prevailing characteristic of walking-dresses, and those of visiting-toilettes, will always be found repeated in gauzy materials, for the ball-dress. Fashion will however hold the scepter, as to the mode.

GREEK-PLAITS.

 The "Greek-plait" is only a box-plait, with about two inches of the outside cut away at the hem ; and the under sides of the fold sloped to the point where they touch. Any box-plaited skirt may be transformed into "Greek-plaits" by the expenditure of little time and trouble.

BUTTONS.

 To sew on a button neatly, and yet firmly, is an accomplishment that few possess ; and it should be highly appreciated in the work-room, or in the home, where it is found. It is not, because it is so difficult a thing to do ; but, because of the general prevalence of a uniform carelessness or indifference toward the little, but necessary essentials, required. And yet, to all who aim even at an approximate perfection, in garment-making, it is an important matter. We can only give a few hints, on the subject ; for we cannot afford to go into details ; on all subjects. We therefore, suggest the following ; simply as a help, viz :
 1. The Covered-Button : To attach this button nicely, make a cover of twist, underneath ; that is, on the under-side. Give one-sixteenth of an inch space, on the cloth, or material. Knot your thread, and place the knot exactly under the neck of the button ; and, at the right-side. Always begin, by putting the needle in, from right to left ; and always, bring it back again. Two stitches are necessary, at first. Now put the needle through the material and the crossed threads, at the under-side

of the button; and bring it through the cloth, close by the first insertion of the needle. Continue this process, seven-times. Do not draw the twist, too closely. Then wind the thread or twist, seven-times around the stitches; and complete the work, by securing it, on the wrong side, by three carefully taken stitches; Be careful not to draw the cloth, or material.

2. The Linen-Button.—The "linen-button" is begun, in the exact same manner, as the "covered-button." You first knot the thread, and hide the knot underneath, and directly under the neck of the button. Take two stitches, as before, from right to left; and then placing the needle up, in the centre of the button, arrange the stitches so as to cross, on the bottom-side. Repeat the stitch, seven-times; now, wind the twist seven-times around the stitches made; underneath the button, and above the material. After winding thus, finish the work on the wrong side, as before; and secure the same, by three carefully-taken stitches.

3. The Eye-Button, is attached to the garment, exactly in the same manner as the "linen-button," only inserting the needle through the "eye;" the rules for which, are given above; and which, is a very simple process.

4. The Woven-Button. — Stitch through the material from the right-side, through to the wrong-side, and draw the knot exactly underneath the centre of the button; this is done by two repeated-stitches, as before. Now using short "back-stitches," form a circular-chain underneath the button. The circle should not exceed one-sixteenth of an inch, in diameter. Insert the needle, and draw the twist out between the material and the button; and wind the twist seven-times around, as before. Then secure the work on the wrong side, as taught above; by three carefully-taken stitches.

5. The Shank-Button.—Knot your thread or twist, as before. Put the needle through the goods, concealing the knot by the work. Two stitches should be given first, to secure the twist. Then insert the needle and twist, through the shank, seven-times. Do not wind the thread around the stitches, as before. But, secure it on the wrong-side, by three stitches, securely made; as taught and given above. These simple rules, will sufficiently cover the necessary requirements.

152 THE ENCYCLOPEDIA OF GARMENT-MAKING,

PLATE XXVII.

Fig 124

Fig 129

Fig 184

Fig 125

Fig 130

Fig 135

Fig 126

Fig 131

Fig 136

Fig 127

Fig 132

Fig 137

Fig 128

Fig 133

Fig 138

International "Tailor-Buttonholes."

International "Tailor-Buttonholes."

PLATE 27.—FIGURES, 124, 125, 126, 127, 128, 129, 130, 131, 132, 133, 134, 135, 136, 137 and 138.

EXPLANATION.

Figure CXXIV, shows the method of working the "Plain-Buttonhole ;" Figure CXXV, shows the method of barring the "Round-Buttonhole ;" Figure CXXVI, represents the method, and the American "Tailor-Buttonhole ;" Figure CXXVII, represents the method, and the Prussian "Tailor-Buttonhole ;" Figure CXXIII, represents the "Double-Stitch-Buttonhole ;" Figure CXXIX, represents the method, and the "Pointed Tailor-Buttonhole ;" Figure CXXX, represents the "Twisted-Stitch-Buttonhole ;" Figure CXXXI, represents the method of making "Knots;" Figure CXXXII, represents the "Knotted Tailor-Buttonhole ;" Figure CXXXIII, represents the "French-Tailor-Buttonhole ;" Figure CXXXIV, represents the "Scotch-Tailor-Buttonhole ;" Figure CXXXV, represents the "English-Tailor-Buttonhole ;" Figure CXXXVI, represents the "Tatted Tailor-Buttonhole ;" Figure CXXXVII, represents the method, and the "Bound Tailor-Buttonhole." All of which, will be comprehended by the mere novice, at a glance; and by combining the various instructions, as given under the subject of "Tailor-Buttonholes."

Tailor-Buttonholes. — In ladies' garments, the custom fixes the position of the buttonholes on the right-hand front, when the garment is upon the form. The opposite of this is true, of gentlemen's garments. And the buttons, of course for ladies, are on the left-hand front. As general rules, we submit the following brief instructions, viz. :

1. Accurately locate the proper " spacings," on the draft ; beginning at the lower extremity of the " Front-waist-depth," and locating the spacings equidistant from each other, up and down from said point ; and from within, toward the front, exactly to the line representing the fit, at the fore-part of the front. The buttonholes on one side, from within to said line ; and the buttons are attached exactly upon said line, directly opposite the buttonholes of the other side. It is best to *trace* the indications of the buttonholes, on the linings, while tracing down the different pieces ; then, it will be an easy matter, simply to set the buttonhole-scissors to the right width, and cut the buttonhole exactly the right width and up to said line, toward the front ; and as indicated by these trace-marks.

2. Cut the buttonhole exactly true with the goods ; or in other words, on a horizontal line with the body. The strain of the buttonhole, must be on the "lengthwise." A width of $\frac{3}{8}$ or $\frac{1}{4}$ of an inch, in front of the line representing the fit, is sufficient to give the necessary strength, to the buttonhole ; at least, in all ordinary goods.

3. In making, or "working" the buttonhole, use the very best quality of "Tailor's Buttonhole-Twist."

4. It is a good plan to stitch, or "encircle" the space, where you wish to cut the buttonhole, before cutting it. This is done, by a double-line of stitching, or "running," as some term it. Two threads of goods are usually left, between the two stitchings. The buttonhole is then cut, between the two inside-rows of stitching.

5. In " working " buttonholes, always work " from left to right." Keep the eye of the needle, very near the buttonhole. The point should be, below the outside row of stitches. Close the stitch very closely to the edges, where the cut is made. In order to accomplish this, turn the twist around the needle, pulling it out, and then drawing it upward ; this brings the stitches near the edge, and, in their proper

place. A short bar is worked across the end, when one side of the buttonhole is completed. This bridges the ends, and gives strength to the buttonhole. And it also unites, or brings together, the entire stitching. Four stitches usually, complete this bar; it, of course, reaches entirely across the width of the buttonhole. Then, work these stitches back; using again, the regular "buttonhole-stitch." Next, you work the other side, in the same manner, as you did the first; and when the side is completed, finish the work by "cross-barring" the other end of the buttonhole; then completing the process, in the regular "buttonhole-stitch;" and then, securely fasten the twist, and insert the needle, and draw the twist underneath the work—hiding the end.

6. The Plain-Buttonhole. — First encircle the space to be cut, by the "double-stitching." Then, having cut the goods the proper size, bar the cut thoroughly, and at once; before the threads of the goods have an opportunity to become displaced. It is often a good plan, in some goods at least, to bar the same before cutting the buttonhole. Be sure you bar it securely, at each end. Always use twist, that is much coarser than the thread of the goods. A buttonhole shows better, and is also firmer, if this is observed. Do not hasten the work too much, and thus spoil its beauty and finish; by taking up too much cloth or goods at one time, in the stitches. They should be very short; and should be exact, and uniform. Let each stitch be drawn firmly, and have a substantial solidity in the work, on the top. The stitch generally called "Buttonhole-stitch," is made by winding the twist only once, around the needle. Be certain to have the stitches close together, and evenly made; so that the barring will be completely hidden, underneath. At the ends, work directly across; and always thoroughly secure the twist, when you have finished; hiding the end, underneath the work.

7. The Round-Buttonhole.—In the "round-buttonhole," the ends are barred in a curved form, instead of an angular form, directly across. It can be worked otherwise, as the "plain-buttonhole;" but in the main, the "chain-stitch" is used, in working the "round-buttonhole." This is positively necessary, in very heavy-weight goods. And indeed in all work, where it is desired to show the work with a heavy or raised appearance. Some work both sides of a buttonhole, first; and after-wards, the ends. But we think this both improper in theory, and imperfect and deficient in practice. Each stitch, in working any buttonhole, should be taken on the same line with that of the cloth or the material, upon which the work is executed; and that, each way, too. First, over-cast the cut, very neatly and closely, all around; and never work a buttonhole, on simply one single thickness of cloth, or material. It will invariably break; and will finally "pull out." This applies to all buttonholes.

8. The Double-Stitch-Buttonhole.—The name of this buttonhole, almost, indicates the process. It is simply made as follows, viz: First, run the double-stitching around the cut; and then bar it, as in other buttonholes. Then work the ordinary "buttonhole-stitch," all around it. Upon this as a foundation, neatly work the "Diagonal-Stitch," over all. This is made by simply taking short diagonal-stitches, from the left-hand toward the right; throwing the loop and drawing the twist neatly and firmly, at the top.

9. The Knotted-Buttonhole.—As its name indicates, this buttonhole is worked in the "knotted-stitch." First, it embraces all the elements and principles of the "plain-buttonhole;" and it is first worked, as any ordinary buttonhole. Upon this foundation the "twist" is laid upward, and drawn through again, underneath the twist or thread of the original work. This is then ornamented with the "knot." The "knot" is made as follows, viz: Pass the twist twice around the needle; and then, put it back through the same insertion, into the cloth; when it is drawn, it forms the "knot." This is repeated, until the work is completed.

10. The Tatted-Buttonhole.—This buttonhole, differs radically from either of the above. First, instead of stitching the twist around the edge of the buttonhole,

we sew a cord to the edge of the cut, using fine "overcast-stitches." Then, it is worked as the "knotted-buttonhole." It brings a similar result; but, in a different way. It gives it a prominence, and boldness of finish, that is much admired by many.

11. The Pointed-Buttonhole.—This is a peculiar stitch; but is very pretty. First, proceed with all the elements and principles as given above, for laying the foundation, viz: Cutting, barring, etc. Then, work it in the ordinary "buttonhole-stitch;" setting the stitches in the cloth or material in such a position, in which the stitches will form points. Each stitch in this buttonhole, is worked off with a second stitch, that is shorter than the first. And it is, of course, worked from left to right. It makes a handsome buttonhole, if it is worked carefully and artistically; and it has many admirers.

12. The Herring-bone-Buttonhole."—The name indicates the process, in this buttonhole. Neither does the "herring-bone-stitch" need any explanation ; since it is known to almost all ; and which, can also be seen in the illustrations. But the "herring-bone" should consist in diagonal "buttonhole-stitches;" one setting the needle from above, below; and the other, in the opposite direction, from below, above ; in the process of which, fasten the edge of the material thoroughly. After you have completed this "herring-bone-stitch," work the inside stitches, nearest the cut, a second time; and do this, with a short "buttonhole-stitch." This completes the entire process. All of which is very simple, and very pretty.

13. The Twisted-Stitch-Buttonhole."—The stitch has given the name; and we might almost say, also the process. First, insert the needle through the material, very close to the edge ; and then, up through the cut. Then wind the twist seven-times around the needle; and draw both the thread and the needle, through said windings. To do this nicely, may require a little experience. Slightly press the windings down, when you pull the needle up. This holds the windings down upon the material. Then, put the needle through the exact same place where it was inserted for the last stitch, and draw it through the material, to the side underneath ; now work all the necessary successive stitches, in the same manner; and continue, until the work is completed.

14. The Piped-Buttonhole."—This buttonhole is made on heavy goods; and where very large buttons are used. We simply apply the elements and principles of "piping," given elsewhere, to the buttonhole. Any one will understand, that it is simply finishing the buttonhole with "piping;" and thus binding it, neatly.

15. Loops.—In thin dresses, such as muslins, grenadines, etc., we sometimes dispense with buttons for closing, at the front; and substitute therefor "loops," when the trimming at the front, consists of bows. They should be worked very neatly, and very closely.

156 THE ENCYCLOPEDIA OF GARMENT-MAKING,

PLATE XXVIII.

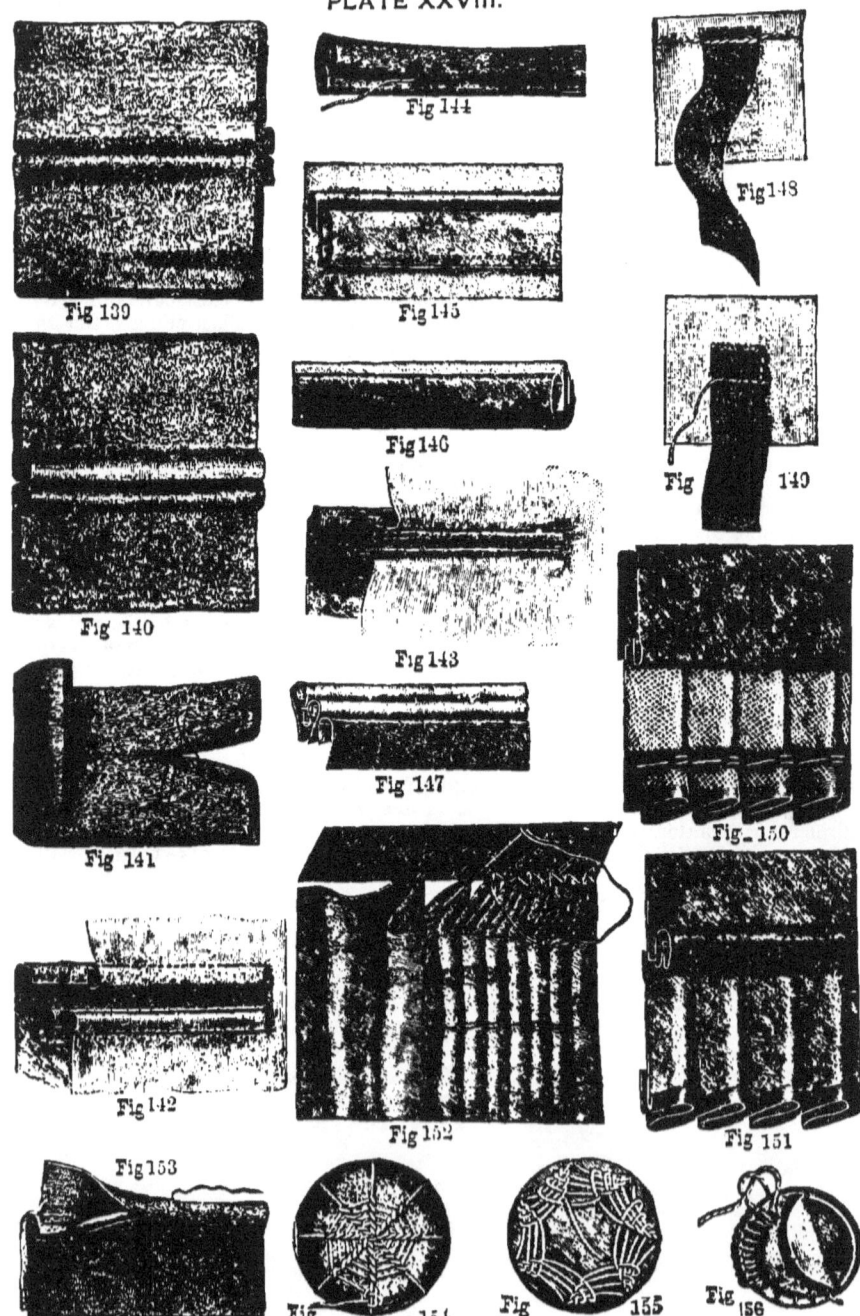

Tailor-Finishing.

PLATE 28.—FIGURES 139, 140, 141, 142, 143, 144, 145, 146, 147, 148, 149, 150, 151, 152, 153, 154, 155, and 156.

Explanatory Instructions.

Figure 139, shows the method and appearance of the right-side of the "American Cloth-Seam." Figure 140, shows the method and appearance of the wrong-side of the "American Cloth-Seam." Figure 141, represents the method and appearance of the "French Cloth-Seam ;" this is sometimes called the "Invisible-Seam," in this country—and is often also, used on thin, and on transparent goods. Figure 142, shows the method of binding, and the appearance, of the "Berlin-Pocket-Finish." Figure 143, shows the method of cording, and the appearance, of the standard "English Pocket-Finish. Figure 144, shows the method and appearance, of the standard "French-Piping." Figure 145, shows the method and appearance, of the standard "French-Fold." Figure 146, shows the method and appearance of the "Berlin-Fold." Figure 147, shows the method and appearance of the "English Double-Cording." Figure 148, shows a process in the American-method of adjusting the tape. Figure 149, shows another process in the American-method of adjusting the tape. Figure 150, shows a process in the French-method of "Kilt-Plaiting." Figure 151, shows another process, in the French-method of "Kilt-Plaiting." Figure 152, shows a process, in the French-method of securing and attaching the "gathering," or "gauging." Figure 153, shows the English-method of attaching the "braid," on the wrong-side. Figure 154, shows the French-method of "Button-Embroidery." Figure 155, shows the English-method of "Button-Embroidery." Figure 156, shows the French-method of "covering" and "garnishing," the circular-wire or "ring." This is executed in linen, or silk; and is often used instead of the button, as well as for adornment. All of the above examples represent simply, "Principles-of-Finish ;" and which are exceedingly simple, and easy of attainment ; and which also, will be comprehended, at a mere glance, when studied in connection with the instructions given, elsewhere in the work, on stitching, embroidery, tailor-finishings, needle-work, etc. We have given the above, because they represent standard-methods ; the principles of which, will ever remain the same ; and which, can not be effected by fashion—as they can be adjusted to every conceivable style.

TAILOR-FINISHINGS.

"Tailor-Finishings," is a broad and extensive term ; and it covers an extensive province, in the art and science of garment-making. Tailors having been the first to give special attention to exactness, in the science of "finish," the various forms of trimming, stitchery, buttonholes, etc., which, by the taste and experience of the world, have been stamped, as the standards of utility, and decoration; these said standards of forms and methods, received the name "Tailor-Finishings." It is an easy matter to detect at once, the difference in the nicety of finish, in favor of the tailor-made suit, when it is compared with the results of the cheap wholesale-processes, which, are applied to ready-made clothing. The ready-made suit, lacks woefully in the fit; but it lacks infinitely more, in the finish! Everywhere, it reveals an utter and wretched disregard for those "little niceties" of finish, which so richly adorn and beautify the tailor-made suit; and which enhances the graceful appearance of the same, on the one wearing it, in the exact same ratio of said difference. It pays abundantly, as all will find out by experience sooner or later, to get a first-class article in goods; and also to have it made up in the very best manner possible. A first-class tailor-made suit at $100, is cheaper than the same class of goods, ready-made at $50! As a help and aid to the novice, we divide the subject into the following divisions, under the heads of the following classifications, viz.:

1. The materials upon which the work is executed;
2. The materials used for garnishing;
3. The process of their execution;
4. The aim and object, of the same.

On the first, we remark as follows, viz.:

   a. They should be chosen with a due regard to the customer, for whom they are intended; as well as the purpose for which they are intended. Complexion, etc., should be considered. Color, also, has a marked effect in enhancing or detracting from appropriateness. Color, also, effects the size. Neither does a plain suit, reflect the same size upon the same person, as a suit of checked or barred goods, does. The best of "tailor-finishings," will not correct a blunder here.

   b. It should be of a texture of sufficient firmness to hold the stitchery, etc. A "crow's-foot," will not long remain a "crow's-foot," in some material!

On the second, we remark as follows, viz.:

   a. They should likewise reflect appropriatness, in color, and durability;
   b. They should blend in the proper shadings, with the original ground-work, of the goods, upon which they are placed;
   c. The cords, linings, twists, etc., should be of the very best quality and grades of silk;
   d. Their object, aim, and purpose should be to strengthen, as well as to garnish.

On the third, we remark as follows, viz.:

   a. The cut and style should reflect mathematical truth, exactness of angles, curves, parallels, symmetries, etc.
   b. The choice of combinations, should reflect appropriateness in size, shape, and color, etc.
   c. The texture of the goods, and the purpose of the garment, etc., will help to decide upon the most proper stitches to be used for the buttonholes, attaching garniture, etc.
   d. The methods of the stitchery, trimming, etc., we have given and treated in other departments of the work; to which, we refer the student—the same being pointed out, in the index.

On the fourth, we remark as follows, viz.:

   a. The main object should be to enhance the appearance of the form, for which the garment is intended; and to hide deficiencies, in both form and material.
   b. This embraces the effects of color, size, and form; each and all of which, have a legitimate claim upon the attention of the artist;
   c. Last but not least, they should lend strength and durability to the garment, whatever its name, form, or style. It is by the careful observation of all these "little details," and the power of their influence, when brought to bear upon the work of the artist in garment-making, that above all other influences, placed the present halo of glory around the lustrous name of Worth—the reputed man-dress-maker of the world.

PLATE XXXI.

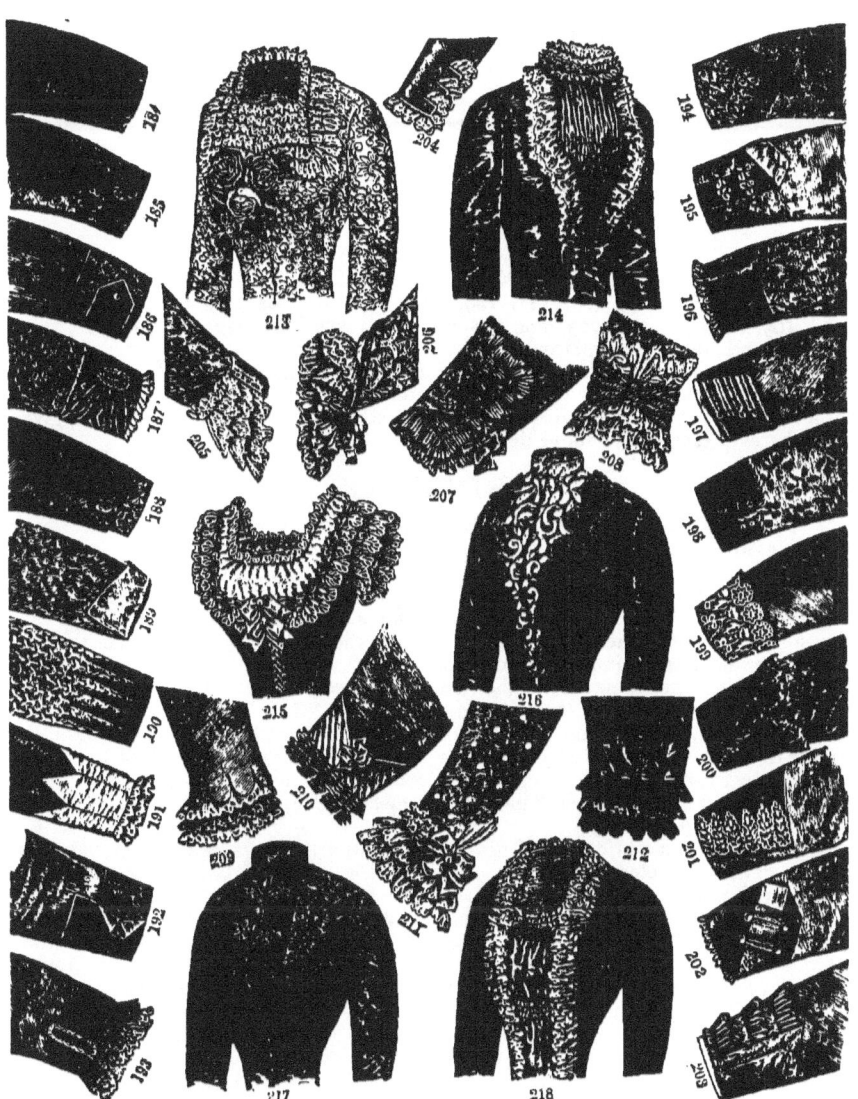

Sleeve and Bodice-Garniture.

Sleeve and Bodice-Garniture.

PLATE 31.—FIGURES 184, 185, 186, 187, 188, 189, 190, 191, 192, 193, 194, 195, 196, 197, 198, 199, 200, 201, 202, 203, 204, 205, 206, 207, 208, 209, 210, 211, 212, 213, 214, 215, 216, 217, and 218.

EXPLANATORY INSTRUCTIONS.

Figures 184, 185, 186, 187, 188, 189, 190, 191, 192, 193, 194, 195, 196, 197, 198, 199, 200, 201, 202, 203, 204, 205, 206, 207, 208, 209, 210, 211, and 212, show various and different standards, in methods of sleeve-garniture. Figures 213, 214, 215, 216, 217, and 218, show various and different standards, in methods of bodice-garniture. Each and all of which can be read and comprehended, by the mere novice, at a single glance; by studying the same, in connection with the instructions given elsewhere, in each of the above departments.

TRIMMINGS.

Trimmings are to garments, what color is to the painting; or what leaves and blossoms are to spring, and flowers to summer. Trimming has its source in the appreciation of, and the desire to emulate the beautiful in nature; whether as displayed in the mingled shades of color along the edges of the over-hanging cloud, or upon the wings and plumage of the most tiny bird or insect. It can best be studied, in her sanctuary; and we receive the inspiration of her gifts, through the endowment and ministrations of good taste.

Trimmings are required for the double purpose of weakening the power of uncomeliness, and enhancing the grace of beauty.

When material is poor, thin, and transparent; when there is something in the figure of the wearer, from which we wish to divert the eye: under all of the above circumstances, the sovereign remedy is found, in trimmings; while rich and costly fabrics, around fair faces and handsome forms, like the fairest of flowers, look most adorned, when unadorned!

Again, certain fabrics should never be used as trimmings, upon other materials. For instance, velveteen cannot be ornamented with velvet bows; but must have faille, or silk, etc., as ornaments. Grenadine cannot be used to trim on washing fabrics; nor is white embroidery appropriate with barege. Poplin looks best when velvet-trimmed, but is not offended when in company with gros de Naples; while turquoise, and faille, can both be made to add lustre, as garnitures, to cashmere, merinos, and all plain woolen goods. Clear fabrics, grenadines, and bareges, etc., look best self-trimmed; while foulard-silk may be used upon the same with good effect. Passementerie, or "gimp," is suitable on all opaque fabrics; and fringes, it should be remembered, are used to edge garments only; and should be placed under, and not upon the fabrics, which they ornament. Velvet bands are utilized to garnish the borders of garments. The effect of trimmings, as well as that of seams, when placed in a horizontal position with the form, have the peculiar effect of apparently producing a "broadening" and "widening" effect; while perpendicular or "lengthwise" seams and trimmings have the opposite effect, viz: of apparently "increasing" the stature, and reducing the width. This principle is of great service to the artist, in properly selecting and recommending the styles of seams and trimmings, which correct any undue proportion, in the direction of either extreme, in any special form. Again, some styles or modes of trimming are adapted to children, and especially to the young; while others belong to old age. Very ludicrous mistakes are often made in violating this principle of adaptation, in the choice and selection of seams and trimmings, by different artists. The most useful trimmings to conceal the ravages of time, perhaps, in the whole catalogue, are the various styles of "Bias-Folds;" which, therefore, naturally cling to old age ; as much so as the vine clings to the trellis. As a few examples in trimming, we give the following, viz:

1. The Plain Gathered-Ruffle.—This is standard, and is the letter "A," in the alphabet of "trimming;" and it is the most simple, common, and, perhaps, we might add, the most useful of all skirt-trimmings. It has done good service for many centuries, and will most likely continue to do so. It is, of course, cut on the exact bias. You can then have your choice out of three distinct and standard methods of finishing the "Flounce" or "gathered-ruffle," viz :

a. Bind the edge by a strip ¾ of an inch wide, stitching it to the right side, and taking up as small a seam as possible; then turning it over, and hemming it down on the wrong side ; and so arranging the stitches in this last process, that the stitches will be put in the first seam, and be invisible on the wrong side.

b. Or again ; you can turn it up on the wrong side, and complete it with a a neat hem, ¼ inch in width.

c. Where the material is not adapted for either of these processes or methods, you can neatly cut a lining of undressed crinoline, or some similar goods. Having done this, sew both together on the bottom edge, and then turn it so that the seam is hidden between the two materials—on the wrong side.

2. "The French-Ruffle.—This "flounce" is very frequently used. It is simple, and is formed as follows, viz : Cut the "flounce" two inches wider than it is to be when finished. It is, of course, cut on the exact bias. Deep, broad points, the desired depth and widths, are then cut in the bottom. The points are sewed all around, neatly, to a lining ; which should be of silk, since it shows slightly. Fold it across the top in very regular and small side-plaits. To do this artistically, arrange it so that each is taken exactly at the upper end of each respective point.

3. The Corded-Ruffle. — This "flounce" has many admirers, and is one among the leading standards. This should also be cut two inches wider than you wish the flounce, when it is completed ; and you should cut it also, on the exact bias. A flounce will not hang true, unless cut exactly on the bias. The extra amount allowed in cutting, is taken up in the "cording" and in the "heading." The "heading" should be turned in double, to the cord. It can be cut in points, or otherwise; and can then be neatly finished by attaching it to a crinoline lining or otherwise silk. Preceding the process of turning, the upper part of the point should be cut; then you can easily turn it, and draw it out. The cord should be inserted, and the gathering done, at one and the same time. The whole process will be found very simple, and makes a handsome standard finish.

4. The English - Ruffle. — This standard "flounce," is a cousin to the "French-Flounce." It differs, however. Cut it the same, in every way as you did the "French-Flounce." Give it the desired shape, as to the points. It differs from the "French-Flounce," in that the points are not lined ; but instead of this finish, they are neatly bound; this is done with a very narrow bias-piece of material.

Arrange and adjust the flounce at the top, in a series of accurately-spaced and neatly-laid box-plaits. Said plaits should be the exact, or of uniform width, of the spacings between them, respectively. The centre of each plait, will thus become the centre of each point, respectively.

5. Bias-Piping.— This useful and standard trimming is of French origin. It is the close friend and companion of old age, rather than that of the young. Because

it is very narrow when completed, do not reason falsely, and conclude to cut it narrow—proportionately. It should, as a general rule, be cut one inch in width. It must be on the exact bias of the goods. Deep seams are required on the inside; which, will take up the surplus width allowed in cutting.

The process of forming it is very simple. Turn the material, and place it in such a manner that the edges are turned on the inside; having the upper, the mere width of a cord above the lower. Then finish it by close, firm, and regularly made invisible stitches.

6. Double-Piping. — A standard, very pretty, and effective trimming in "piping," is formed by showing two different colors in the same "piping." Black silks are sometimes thus decorated, by heading the ruffle, with a slight showing of some proper shade in a different-colored silk. This is termed "double-piping." The process is very simple. First, cut the bias-pieces as usual ; attach them, by the ordinary seam ; they are then opened and turned over. This, when adjusted, shows to good advantage, if good taste is used in the selection of shading in the colors.

7. Kilt-Plaiting.—This is a standard, very popular, and stylish plait, if properly formed and artistically planned. It is sometimes termed "side-plaiting." It should invariably be formed on a straight ruffle. There are two general forms of making the "kilt-plaiting," viz :

a. First, it may be lined entire.
b. Second, it may be jointly "bound and hemmed."

It is most standard, among connoisseurs in the art of dress-making, to line it entire; and finish it with the "invisible-stitch." Proceed as follows, viz :

a. First, design the width of the plaits.
b. Second, lay the first carefully.
c. Third, then baste it thoroughly into that position.
d. Fourth, proceed with the second, laying it exactly as the first, and the exact same width; and baste it in said position.
e. Fifth, repeat this process until the work is completed.
f. Sixth, after laying them in the same folds, baste them thoroughly across the lower edge.
g. Seventh, press them thoroughly on the wrong side.
h. Eighth, now secure them thoroughly, by joining them with strong cotton, before removing the former basting. This completes the work ready for adjusting and finishing.

8. Double-Corded-Piping.—A handsome standard design on some groundwork, is frequently made in the "double-piping," which we have already described above, as follows, viz :

a. First, cut the two materials on the bias, as usual.
b. Second, attach them by the ordinary seam.
c. These are not turned and hemmed on the wrong side, as for the "double-piping" described under the sixth paragraph above ; but instead of this process, it is stitched near the bottom-edge, with silk of the same color as that used in the cording. Into the upper, is inserted a medium-sized cord.

9. Double-Cording.— A very popular standard, and very useful trimming, is termed "double-cording." It is formed as follows, viz :

a. The standard width of the bias-strip is 1½ inches. Cut said strip exactly on the bias.

b. Stitch one cord in this, in the usual way. A wide seam should be left at the lower edge.

c. Insert a second cord, in like manner.

d. Turn this last cord up, stitching along the edge, and connecting the seams of both cords.

e. Now run another stitching near the second cord, in such a manner that when reversed, the seam is not seen on the wrong-side.

10. The French-Fold. — The "French-Fold," another very standard trimming, is usually made out of the material of the dress, edged with satin. It is very simple, and is largely used in all kinds of dress-trimming. Proceed as follows, viz:

a. Cut the strip out of the material, exactly on the bias, $3\frac{1}{4}$ inches wide.

b. Cut the strip of satin, on the bias, 1 inch wide.

c. Fold the satin on the centre-line, and stitch it to the edge of the fold.

d. Then reverse the seam, and finish it by hemming it upon the other edge, on the wrong-side.

164 THE ENCYCLOPEDIA OF GARMENT-MAKING,

PLATE XXXII.

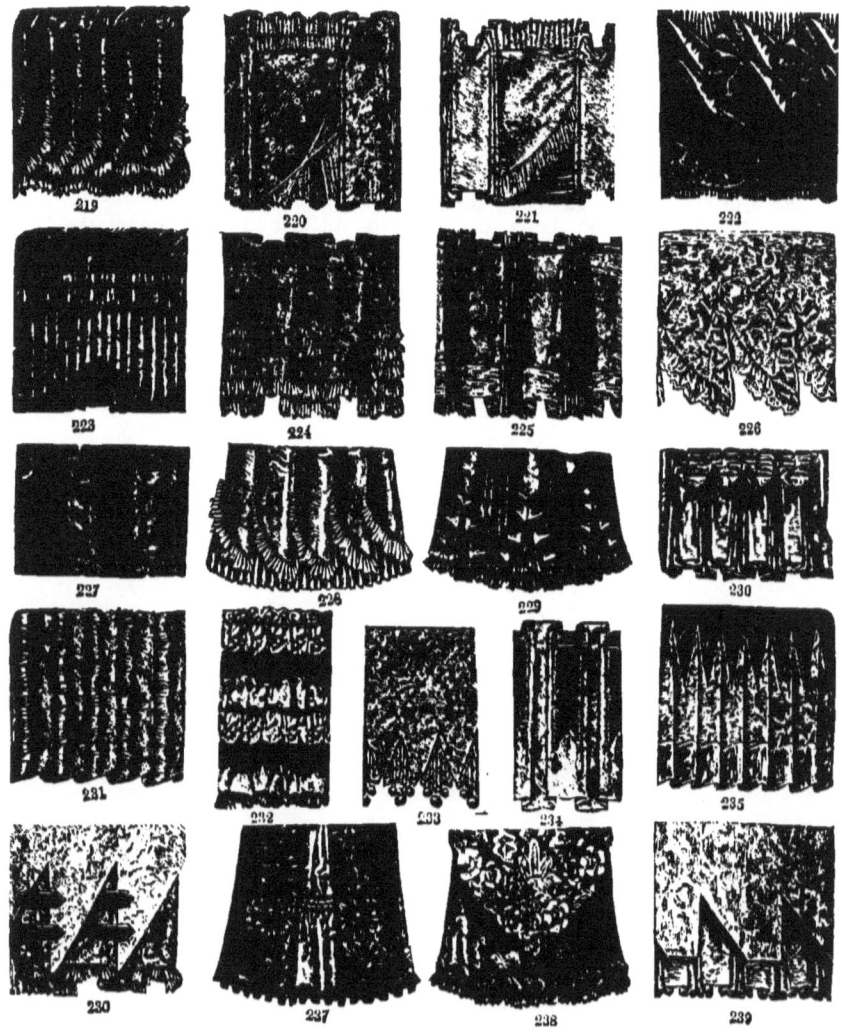

International Standard Skirt-Trimmings.

International Standard Skirt-Trimmings.

PLATE XXXII.—Figures 219 and 239.

Explanatory Instructions.

Figures 219, 220, 221, 222, 223, 224, 225, 226, 227, 228, 229, 230, 231, 232, 233, 234, 235, 236, 237, 238, and 239, show twenty-one different standard-methods of skirt-trimmings. Each and all of which, explain themselves; and which can be accurately read, and directly reproduced, by even the most inexperienced; by simply studying them a few moments, jointly and in connection with the instructions under "Copying," "Tailor-Finishings," "Trimming," etc.

11. Quadruple-Bias. — This has, in recent years, assumed a place among the standard-trimmings. It is sometimes termed "Four-Fold-Bias"; because, it is folded four-times. Its name "quadruple-bias" comes from the same reason. That it folds four times, should be remembered in "cutting-out" the material; which, of course, is on the exact bias. Baste each fold down and finish easily, but firmly and securely; and, on the wrong-side of the goods. Then removing the basting-threads, continue this process, until all the folds are stitched on. Finish or fasten the top-fold, securely, by means of the "slip-stitch."

12. Rouleaux.—These were once very popular, and much used as garnitures. Especially so, for garnishing mantles; and for edging skirts, and princess-dresses. The process is so simple and well-known to all, that we deem it unnecessary to give space for anything more than merely calling attention to the same.

13. Plaited-Belts. — A handsome and much-admired finish, in the form of "Plaited-Belts," is often used, in dress-making, and for loose garments. Proceed as follows, viz.:

    a. Cut the material, twice the desired width of the belt.
    b. Cut it thus, on the exact bias of the goods.
    c. Using but a narrow seam, stitch the material to the wrong-side of the lower part of said belt.
    d. Having done this neatly, reverse the same; by turning it back, and stitching the plaits to the same, near the edge on the lower-side. This should be performed, with great accuracy and care. Fold the top-edge over, and secure it by the finish of hemming; which, of course, is done on the wrong-side.

14. Flounce-Heading.—This pretty finish, is of great service on all kinds of "flounces." It is made in different ways. But perhaps the most standard-way is, to make it exactly on the same principle, as that described in the above instructions, on making the "Plaited-Belt." To which, we refer the student.

15. The Tape.—Everywhere, among all nations, we find this form of finish, highly prized and appreciated. It is standard, both as to time and place; and will likely ever remain so. It is ornamental; but it is, also, very useful. To attach it nicely, proceed as follows, viz.:

    a. Place the top near the hem, on the wrong-side.
    b. Taking the stitches from the right-side, stitch it down securely, in that position.
    c. Fold the tape over the seam, so that it laps over the edge.

d. Now finish the process by stitching it down firmly, on the right-side.

N. B.—Another Process.—Proceed as follows, viz.:

a. Place the tape on the material, as taught above.
b. By means of the "back-stitch," attach it to the material.
c. Fold the tape over the seam first made, and stitch it down firmly, near the edges.

16. Fringe.—A beautiful means of finish can be made as follows, viz.:

a. Ravel two pieces of gros-grain, of equal length, to the desired depth.
b. Secure the former under the edge of the latter.
c. Give this a finish on the under-side, by a neat facing, or cover.
d. Dividing the strands into equal sizes, link every second pair from underneath, by means of "buttonhole-stitches," of the same color of silk.

The process is not difficult; if the following is observed, viz.: First working the first strand, the desired distance down from the goods, with two stitches; and then joining half of this, with half of the next, by two stitches—twice as far down, as that of the first junction; then, coming back to the first distance from the material again, and joining half of the first with that of half of the second, as before—leaving the silk with which you work, on the under-side. Continue this process, until the entire work is completed. It is simple, and very beautiful; and admits of many changes, and as many variations.

17. The Berlin-Fringe.—This is very popular, and is made as follows, viz.:

a. Proceed exactly as for the above fringe, in cutting, and joining, and facing the two pieces.
b. Form an equal division of strands, as in the above.
c. Join them in pairs, alternately, by means of the "looped-knot."

This process is likewise, very simple; and the mere novice can "catch the idea," by a little thought and practice.

The first, is brought around and over the top of the second, and then back through its fold or loop; the upper strand is then drawn on the outside; and to complete the process, the two are joined together, by being wound a few times with silk twist—of the same color.

18. Bows.—"Bows" are to trimming, trousseaux, and evening dresses, what stars are to the firmament at night, or dew-drops to the rosebud or flower in the early dawn. They are the "finishing-touches," of the true artist's brush! And we might add, that as the dew-drop—mirror-like, reflects the face of Nature; and through Nature reflects the infinite conceptions of beauty and perfection of Him who is the Author of Nature; in like manner, we can invariably read the degree of scholarship and education of taste attained by the modiste, from these instructive characters; which she, perhaps unconsciously, has placed upon her work! There is something about its secret, for which it is impossible to give a law or rule, that will sufficiently cover the ground. It is like music, or painting—natural to some; almost, impossible, to others! And yet, we do not like the word "impossible," in anything. Neither Demosthenes nor Cicero, were born orators! Much can be accomplished by industry, patience, observation, comparison, reflection, and a determined will—to succeed.

The very carelessness with which Nature has placed the opening rosebud, nestling half-concealed within the green and shaded leaves, lends it richness of lustre; and makes it seem, because of that, only the more charmingly beautiful.

To describe all the styles of "bows," would be too great a task. To give simply a catalogue of the various names, would more likely confuse than help the inexperienced. Suffice it to say, in conclusion: All can attain progress, at least; and all can improve in this art, as they can in their penmanship! But while all can become skilful with the pen, it is given to some, to excel! Observe, think, and practice; and be industrious, patient, determined, and hopeful; and success will crown your efforts.

19. Knife-Plaiting.—This element of trimming is much used, in the art of garment-making; and it is one of the principle standards—in "methods of finish." The principle which underlies the process, is exactly the same as that which underlies all plaiting. The plaitings however, are much smaller, and are generally laid in regular successive widths, in one way. This, is called "Plain-Knife-Plaiting." It is sometimes varied, by laying a different number of a size, in alternate places; this, is then termed "Cluster-Knife-Plaiting." The plaiting in general, takes its name from the form, having the appearance of a knife, over which, and into which it is laid. The "Cluster-Plaiting," received its name from the plaits having been arranged in "clusters," at equidistant spacings. The process is so simple, and so generally and well-known, that we deem it of insufficient importance, to treat it by an extended explanation.

MOURNING-DRESSES.

Personally, we both prefer to express our sorrow and see that of others expressed, by "draping the heart," rather than the body. But in all ages and places, it has been a universal custom to give expression to "sorrow for the dead," by draping the body, for a limited time, in outward symbols and tokens of grief. And as this naturally comes within our province, we give it a passing notice. The "mode," finds expression in all the ten-thousand variations, and all possible degrees; from the mere simple and plain "band-of-crape" around the arm, to that of the most elaborate, extensive, and most costly costumes, worn. It does not always express sorrow, in the exact ratio of its form, and its extent—only as measured by the false reed of "public custom!" But the decrees of custom are merciless, and do not admit of discussion; but demand instead, a quiet and submissive acquiescence; and our inclinations must very often—however unwillingly, become for a time, its slaves! But it is possible even for the slave, to be superior to his master. And we therefore, simply point out how, most expressively and most gracefully, to adorn and wear its chains! And we therefore, submit the following, viz :

1. The Dress.—The dress should be,

   a. Grosgrain-silk, drap du nord, satin-merveilleux, paramatta, barathea, bombazine, or some one of the like-materials, in deep-black ;

   b. Cut plain "demi-promenade," in the skirt; and with long "jacket-bodice," or otherwise, "en princesse-form," in the body-part ;

   c. Almost entirely, at least very heavily, "trimmed" in either "Albert-crêpe," "Rainproof-crêpe," or otherwise, in "Rainproof-crêpe-cloth," etc. ;

   d. Without lace or ribbons; unless very sparingly used, and these of the highest grades and best qualities made; and which of course, are only in the heaviest shades of black.

1. The Wrap.—The wrap should be,

   a. "Paramatta-cloth," "Cypress-crêpe-cloth," "Crêpe-trimmed-cloth," "Cloak-silk" or "cloak-cloths" of various names and shades, in deep black ;

   b. Cut "Dolman-sleeved," for elderly ladies ; and in long "jacket-form" or "paletôt-form" for younger ladies ;

   c. For heavy winter-wear, dark-seal-skin, astrachan, velvets, or cloths;

   d. For light summer-wear, simply of silk, "crêpe-silk," or lace; all of which, must be "crêpe-trimmed ;" or otherwise, the simple crêpe is used ;

3. The Bonnet.—The covering for the head is,

   a. Silk-crêpe, founded or built upon, what is generally known as the "widows-cap."

   b. "Bonnet-shaped;" hats, not generally being allowable ;

   c. "Trimmed" with a "crêpe-veil" of medium length, at the front; and with a deep, heavy, and long "crêpe-veil," at the back.

4. The Cap.—The cap is,
   a. Tarlatan, lisse, or tulle, etc ;
   b. Generally, in the form of what is known as the "Marie Stewart"-shape ;
   c. "Trimmed" with long "streamers."
5. Supplementary Requisites.—These in the main, are are as follows, viz:
   a. Ornaments—only in jet ;
   b. Collars—linen, lawn, or muslin; edged in deep-black;
   c. Cuffs—the same, as the collars ;
   d. Gloves—black kid.

DRAPING.

When we stop to think that mathematics fixes the perspective of the Artist, garnishes Architecture, and brings harmony and musical cadence out of discord ; we certainly should not be surprised to find that "Draping," claims a legitimate birthright and inheritance, in the wedlock of Art and Science. In a preëminent sense, "Draping" is an art; but it is also, more than an art. It is a science; as much so as any or all of the generally accepted sciences, recognized and honored throughout the civilized world. It has an extensive province; and as a subject, for exhaustive discussion, would lead us not only through the department of garment-making, but would also lead us through the provinces of painting, modeling, and sculpture, etc. Many excellent artists with the brush, and many excellent paintings, have been shipwrecked on its dangerous shoals. For if the draperies of a painting, or a work of art in sculpture, are not mathematically symmetrical, and do not reflect true grace and real elegance, the entire work is a complete failure. In each of these provinces, the drapery is all-important. But we have, in a special manner, to do with the subject only, as it effects the province of garment-making. And to this field, we now turn our attention. In a general way, and for the purpose of simply aiding the inexperienced student to a rudimentary knowledge of the process of draping in garment-making, we submit the following general rule: For instance, suppose we have taken the measures, completed the draft, and have cut the linings; and afterwards have formed the "polonaise-draft," upon the silk ; according to the instructions, as given and illustrated in the engraving; and, that the polonaise is basted, and partially stitched; and that it is now ready, for draping. Then proceed in the following order, viz :

   a. First, place the entire polonaise length-wise, upon the cutting-table ; drawing it down neatly and smoothly, in a length-wise manner.

   b. Second, suppose the design to call for "medium-high-drapery" at the back, and a series of "shell-puffs," at each side ; and with the points ⅔ ways from the waist, down, each side.

   c. Third, suppose these "shell-puffs" to be in alternate positions, as to their height and depth ; instead of each row or series, being on the same horizontal line.

   d. Fourth, notice how far down below the waist, the polonaise is drawn to its closest degree, toward the back of the form—on the given design ; which is usually, on the "hip-line."

   e. Fifth, now mark these points, equidistant from the "waist-line" at each side of the polonaise, with the "chalk-pencil ;" or otherwise with pins. This should be done very accurately; and only, by actual measurement.

   f. Sixth, now "sew-on" or attach the elastic, at each respective side ; which, is afterwards to be drawn to any desired degree of closeness, or ease of fit—when upon the draping form.

   g. Seventh, now locate the desired and required points, on the exact centre-fold at the back ; either, only below, or otherwise, above and below said line—upon which the first points were located ; and as the design or plate calls for ; and where the intervals occur, that secure and form the "shell-puffs." These distances are usually, from 10 to 14 inches; but often, only from 4 to 6 inches.

h. Eighth, then on a direct perpendicular line, half-way between, or equidistant from the centre-fold at the back and the lines at each side, and upon which the elastic is adjusted, locate points in a similar manner, as above; and also, of equal and uniform-distances apart. This is done either on the same horizontal line, with the former locations, or on alternate heights, and on different and alternate horizontal lines; and as the design may call for—in the location of said "shell-pouffs." Locate both these points and those on the centre-fold, at the desired or required depth, shown in the design.

i. Ninth, if it is desired to continue the "shell-puffs," locate the points on a slightly curving-line.

j. Tenth, if it is desired to effect the opposite from the last result, locate the points between the former.

k. Eleventh, Now fasten "loopings," by means of tape, at each respective point.

l. Twelfth, then run through these "loopings" a "runner,' by means of another tape; which can be arranged so as to be drawn at will, and to any degree of closeness desired for the "Shell-puffs."

m. Thirteenth, if a double-series of "shell-puffs" is desired, simply use a second "runner" of tape.

n. Fourteenth, then cut the tape, the desired lengths; and secure each to its respective given point, on the centre-line of the back.

o. Fifteenth, tapes should be adjusted at the side-seams of the "skirt-part," to hold it back, firmly; and which, should be neatly and firmly joined together; and which should also, be concealed under the draperies at the back, when the work is completed.

p. Sixteenth, now place the polonaise upon the "draping-form;" which, should be arranged to the proper sizes at the bust, and the waist, and at the hips; and which should also, be arranged to the proper degree of the length, in the skirt.

q. Seventeenth, now draw and secure the elastics to the proper degree of closeness, and at the proper points; and arrange and secure everything as arranged for, when on the cutting-table; and study to bring everything carefully, symmetrically, artistically, and gracefully into the position desired, or otherwise, called for in the design. Although the process of draping, as described, at first thought may seem intricate and somewhat difficult, to such as are entirely inexperienced; yet it is only apparently so, and not in reality. Indeed, the very opposite is true. For it is an exceedingly simple process; as will be found by all, who once have made the attempt. Things entirely new to us, always seem more difficult in the printed instructions, than when we come to put them into practice. It apparently comes to one then, by one thing suggesting another. Begin on prints, or something of no great cost or value, at first. If, you "spoiled" it entirely, the loss would be atoned for, in the merit of the effort! But you certainly, cannot spoil it entirely. And if it does not just come out exactly right in every respect, it will most likely have some attributes of grace; and should it be entirely original, and possess the above qualities, it would be all the better for that ! The object of this entire work is, to arm the student with correct scientific-principles; and then to create within the student, a well-founded self-reliance. To aid the student, we classify the various styles of drapery, as to "position," as follows, viz.:

  a. High-Drapery;
  b. Low-Drapery;
  c. Medium-high-drapery;
  d. Front-Drapery;
  e. Back-Drapery; and
  f. Side-Drapery.

In deciding upon the style of drapery, suitable for any particular given form, much depends upon the education and skill of the artist in the choice as to location; and in so arranging it, that its position and location will not detract from the symmetry of the given form; but on the contrary, will enhance the same. It is both ludicrous and wonderfully interesting to the thoughtful observer, in noticing how vastly different the same person appears on coming forth from the establishments of different artists. The true artist will study not to detract from any grace or symmetry of form, that may exist; and will study to create it, where it is lacking. And yet, this must be done in such a manner, as not to leave fashion entirely out of consideration; and thus highly offend both fashion, the customer, and the public! But after all, correct art and correct taste are always "in fashion;" for fashion is their slave! and hence, fashion should be the artist's servant; and should therefore minister to his or her wants.

As to the "kinds" of drapery, they come *a la Franciase*, as,—

a. "*Boufante-drapery*"—which is known to all.
b. "*En Biais-drapery*"—On the cross, or diagonal folds.
c. "*En Châle-drapery*"—Resembling a shawl, or shawl-like folds.
d. "*En Coquille-drapery*"—Folded backward and forward, in zig-zag shape, forming shell-points.
e. "*En Echelle-drapery*"—Representing a ladder-like appearance, in uniform interval folds.
f. "*Pannier-drapery*"—Representing panniers, at the side.
g. "*Flots Coques-drapery*"—With quantities of lace or ribbon looped, and so arranged as to fall over each other.
h. "*En Eventail-drapery*"—Having the folds so as to show an apparent centre, giving it a fan-like appearance.
i. "*Cascade-drapery*"—Representing in arrangement of lace, or in the material, the appearance of the cascade, from which it derives its name.

This catalogue of "kinds" is by no means exhaustive. But we have first given these as aids to help the student to comprehend and formulate the subject; and secondly, because they represent at least the principal standards. Study these "classifications of position and kind," as you would the fundamental principles of arithmetic, if you aimed at becoming a mathematician; for you will ever need them in reading the productions of the fashion-journals: and in the solution of every problem, however far-advanced in the study and practice of the art. We do not abandon the laws and principles of multiplication, addition, subtraction, or division, when we advance to astronomy, in mathematics! Neither will you advance to a position in the art and science of draping, where you will not find use for the above classifications; and for the principles they embrace.

As new conceptions are born, they soon take form in some way, and thus create Fashion; and for which, new names are coined, expressing the leading features of the same. But the strength of the true artist lies, in the power of the inventions of combination, variation, and adaptation; and out of every new standard shade of thought, to create an almost infinite variety of styles; and at the same time, remain both in the line of correct art, and that of good taste; and yet, so as not only to honor Fashion, but to lead the same!

In this department, more particularly than in any other, will be displayed the skill and taste, as well as the true education of the artist. To be a copyist, is to attain an average-success; but it is something far different from this, to be an artist—in draping! Nature comes to us each season, with the same leaves, the same buds, and the same flowers; but she never arranges them just exactly the same, each time. Above all, strive to educate the perception and appreciation of correct taste; then learn to think, decide, and act upon your own best judgment, taste, and skill; making fashion, minister to your wants. Do not allow yourself, to fall into the habit of sameness. The work of some artists is ever and always the same; and their customers are known on the streets, by the uniformity of sameness in the cut and style of their dress.

"In accordance with fashion," does not mean sameness. The "fashion" of the countless millions of pebbles along the lake or sea-shore, is the same; and yet, no two are just the same! In all this world of beautiful leaves and flowers, you can not find two, just exactly alike! In nature, you will find the true "Fashion-Journal," after all; and there alone, will be found the true dictionary of both science and art. In the arching rain-bow, and upon leaves and flowers, will be read the true law of the shading and blending of colors; on the dark and shadowed folds of the evening clouds, will be suggested new and befitting methods of arranging the emblems of sorrow and grief—in crêpe; while in the overhanging mists of early dawn veiling the blooming orchards of spring, will be born new ideas as to the most appropriate arrangement of the glad tokens of the orange blossoms, and of the bridal-veil.

STAMPING.

Stamping is a process of very ancient origin. We find traces of its history as far back as the times of the Egyptians. At least in records of the strongest possible inference, as to the fact of its existence. The very fact of the existence and the attainment of the high degree of perfection possessed by the Oriental nations in "embroidery," would prove, almost to a demonstration, the existence of stamping; as well as a practical knowledge of the same. Stamping is the very "corner-stone" upon which the whole monument of "art needle-work" rests. Upon the wall constructed out of the elements and principles of the art of stamping, rests the whole superstructure of "embroidery." Embroidery had its birth in the conception to emulate the beautiful in nature, through the reproduction of the same by and through the needle and silk; just as the painter would, through his brush and paint. The needle and stamping are simply instruments, or means of conveying and placing the silken cord, in the art of embroidery; just as the brush and the instruments in which the colors are mixed, are to the art of painting. Anything that will perfectly answer all purposes to this end, come legitimately under this head, and within the province of stamping. Notwithstanding the futile attempts to cover this simple art by the gauzy veil of "secret-mysteries," for mere mercenary purposes, it is an exceedingly plain and simple process. As embroidery is most highly cultivated among the French, in modern times, we owe much to them for late improvements in the art and methods of stamping.

Stamping, naturally divides itself. under the following simple classifications, viz.:
1. The materials upon which the work is performed.
2. The materials used for executing the work.
3. The instruments used for their application.

On the first, we simply remark that, all materials which have a surface that is sufficiently even and plain, to show either the liquids or powders, and upon which embroidery is placed; are, legitimate materials for stamping.

On the second, we remark that, the theories as to what are the proper compositions to use, are almost as numerous as the artists performing the work. Each, of course, claiming their particular and special theory as the *ne plus ultra*, in the art and science of stamping. To quote all, would fill a volume in itself; and to quote a few, to the exclusion of the many, would in the estimation of some, make us a partialist; and thus, we might highly offend. Suffice it to say, that the matter of the special chemistry in the composition of the materials, is one of but little concern to the practical student, in any way. For all the materials, for all methods and branches of the art of stamping can be had ready-prepared, and in a neat form; in any of the art-stores, in any of our county-seats or cities. You can simply ask for the liquid, called "stamping-liquid," or the powder-form of the material used, which is termed

"stamping-powder;" you will in all probability get a satisfactory result; as all the compounds generally sold by the trade, are in the main, good; and will answer all purposes. So much for the "material" upon which it is executed, and the "compounds" used. As to the means of applying and transfering the powders and liquids, they are equally simple. The great general divisions of the "methods" are the following, viz. :
    a. The English, or the "damp-process;"
    b. The French, or the "dry-process."
    The French, or the "dry process," by means of powders, is now however, almost becoming universal. The liquid, being mostly confined to the color of white; which, upon certain qualities and textures, when applied by the "oil-processes," is deemed most practical.

    The Requisites.—The requisites therefore, will be,—

    1. The proper material upon which the work can be produced.
    2. The various colors desired in the "dry-process" of prepared "stamping-powders."
    3. The various colors desired in the "damp-process," of the prepared "tube-stamping oils," or "tube stamping-paints."
    4. The instrument in which to deposit the contents of the "tubes," so as to be easy of access in the method of application.
    5. The ivory "pallet-knife" in the "damp-process" for spreading the paint.
    6. The "pad," for the "dry-process;" and last but not least,
    7. The "stencils," or the patterns.

    Having all these, the process will almost suggest itself. The "stencils" or patterns, formerly were made out of soft metal. The modern "stencil," comes to us in fine "French-stamping-paper." All patterns, from the simplest buds, vines, and flowers, to the most complicated designs and combinations seen in the richest Oriental laces, can now be purchased at a very trifling expense, at any first-class art-store.

    The colors most generally used in the "powder-form" are black, blue, and white.

    The material is then evenly and smoothly placed upon a plain, flat-surface; stretched somewhat, and secured in said position. Upon this, the "stencils," either as one entire whole, or in parts, and in their proper combinations, are carefully placed. Care should be taken to have everything true to the goods, and accurate as to its horizontal and perpendicular position; and also as to its combination with the design in general, if more than one "stencil-plate" is used. You then simply apply the powders by means of the "pad," rubbing over all the perforations thoroughly. Be sure you do not omit part of the perforated-indications. All of which is very simple, easily done, and quickly performed. On removing the "stencil-paper," or the perforated-design, you will see all the outlines beautifully and elegantly outlined, ready for the needle and the silk of the artist in the science of embroidery. In the "damp-process" the "oil-bath" is used. Then the "stencil" is placed upon the material, and the "pallet-knife" used to spread the paint, which marks the outlines of the completed work. See that you have all the necessary requisites. Then select your design in the proper stencils or perforated-papers, and begin on simple and plain work at first; and experience will soon make you proficient, and superior to all possible emergencies. Furthermore, there are many beautiful designs now, ready-prepared on paper, in "copying-ink," and which the student can easily transfer upon the material, by simply placing the sheet, upon which the design appears, on the material; and then pressing over the top of said paper with a hot iron. This transfers the impression directly upon the material; and completes it, ready for the work of the needle and the silk of the artist in embroidery. All of which the most inexperienced can easily and quickly attain, as given and explained above.

Embroidery.

Everything in embroidery, depends upon the proper foundation being laid, as to the design; its arrangement, and the choice of colors, that shall become the medium or language to express the conception in the artist's mind. If deficient here, it must end in failure. This of course requires practice, observation in nature, and a true education of both head and heart, in the artist. Good suggestions sometimes come to us, from woven work; the correct drawings of the various animals, birds, flowers, etc. You cannot be too natural. He who always follows nature, in art, will always stand the crucial test of the severest critics. We want both form and color, in embroidery; and it should reflect truth, in order to be beautiful. The pottery of every age, furnishes good patterns; but which, advanced thought should be able to improve. This is also true of the professional work of painters, both ancient and modern. The patterns can be stamped upon the material, outlined by pencil or colored inks, or transferred by "pouncing" and "tracing-papers."

The original meaning of the term is "to border," or "to decorate." It is of very ancient origin ; and, was much used among the Egyptians, Chaldeans, and all oriental nations. It is preëminently "woman's work." It clings to her hands, as the vine clings to the trellis. It is not alone, the art of bordering in stitches; it is the fine art of stitching—it is A MODE OF EXPRESSION. It is "painting ;" and the needle and the silk, are the paint and "brushes!" It is "music;" and they are the "key-board." It commands the respect and admiration of all cultivated people; and creates demands for new fabrics, and new material. Tapestry, as embroidery, also uses the needle for expression. But there is a difference between them. Embroidery employs many kinds of fabrics and every variety of needle-work ; tapestry has only one fabric, as a "backing;" and the work is "one complete fabric," with a uniform surface. It is stitching into a woven fabric of new threads, that pass under the warp and over the filling. In tapestry, the simple "running-stitch" becomes the basis of the most complicated, and most artistic and beautiful effects. Whatever can be done with the needle in embroidery, in the main, is useful as well as ornamental. Mere ornamental stitching in colors, is something less than embroidery. It is "the art," "the design," "the drawing," and "blending of shades and colors," that fills the meaning of the term embroidery. Again, we lead the student to nature. Do anything ; but get your effect, and reproduce nature, with the needle and the silk. Begin on table linen, or on some other plain material. Choose simple things, natural objects, sprays of foliage, or outlines of flowers, ferns, or mosses. Then insects, colored shells, etc. Be true to nature. Simplicity in embroidery, as in all things, is the chief thing, until we grow more experienced ; when it will seem a very easy thing to copy nature directly, in some splendid curtain, rich with portraits of a thousand different flowers and buds and blossoms.

Materials.—The "materials" are, various kinds of canvas-cloth, sateen, plush, felt, satin, and plain cloth, etc. Crewel is the most popular, excepting silk, for "decorative needle-work." It is a soft brilliant finished wool, in all shades. "Chenille-embroidery," and "Arasene," are on the increase, just now. The best is imported. But the American goods has the richest lustre. But this is used for coarse work. It cannot be made to show the artistic effect of filoselle, and other fine "silk-threads." Silver and gold-tinsel, are effective for outline work on velvet or plush ; but the work is transitory ; being effected by light and air. Decca-floss, is a slack-twist glossy-silk. It is for pongee-silks, and transparent canvas. For small pieces of work you can use, as frames, a small hoop. Larger pieces, the "Patent Embroidery-Frames." Needles are all sizes. There are many kinds of "embroidery silk," in the market; some is rough-finish silk; others smooth, soft-silk. The former, has a stiff and wiry appearance ; and it is not suitable to fine work.

PLATE XXIX.

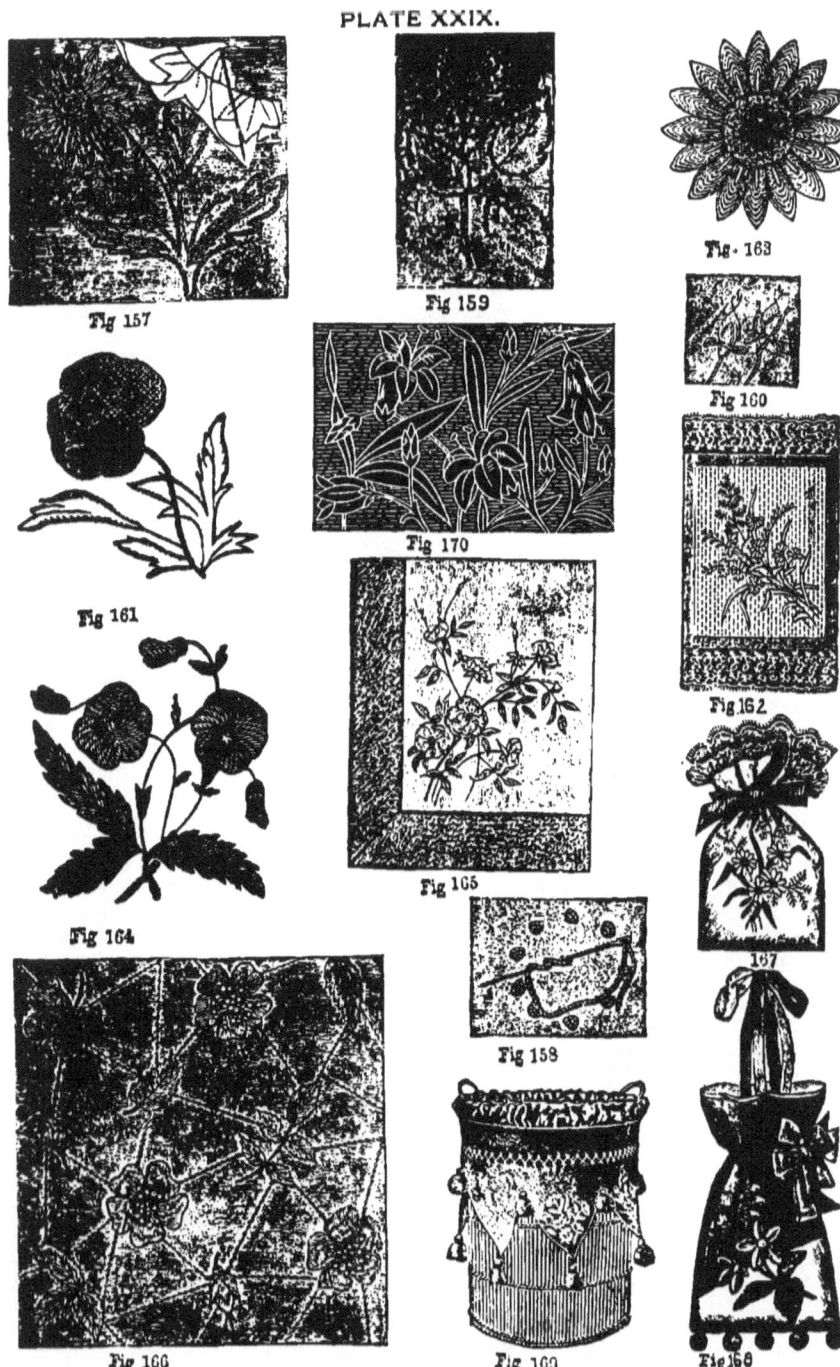

International Embroidery-Designs.

PLATE 29.—FIGURES 157, 158, 159, 160, 161, 162, 163, 164, 165, 166, 167, 168, 169, 170.

EXPLANATION.

Figure 157, represents the process of the plain "outline-stitch," in embroidery. Figure 158, shows the "French-knot." Figure 159, shows the "double-outline," or "skeleton-stitch." Figure 160, shows the "darning-stitch." Figure 161, shows the "Queen Anne-Stitch." Figure 162, shows the "transparent-tidy." Figure 163, shows the "rick-rack-daisy." Figure 164, shows the "French-Pansy." Figure 165, shows a sample of "ribbon-work." Figure 166, shows a "square," for a "silk-quilt." Figure 167, shows a "satchet-design." Figure 168, shows a "silk-work-bag-design." Figure 169, shows a design for a "scrap-basket." Figure 170, shows a design for a "table-spread." All of which are simple and very easy of attainment, when studied in connection with the instructions in embroidery and art-needle-work, as given and taught elsewhere.

Embroidery-Stitches.
INSTRUCTIONS.

Stamens and veins.—In making the outlines of these, in leaves, use the "running-stitch ;" but this, is never used for filling or shading.

Tacking the edges.—In appliqué-work, or in all work indeed, use the "chain-stitch" for securing the edges. It makes too coarse a line for "outlining." It is made simply by a diagonal-insertion of the needle, and folding the thread once over, and around underneath the front of the needle.

"The Tracing-Stitch."—This is useful in appliqué-embroidery, and is worked by first laying down a line of filoselle or embroidery-silk. Secure this with a thread of another color, by bringing it up from the back of the material on one side of the embroidery silk, and then carrying it back on the other side. The stitches must be, equidistant in length. Gold-cords can be fastened down in the same way, using fine sewing-silk to fasten them. When the outline is finished, a small hole should be pierced, and the cord cut off and passed through to the other side, where it is fastened.

"The Skeleton-Stitch."—This is sometimes called the "double-outline-stitch," and is done in the regular "embroidery-stitch," making every second or third stitch longer than the others; after which, the veins of the leaf or design are traced, in the "single outline-stitch."

"The French-Knot."—This stitch is used for the centre of flowers, and for making a "raised-foundation" for such flowers as need it. It is made by taking a "back-stitch," passing the silk twice around the needle, and drawing the latter through; and at the same time, holding the coil down in place.

"The Darning-Stitch."—This simple stitch is used in many ways. A design can be outlined, and the interior darned; or the interior can be left blank, and the back-ground darned. The effect can be changed by using foloselle embroidery-silk, or etching-silk; each, giving a different appearance to the work. This stitch can be successfully used for making apples, cherries, oranges, etc.; and they show to a very fine advantage.

"The Queen Anne-Stitch."—This stitch is sometimes termed the "weaving-stitch." Suppose a spray of foliage, or a flower to be formed; first, outline the flower either in etching, embroidery, split, or filling-silk. If on fine, closely-woven material, Japan-etching-silk is best. If on pongee or coarse linen, embroidery-silk shows to a better advantage.

In weaving, cover each petal with parallel-stitches, extending from one outline to another, leaving very small space between each. Cross these at right angles, in the regular "darning-stitch." The effect may be varied by changing the angle, at which the silk crosses. A knowledge of colors, and the art of drawing, aids largely to success in embroidery. A love of flowers should be cultivated; and the student should observe them closely.

"The Plush-Stitch."—This stitch is best adapted to making such flowers as the golden rod, Russian Snow-flower, Coxcomb, Sumac, Marigold, and others of a similar nature. First, fill in the flower with "French-Knots" of the leading color; then using filling-silk, pass the needle through from the back; then take a double strand of filling-silk, pass the fine thread over it, and through the work near the original place where the needle came up. As the split is tightened, the double-silk will naturally fall into place. Cut the double-silk, the length best adapted to the height of the flower. Repeat this stitch, until the flower is sufficiently covered to appear well. Do not crowd the stitches too much, unless you wish it to look very heavy. By varying the size of the "French-Knot," which forms the ground-work of the flower, its surface can be raised more or less, as may be desired.

"The Kensington-Stitch."—This stitch derives its name from the celebrated art school at South-Kensington, England. It is not, as is generally supposed, simply a stitch of itself, but is a "plan of shading" and "blending"-in of colors, according to the principles of art, by using a "combination of stitches" to secure artistic effect. By this stitch or plan only, are we able to achieve success in the embroidery of natural flowers and natural colors. The stitch is taken slightly diagonal, toward the last stitch. Commence the work on the stem of the design, using the "outline-stitch;" the stem made, commence on the leaf at the centre-line, and at the lower part, giving the needle the slant upwards, on the angle of the natural veins in the leaf; the stitches must be proportioned in length, to the size of the leaf. If the leaf is a tiny-one, one shade of the leaf-color is sufficient; in which case, take the stitch from "outline to centre;" but in larger leaves, where two or more shades are required to fill the leaf, proportion the stitch according to the number of shades used. These stitches must be of irregular lengths, where they are to join and blend with the next shade, so as to more perfectly blend in the shades and in the colors. In making the flower, commence on the outside-edge of the petals, etching up from the centre or circle of the flower; and proportion the length, as in the leaf, shading down toward the centre, with darker shades of the flower's color; and according to good taste and the correct principles of art. Thus, in this combination, we use the "outline-stitch," the "satin-stitch," the "appliquéd-stitch," in order to complete the work. By this process, nearly all the material is brought on the face of the work, without the waste there is in the "satin-stitch," which leaves as much on the wrong side as on the face of the work; and the "French knot-stitch," which is used to represent the seeds in the centre, and also when necessary, on the ends of the stamens.

"The Herring-Bone-Stitch," is used for plain, simple work on linen, etc.; and is so well known, that we will not occupy the space required for a description. By inserting the needle diagonally, instead of straight, and with a double branch instead of single, it becomes the "coral-stitch." The "Arasene," the "Grecian," the "Persian," the "Tent," the "Star," and the "Feather-Stitch," each and all, deserve a special treatise; but as we have given in the main what is embraced in all of them, we are reminded by the matter that claims place in the work to close, by giving some examples in actual work.

UTILITY AND DECORATIVE ART. 177

PLATE XXX.

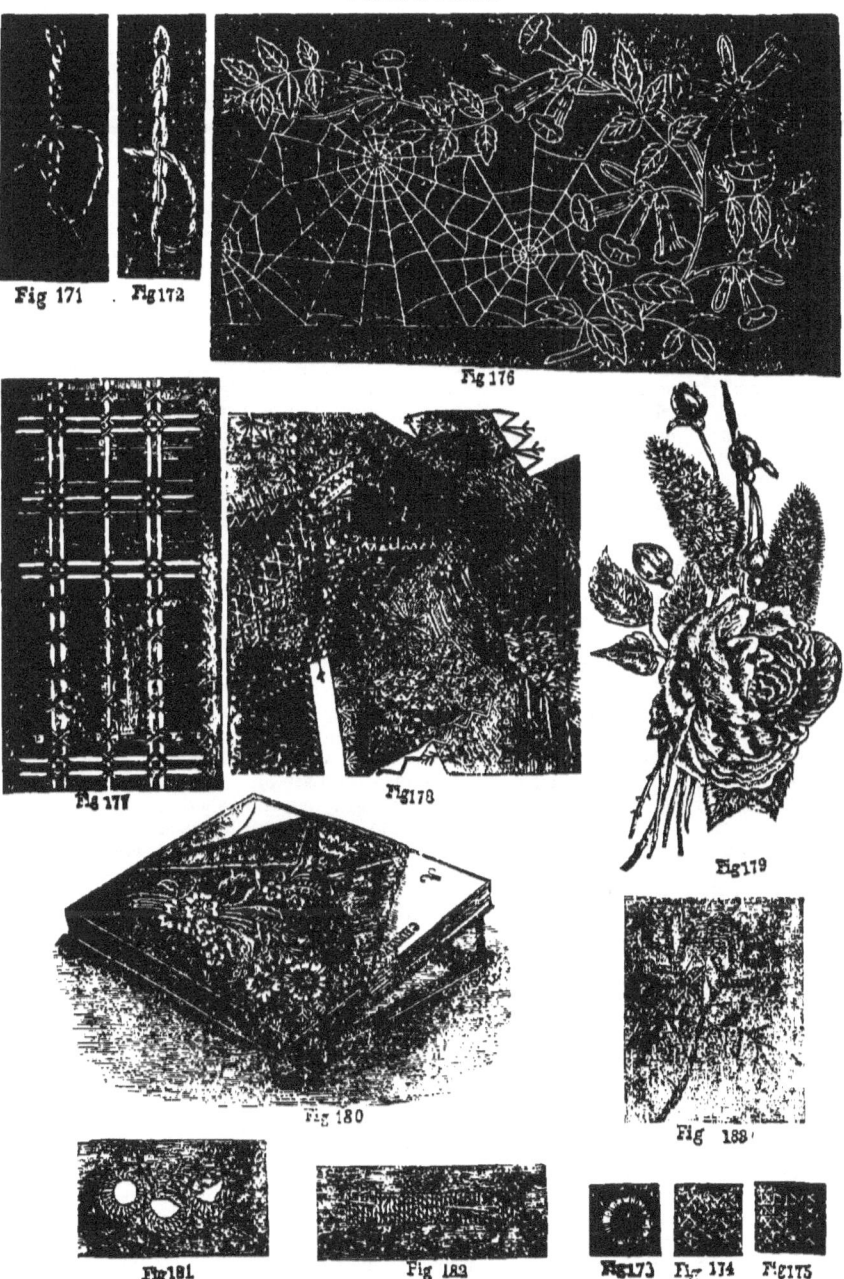

International Embroidery and Lace-Work.

PLATE XXX.—Figures 171-183.

EXPLANATON.

Figures 171 and 172 represent two methods of the "chain-stitch;" figures 173, 174, and 175, represent three different styles of "point-lace;" figure 176, shows a design for "Lamberquin-work;" figure 177, shows a fancy design for a "waist-belt;" figure 178 shows a design of "crazy-work;" figure 179, shows a "cluster of gauze-flowers and buds," for a toilet; figure 180, shows a Morocco and satin-lined "envelope-holder and "ebony-stand;" figure 181, shows "stitches for eyelet-holes;" figure 182, shows a "bordering-stitch;" and figure 183, shows the "stem-stitch." All of which explain themselves, in connection with the instructions on "Embroidery" and "Lace-Work" given elsewhere.

The Rick-Rack-Daisy.—Crochet the centre with yellow "embroidery-silk;" and gather in the rick-rack-braid about it nicely, imitating the flower as accurately as possible. Or it may be made in still another manner, and as follows, viz: Make the flower of felt; the white petals or points being connected where, and at the points, they come towards the yellow-centre. This is beautiful, if it is nicely formed.

The Wild-Rose.—Embroider the centre of the wild-rose with dark olive chenille; and then, make "knots" of yellow brown floss, with stitches of a paler shade, for the stamens. This represents it beautifully.

The Snow-Ball.—Make the foundation in "double-French-knot," in double-crewel; so as to bring it out, in bold relief. Attach to this, very narrow silk ribbon; which should be cut in pieces, ¾ of an inch long, and pointed at the ends. These are then crossed, and fastened with gold-silk. If small pieces of ribbon are closely put in, and the ends are frayed, the effect produced is very natural.

The Wild-Clematis.—This is much more easily accomplished, than the "snow-ball." Instead of using silk-ribbon, arasene is cut and fastened down by silk stitches; which, imitates perfectly, that fuzzy effect of the wild flower; and is charmingly beautiful, if nicely executed.

The Wild-Cucumber.—This luxurious and beautiful vine, with its delicate feathery sprays, is imitated by couching down white arasene and chenille; thus, throwing the spray into marked relief. It is exceedingly simple, easy of attainment, and very pretty.

The Transparent-Tidy.—Make the body of transparent-silk; hem the edges 1¾ inches all around, and stamp it in some pretty design; which, will look well worked in the plain "outline-stitch." Fill in all the space between the outside-edge of the design and the border, with silk. This "filling-in" is done in the regular "darning-stitch," leaving three canvas-threads between each silk-thread. Line the whole with bright gold, satin, or silesia ; and then, trim all with lace.

"Crazy" Patch-Work.—"Crazy"-work, or perhaps using the better term, "Kaleidascope-work," most likely received its "crazy" name, from the chance-method of its construction. It has its side of beauty, however. Its supreme violation of all pre-arranged order, makes it interesting; and to some minds, to appear very beautiful. What in other things would seem disorder, is order here. Some very handsome work is done in black silk-pieces. Have your pieces nicely pressed out, and place them on new paper-cambric; then press them in their position thus, before you begin your stitches, viz: "Over-cast" all the edges, so they will not fray-out. The stitches are "crazy" too, and you can do as you please in the choice of the same; and you need not fear of offending any one, but yourself! Among the stitches we see the "cat-stitch," "buttonhole" and "feather-stitch ;" the "coral," "loop-stitch," "knot," "star-stitch," and the "double-feather," "diagonal," and "zigzag-stitches." And as their names indicate the process, we will simply submit them. There are a variety of others; but you can do elegant "crazy"-work, with these, and with their combinations with each other. If you fall short of a sufficient variety, draw on the "chain-satin," "French-knot," "border," "brier," "cross," and "point-russe" and the "crow's-foot" stitches. All colors are used. Principally, red, pink, lavender, blue, yellow, and white; and five shades of green "embroidery-silk" are used. Roses and figures can be cut out of brocade, and applied; and these can again be ornamented by needle and stitchery. You can also mix cream, white, black, olive, cardinal, garnet, bottle-green, sky-blue, moss-green, gendarme-blue, silver-gray, golden-green, lavender, pink, and anything else you wish; and can all the time be in good style and good taste! You can embroider in the centres, flowers, elephants, birds, plants, fishes, or anything you may fancy; and then again, work sprays of blue forget-me-nots, intermingling with blossoming vines and white rose-buds, upon the strips. Give solidity and security to your work; and finish with care and artistic neatness. And if true to this one great law, you can do "sensible" "crazy-work."

Crocheting.

Chochet, is a French term; and its literal meaning is, a hook; to draw; a species of knitting, performed by means of a small hook. The materials are fancy worsted, cotton, or silk. But indeed it applies to all materials that can be "hooked," and "knitted" by hooking. In its province are embraced tidies, lambrequins, edgings, cushions, bibs, jackets, shawls, stockings, gloves, nettings, and all the many things that are made by "hooking, and drawing" the threads through. The stitches, as in embroidery and other work, are various; and we might add, almost endless; but fortunately they are very simple, and are nearly all well-known to all, and are very easy of attainment. Furthermore, any new stitch can immediately be attained, by the simple process of unravelling a stitch or too, and noticing accurately, its construction. The simplest stitch to begin with is the "chain-stitch;" and the simplest kind, is "chain-crocheting." In crocheting a "chain," simply draw one loop through the other, until the proper quantity is obtained; this constitutes the foundation. Take the needle in the same manner in which you would hold a pen; then, pass it through the loop of the "chain," and on the needle place the silk or wool, which must be drawn through the loop; continuing back and forth, which produces a kind of a ribbed-stripe. This is considered the most simple, for the first practice. Hold the work between the forefinger and thumb; the silk or wool is then to be passed over the forefinger, under the second, and over the third. An ivory needle is best for beginners, but in progressing, a steel-one is preferable; rendering the stitches more even, and regular.

1. Chain-Stitch.—Draw the thread through the loop on the needle.
2. Single-Crochet.—Keep one loop on the needle; put the needle through the upper edge of the chain-stitch, and draw the thread through the chain-stitch and the loop on the needle, at the same time.
3. Double-Crochet.—Insert your needle into the upper edge of the chain-stitch on the work, and draw the thread through the work; then, through the two loops on the needle.
4. Long-Crochet.—Catch the thread around the needle, before you insert it in the work; draw the thread through the work and then through one loop, then through two loops, then through two loops remaining on the needle.

5. **Double-Long-Crochet.**—Catch the thread around the needle twice before you insert it in the work, then draw the thread through the work, then through one loop, then through two loops successively until you have drawn the thread through all the loops on the needle.

6. **Treble-Long-Crochet.**—The same as double-long-crochet, with the simple difference of the thread being thrown around the needle three-times instead of twice.

7. **Open-Crochet.**—Catch the wool around the needle, before you insert the needle into the work; draw the thread through the work, then through one loop, then through two loops, then through one loop.

Spanish-Edging.—Cast on nine stitches, knit across plain. 1. Slip one, knit two, over, narrow, knit one, over twice, narrow, knit one. 2. Knit two, knit one, purl one, knit three, over, narrow, knit one. 3. Slip one, knit two, over, narrow, knit five. 4. Knit seven, over, narrow, knit one. 5. Slip one, knit two, over, narrow, knit one, over twice, narrow, over twice, narrow. 6. Knit one, knit one, purl one, knit one, knit one, purl one, knit three, over, narrow, knit one. 7. Slip one, knit two, over, narrow, knit seven. 8. Bind off three, knit five, over, narrow, knit one. This finishes one scallop.

Narrow Scotch-Edging.—Four chain, three double-crochets, two chain, three double-crochets (which makes a shell), in first chain, turn; shell in shell, one double-crochet in top of last double-crochet of shell, two chain, turn; shell in last shell, double-crochet, two chain in three chain until you have six double-crochets with five spaces between; in which, put one single-crochet, two double-crochets and one single crochet, which makes five tiny scallops, one chain, shell in shell, and proceed as before, remembering to fasten the last of the six double-crochets into the one chain before making the little scallops. This can be made wider by increasing the number of stitches in the foundation chain three stitches for every shell desired.

Prussian-Edging.—Materials—No. 40 cotton and steel crochet hook. Make a foundation of nine chain-stitches. 1. Four chain, fasten in ninth chain of foundation, four chain, miss four, short crochet in foundation, four chain, short crochet in same stitch, four chain, miss four, short crochet in end of foundation, four chain, short crochet in same stitch, turn work. 2. Three chain, four double-crochets in first loop, four double-crochets in second loop, one double-crochet in end loop, three chain, one double-crochet in same loop, turn work. 3. Four chain, short crochet in first stitch, four chain, miss four, short crochet between two double-crochets, four chain, short crochet in same stitch, four chain, short crochet between two clusters of double-crochet, four chain, short crochet in same stitch.

There will be a little loop between each cluster of double-crochet, and one at the upper edge of your work for the double-crochet of the next row to be worked in. Turn work, and repeat from second row.

New Berlin-Stitch.—Conceive a canvas to be worked with twenty-six stars, having eight points; and these points so arranged that one set of stars have their centre, where those worked in a darker shade meet, at their extreme points. Each of the sections of these lighter stars, consists of four slanting-stitches for each side; in all, eight stitches for the whole section; the intervals being filled up with dark-green wool. Worked on very fine canvas, this pattern can be used for slippers, travelling-bags, etc. On middle-sized canvas, it will serve for footstools, cushions, etc. Lastly, thick canvas should be chosen for a rug.

Russian Carriage-Rug.—Set up twenty stitches, knitting six in plain. Then knit and purl four alternately, until the twelfth row; then slip four stitches, on the third pin. Keep these at the right-hand side. Then purl four stitches, from the second pin of the left hand, upon the right-hand pin; and leave the third pointed-pin down over the right side; purl four stitches on the third pin, purling four again and knitting four. Knit the next row plain. Knit four, purl twelve, and then knit four again. Then knit eight; and turning, purl four alternately until the completion of eleven rows. Transfer these on the pointed-pin, and let the same hang down along the right-side. Transfer four stitches from the left to the right-pin; then transfer by knitting four from double-pointed pin. Knit eleven rows. Then knit, and again

UTILITY AND DECORATIVE ART.

purl twelve; then knit four. Knit plain for next row, and repeat from seventh row until the completion of 1½ yards. Then knit six plain rows and cast off. Select any colors, and knit five strips. For the fringe, proceed as follows, viz: Form twenty-four stitches; double this, inserting needle in a stitch of the main work, and drawing the ends of the zephyr through the loop. Secure the work firmly, as you proceed thus, until finished.

Mignardise-Work.—Materials—smallest size of mignardise; cotton, No. 26; crochet hook, No. 20. Bell gauge. Begin with second loop; work five, in chain; then pass into second loop; chain five, and repeat the process. Then join this to the first stitch of the chain in five; chain five, and pass into centre stitch of chain in five; pass back into same stitch, chain five, and pass into next five; then chain five, and repeat the process from *. Attach the thread through centre-stitch of first chain in five; * chain in six, and one in single, through third stitch from the crochet-hook; chain three, pass into centre of next in five chain, passing between the points in the preceding row; then repeat the process from *. This makes a very handsome design in this once very popular and highly-prized work.

The American-Tidy.—This choice and highly prized tidy, among American ladies, is very simple, and as follows, viz: k l, n, tto, k l; then repeat this process, and purl the next row. N, tto, k two; and again repeat this process. Now purl the next row. Any desired number of stitches that may be divided by the figure four, may be used as a basis for the work; and the process as given for these first four rows can be repeated, the required number of times, until the tidy is the required or desired size. It is very handsome.

The Louise-Collar.—Arrange thirty stitches, and knit back in plain stitches. Now purl four stitches less than the number taken in first row, and knit eight less than this last number, in plain stitch; o, n, o, n, o, n; and then knit two, in plain. Now purl twenty-eight, turn, and knit twenty-eight plain. Next row plain to thirty-two; four plain in next row, purling twenty-eight. Knit two, o, n, o, n, o, n; knit twenty plain, turning the work. Purl twenty-eight, knitting thirty-two plain; and commence again, at first row. The "edging," is made as follows: Take on nine stitches, knitting across plain. Then purl seven, and turn the work; slip one, o, n, o, n, and knit two. Purl seven, and turning the work knit seven plain; and knit nine in next row, plain. Knit two, o, n, o, n; and turning the work, purl seven. Now complete, by crocheting nine, plain.

French-Edging.—Arrange fourteen stitches in plain; and then thread over, knit one, turn, knit two, narrow second time, knit two, turn, narrow, turn, narrow, turn narrow, and knit one. Seam across every second row, beginning at the second. Turn, knit three, turn, knit one, narrow twice, knit one, turn, narrow, turn, narrow, knit one. Turn, knit five, turn, narrow twice, turn, narrow, turn, narrow, knit one. Turn, knit three, narrow, knit two, turn, narrow, turn, narrow, turn, narrow, and finish by knitting one. Which, makes a very handsome design.

Clover-leaf Crochet-Stitch.—Form the desired length in chain; turning, make eight "double-crochet," and a chain of four more than the "double-crochet." Now reversing, form a single crochet in seven-stitch; this, will leave six stitches for the stem. In this, three loops are made thus, viz: Crochet one single, chain four, one in single, chain four, and join the least at the lower part, where the loop was made in single-crochet. Now reverse the work, and pass the thread under, and within the first loop; form one single, three double, and secure in centre of eight double-crochet, on the chain. Chain three in double, passing the hook through first, double-crochet in chain; and then draw thread through, forming six double-crochet and one in single, in said loop. Form one in single, twelve double, and one single—in next loop. Then again, eight single, over stem; and now double-crochet along chain, and form twelve chain. Form second leaf, same as first. Secure first, at root of stem; and the second, in the seventh stitch of the third part of the first leaf. This is very simple, after a little practice; and is very handsome.

Spanish Wheat-ear-Stitch.—Form, and take up nine plain stitches. Slip one, knit one, turn over twice; join two, by purling; k two; turn over, k one, turn over twice, joining two—by purling. Turn over twice, and join two by purling, and knit four; turn over twice, and join two by purling, knitting two. Slip one, knit one, turn over twice, purling two together and knitting three; then turn over, knit one, turn thread over twice, purling two together. Turn over twice, purling two together and knitting five; turn thread over twice, purling two together and knitting two. Slip one, knit one, turn thread over twice; then again, purling two together knit four. Knit one, turn thread over twice, and join two, by purling. Turn thread over twice, and join two, by purling; then turn thread over twice, purl two together, and knit six. Turn thread over twice, purling two together knit two. Slip one, knit one, turn thread over twice, purling two together knit six. Thread over twice and purl two together. Thread over twice, purl five together, and knit three; thread over twice, and purling two together knit two. Now repeat this process, beginning again, at the first row. When the last two are joined by purling, at the end of the first and each alternate row, the loop following the same is dropped.

The American Raspberry-Stitch.—Using any multiple of four, and increasing it by two, take up the desired number of stitches. Example: say $20 \div 2 = 22$ stitches, which we now suppose to be the desired amount required. Purl across; and form each alternate row, the same as first. Knit first stitch of second row; knit, purl, and knit second stitch, combining the three stitches. Purl next three together, then knit, purl, knit, forming three in one; purl next three together again, and continue to the end. In beginning the fourth row, knit the first, purl next three into one; now knit, purl, knit next three before you slip, and again make three out of one. Repeat this process to the end.

Mittens in French "Shell-Stitch."—Take up twenty-four stitches on each of two needles, and sixteen on the third. Then knit, thread over, and continue this process to the fourteenth; then seam one, and form next seven into thirteen, the same as you did the first seven. Then seam one, and continue this process entirely around. Slip, knit, and slip the first over it; then knit nine through loops first made; narrow two, seam one, and continue thus, around. Form the third row as second, having but seven stitches instead of nine. In the fourth row, knit the five the same. Commence again at first row; knit twelve rows of "shells;" then reserve three rows of shells for back, with seam at each side. Form a braid of the centre by converting seven into twelve. After the work of the first row of shells, slip the first four stitches of the braid upon a different needle, keeping them at the outside while knitting the next four. Then knit the four left; then the last four; then plain across the braid, until having formed the next shell. Then knit the first four, and again slip the next four on another needle; and keep them on the inside while knitting the next four. Then begin again by knitting the four you left, and continue this process every other time you slip the first four, and the second four. Between the "shell" and the thumb of the mitten, knit seven plain stitches; and, of course, all the rest of the mitten in plain stitches.

UTILITY AND DECORATIVE ART. 183

PLATE XXXIII.

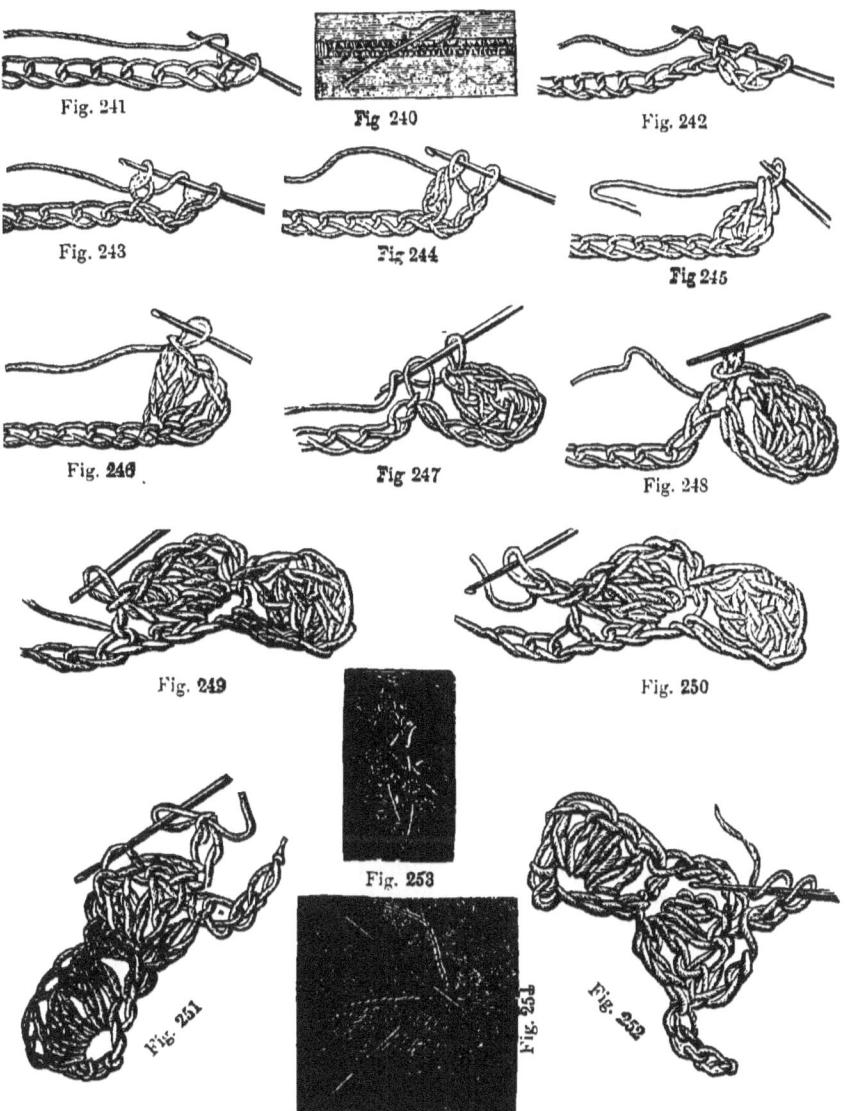

International Embroidery and Crochet-Stitches.

International Embroidery and Crochet-Stitches.

PLATE XXXIII.

FIGURES 240, 241, 242, 243, 244, 245, 246, 247, 248, 249, 250, 251, 252, 253, and 254.

Figure 240, represents the position of the needle, and the method of "hem-stitching;" figures 241, 242, 243, 244, 245, 246, 247, 248, 249, 250, 251, and 252, respectively represent and illustrate in successive details, a series of "crazy-stitches," in crocheting; figure 253, shows a handsome design for either painting or embroidery; figure 254, shows in detail, a "buttonhole-embroidery" for appliqué-flower.

"Wavy-Braid"-Crochet.—Procure a piece of "wavy-braid," twice as long as desired when completed. Work on this a row of *point de Bruxêlles*. Now double the braid, and having the two ends meet, sew the points together. Make a chain of nine, and loop this into the end of the braid, and continue to the end. Chain six for the second row; and loop the same into the centre, of preceeding row. Chain four; loop again, on the same; and continue thus, working to the end. This finishes the crochet. Finish the other edge of the braid with another row of *point de Bruxêlles*. A pretty insertion is also made, by omitting the last row and repeating the crochet to match with the other side. The two, can be thus arranged to accompany each other.

The "Forget-Me-Not."—Use a small crochet-hook. It should be worked in a light, but bright color. First form a small ring of wire or leto, lightly twisted; and out of one end, forming the stem or stalk. Forming a loop on the crochet with "split"-zephyr, form a plain stitch on said ring and make a chain in five stitches, securing this on the ring by one plain stitch. Proceed until you have five small loops, each forming a petal. The ring is then twisted, closely. Thread a needle with a light shade of green silk. At the end of the same, form a small tuft of light yellow. Pass this through the centre of the flower, which will so arrange itself, that this tuft will serve as the stamen of the flower. Now twist the cord around the stamen of the flower. For the leaf, form a chain in a soft tint of a yellow shade of green, of nine stitches. Around this chain is worked a row of rather long stitches, having a wire in the edge. The shade of this latter should be, in a deeper green. Another row of double-crochet is worked in a shade of deeper green, which also has a wire in the edge. These wires are then all twisted, and covered with a shade of deeper green. This serves as the stalk.

The Laburnum.—For the flower, proceed as follows, viz: In a bright shade of yellow "split"-wool, form seven long stitches in "double-crochet," upon a wire—alternating, in one plain and one double. The wire for the first two petals is put in at the base, and not in the edge of the stitches; therefore, the needle should be passed back of the wire, at the first pair of loops in the long stitches, just described. The wire should be covered, about ⅔ of the distance occupied by the stitches. This wire is folded at the back of the work, and the two ends secured by twisting the zephyr around them jointly, fastening it by the "loop-stitch." Repeat this process, and form the second petal. For the chief-petal, make a chain of five stitches, in a deeper shade of unsplit-zephyr. Then, having a wire in the edge, form in a lighter shade, one stitch of "double-crochet," two long stitches in the next two loops, and three long stitches in the next, and one of "double-crochet" in the loop at the top; three long stitches in the next, on the other side; two long stitches in the two succeeding loops, and one of "double-crochet" in the next; which process, completes the entire petal. Now join the two first petals with a light yellowish-green in "split"-material. Over these place the chief-petal, securing them firmly together. Finish neatly by cutting off the wire, and covering the same with the silken cord. For the leaf, form a chain of nine stitches, having one long stitch in each looping; into which, at each edge, a wire is neatly inserted; and the tripple-parts of the leaf, are then secured by giving the wires the twist, as explained before. Each part of said leaf having however, a separate stem ⅜ of an inch in length. The flower is naturally of the "drooping"-kind; and it should therefore, be mounted in that form.

The Snow-Drop.—The best material for working this beautiful flower is white silk, in "half-twist."

The flower contains three small petals and three larger ones, making six in all. Form four stitches, and then arrange the work so that you will have knit and purled in alternation six rows, in plain. Then knit and purl the next succeeding rows, until you come to the seventh row; in doing which, increase one stitch at the commencement of the first and second, and the fifth and sixth rows. Now without increasing, knit and purl successively, the next eight rows. Then, by dropping one at the beginning of each row, cast all off until you have but four left.

Knit the small petals, as three in one. The first row is purled; then form one stitch, knit two, and alternate thus to the end of said row. Purl the rows back. Then form one stitch, knitting three; and alternate thus to the end of row. Continue this system of increase, to the place where seven will be knitted between each. Now purl one row, and knit one plain; and then again, purl one row. Next, knit eight stitches, turning back and purling them. Then knit four, purling them back; and then break your thread so that you have a good length, from the end to the work.

Now secure the four stitches; and by means of the needle, pass the cord through the work for several stitches, and to the point in front of the four stitches left. Knit these four stitches, purling them back; and secure and fasten them as in the former. Now coming down to the next stitch, knit eight successive stitches; and then proceed as before. And the same in the next following. Insert or sew a wire form about the edge of the top, with "split"-material. Touch and brighten up with embroidery, in "heart-shaped forms," to simulate the natural flower. Then placing the larger petals over these, fold the stem with a rich color in green. Make the buds and flower, of either silk or wool; and as fancy and circumstances dictate. Cast from six to twelve stitches on, for the smallest. Now knit and purl a sufficiently large piece to cover the bud; which, is secured by a wire. The larger buds require from ten to fifteen stitches, at the top. Shape the leaves similar to the natural leaf. Begin with six stitches, knitting and purling alternately, until a sufficient depth is reached; then at each side, join the stitches; and also neatly join a very tiny wire around the edge, by means of stitching. This completes the elements and principles of the snow-drop.

The Lupine.—The lupine is a member of the "pea-family;" and it, along with all other small flowers of this character, can be worked according to the rule given above for working the "Laburnum," to which we refer the student.

Grecian-Netting.—For ordinary work, use cottton number 2; a small mesh, ¼ inch wide; and a larger mesh, ½ inch wide. For fine work, best fine silk, and meshes, Nos. 8 and 16. Using the larger mesh, take any even number of stitches. Small mesh: Bring the silk around the fingers as in ordinary netting, and pass the finger through the finger-loop, into the first stitch. By means of the point of the needle, bring the second stitch through the first. Withdrawing the instrument, repeat the process of drawing the first stitch through the second. Now complete the stitch. The next requisite will be a small loop, which apparently, falls across the "twisted-stitches;" and then, repeat the process. For large mesh, a plain row. For small mesh, the same as before; only, that the first and last stitches are netted plain.

Grecian-Border-Stitch.—This is used for heavy work, generally; and it can be of any desired width. Cotton No. 8. Take this triple-fold. A pair of the larger-sized meshes. Net two plain rows upon the foundation. Net four plain in second row, winding the cord thrice around the mesh, and putting the needle into the stitch each time, without netting the same; then knotting the stitch securely, and passing the needle around said "three-loops," without throwing the cord over the mesh; and securing it as the "buttonhole-stitch" is secured, in edging embroidery. All patterns can be adopted.

Begin the "fringe" with the larger mesh, netting four stitches by alternation, into each loop; and then securing, as in fancy stitchery. Net plain for second row, using the smaller mesh. Then using the larger mesh, net third row plain. For the fourth row, use the larger mesh also; and take up four loops of the last row, and form them into one; net the fifth row the same as you did the first; and the sixth as you did the second.

This ancient and very popular netting, is exceedingly simple and beautiful. A little practice will soon make it plain.

Geometrical—Netting. —This ancient but recently popularized "netting," is sometimes termed "Honey-Comb-Netting." Suppose the materials to be, cotton No. 2; a plain flat mesh, three-sixteenths of an inch wide. For the first row, take any even number of stitches, and net plain. For second row, reverse the order of the "netting," by forming the second loop first, and the first, second. Net the entire row, in this manner. Net the third row, plain. In fourth row, take the first stitch, and then the second, first; and net plain, across. The fifth row is entirely plain. And finish, by forming the sixth row the same as you did the second.

Persian-Netting.—This very ancient and pretty "netting," has formed the finest purses, laces, and shawls of Persian ladies, for many centuries. It is used in both fine, and coarse work. It is netting on the principle of "double," "treble," and "quadruple-netting," in an equal number of colors. It is very simple, and yet, can be made exceedingly intricate and complicated.

For "double-netting" in two colors, proceed as follows, viz: Net a foundation, fastening both colors; and then net the "white-stitch," throwing the needle over the left hand, upon the table. Taking up the needle with the color, net one stitch; and throw the needle on the left-hand-side of the position of the needle of white. The pattern depends, upon this arrangement; otherwise, there would be two separate nets—one of each color. Net the white into the white, and the color into the color, throwing the instrument in the exact same manner as before, for the next row. Now repeat this process, by alternations, until completed. Its variations are infinite; and, commensurate with the Arithmetical and Geometrical progression of the numbers and combinations used, in the stitches and colors.

Lace-Work.

The term "Lace," comes from the Latin Laqueus; Italian, Laccio; Spanish, Lazo; French, Lacet. Its various meanings come to us as follows, viz: "A noose;" "a snare;" "a gin;" "a trap;" "a net;" "a delicate ornamental net-work; the meshes of which, are formed by plaiting together threads or cords of linen, cotton, silk, or other material." Shakespeare grasped its figurative meaning in that expressive sentence "Here lay Duncan, his silver skin *laced* with his golden blood." It is of very ancient origin; and it has always been one of the chief ornaments, in woman's dress. It was prized above all earthly gifts, among ancient queens, save the precious stones that burned and flashed fire from among its tangled meshes, as they lay upon hearts that too often burned with the consuming fires of false ambitions and merciless cruelties.

The kinds and their names, are almost innumerable; and they represent every degree and shade of enlightenment and refinement, in civilization and culture. Their values are not based upon positive worth, but are measured by the ratio of expense in the labor and time, necessarily consumed in their production. From Brussels, come our principal laces; and the finest, are hand-made. There are various grades of "Brussels-Lace." The majority of articles termed "Brussels-plait," come within the great general province of "Pointe-á-l' aiguille." These resemble somewhat the "Honiton-applique;" in which however, tne flowers are formed separately on the pillow; and are attached thereafter, to a "made-formation." Two kinds of "net," are also used. The best "résean," is formed by bobbins, on the pillow. It has good imitations however, made by machinery. The best "Brussels-Lace" contains only, the finest possible thread of flax; the best quality of which, comes from Hal, and from Rebecque. The finest thread is hand-spun, in damp and gloomy cellars; the tenuity being so great, that this process can not be accomplished in open light and dry air. Belgian "thread-spinners" are apparently well-paid; but they might be said to almost literally give themselves "a living sacrifice," for the gift and production of their hands. Every inch is accurately examined, as it is drawn from the distaff, and is closely scrutinized, as to any possible imperfections and inequalities; which, are immediately and carefully corrected. The thread is held in front of a dark background, while being formed; and the scene of labor is often so arranged that but a single ray of light is left to cheer the lonely place; and this is made to fall directly and accurately upon the forming-thread! Each department of lace-making is under the supervision of thoroughly-trained specialists.

The "platteuses," make only flowers, and form patterns for the ground-work, on the pillow. Some work them in point, by means of the needle; and which, when joined to the net, forms the "Brussels-appliqué."

The "striqueuses," join the flowers to the net. The "Attacheuses," join the different parts of a pattern to simulate a whole so perfectly, that it often deceives an expert to detect their joinings.

The "faiseuses de point-a-l' aiguille," shows her taste and skill by forming the floral-patterns in relief, upon the plain flat surface; and under her trained hand we see, the plain lace budding and blossoming into all possible variations of clusters and wreaths of grasses, leaves, and flowers; shadow and shades, being reflected by the transparency of the texture. The embroidery is often so fine and perfect, that a needle would pass through the meshes, only with considerable difficulty. It comes to us as "Guipure," "Maltese," "Pillow," "Saxony," "Point d' Alençon" and countless other names; which, are naturally drawn from the circumstance of place, style, or quality. As examples of "method," we give the following, viz:

Point d' Alençon-Lace: Have the design first engraved on copper. This is then printed from the plate, in divisions, on pieces of parchment. Number these, in their order and, then prick holes along the outlines of the flowers.

Placing upon this a piece of course linen, sew with a tracing-thread, by fine stitches, uniting them. Holding two flat threads beneath the thumb of the left-hand, guide them and fix them along the edges of the pattern, by very delicate stitches passing through the punctures of the parchment. This completes the skeleton. Next, fill up the centres of the flowers.

By means of a long fine needle and very fine thread, work in "knotted-stitch" from left to right; until, by successive rows of the same, all the interstices are filled. Form the plain ground by first laying one thread across the pattern, and then completing its formation by a system of the regular intersections; this forms the delicate meshes.

Spaces can be reserved for any desired fancy-stitches. These are sometimes termed the "modes." Ornament it to fancy, and to design, by stitches of "embroidery-in-relief;" which are known, as the *brodé*. Then unite the sections, so that their junctions are not perceptible. This forms the celebrated "point d'e Alençon-lace," sometimes quoted "point de Venise-lace;" and which is one, at least, of the most costly laces of the age.

Brussels-Point-Lace.—This pretty and easily-formed lace is simply a succession of rows of "Brussels'-edge," which is worked backwards and forwards; and which is so simple, that it is understood by all—needing no formal instructions.

Scotch-Lace.—Form seven stitches. For first row: three plain, over, narrow, over twice, two plain. Second row: three plain, purl one, two plain, over, narrow, one plain. Third row: slip one, two plain, over, narrow, four plain. Fourth row:

bind off two, three plain, over, narrow, one plain. Now repeat from the first row.

Berlin-Insertion.—Form thirteen stitches. Knit across plain, slip one, knit one, thread over twice, seam two together, knit one, thread over, narrow, knit two, thread over twice, seam two together, knit two. Then knit all the rows the same.

Venetian-Lace.—This is simply a pretty net-work of successive rows of Venetian-edge, which should all be worked one way—from left to right.

Sorrento-Lace.—This is the same as sorrento-edging; and it is simply worked on the principle of "netting backwards and forwards."

Spanish-Point-Lace.—Secure the desired number of threads for the design, to give the lace a raised or heavy appearance; then work from one side to the other, repeatedly, with "buttonhole-stitches." Finish the edge to taste or design.

English-Point-Lace.—Fill any desired given space with threads crossed at right angles with each other, and at the distance of ⅛ of an inch apart. Secure your thread to the braid, forming the "spots," by passing the needle around the thread until you reach a point where the first threads cross; then insert the needle in a manner, under and over said "cross," until you have a sufficiently large space. Then arrange the next the same way; continuing this process until all are completed. The pattern is at your option.

Open English-Point-Lace.—This is made exactly the same as the preceding, with these exceptions, viz: four threads cross each other, only; and the "spots" are only located upon the "cross." This effect can still be more enhanced in beauty, if the horizontal and upper right-threads are taken several sizes coarser than the other two threads.

Spanish-Bar-Lace.—A very fine lace is formed by making "bars" in filling up any space between two pieces of braid, which are edged either with Venetian or Brussels-edging. The method is as follows, viz: Pass the thread from side to side, through pairs of opposite loops; but the process demands the needle to pass from the under-side of one stitch and the upper-side of the other.

French-Rosette-Lace.—Form this in from four to eight threads, according to the space and the desired effect.

First, form "sorrento-bars;" then bridging the space from side to side, pass the thread under and over the alternate threads, until the proper size is reached—beginning at the centre. Now secure the work by taking the needle around the last bar, and then into the braid; and complete the process by four "buttonhole-stitches."

Portugese-Lace.—This is formed on "sheet-linen," which is best when old. Draw an even number of threads, each way; then join the cotton by means of the needle, to one of the bars; and then carry it on to the next, drawing it up tightly, leaving the cotton of sufficient length not to draw the lace. Complete the entire work one way, first; then the opposite way; and so arranging, that each thread will cross the centre of an open space. Pass the thread over each former thread, and secure it at the cross with a firm stitch; then form a stitch where it meets the bar, and use "cross-stitches" over the linen. The design can be varied, and a plain linen band can be used with raised and profuse embroidery in leaves, buds, wreaths, scrolls, etc., and having a "dotted ground-work."

"Barred-muslin" is sometimes used; the thin part being cut out, and the edges overcast, and then completed as before. This facilitates the work. It is nice for altar-cloths, and spreads for infants' cots, etc.

N. B.—For both crochet-work and lace-work, the student will find the following the "key" to our abbreviations, viz:

K,—knit; o,—over; t o,—thread over; n,—narrow; s,—slip; p,—purl; k 2,—knit two stitches; p 3,—purl three stitches; s 1,—slip one stitch; t o t, or t o 2,—thread over twice; p t t, or p 2 t,—purl two stitches together; s t t, or s 2 t,—slip two stitches together; and k 1,—knit loop or last, etc.

Torchon-Lace.—O two, purl two together, knit three, n, o, n, o, k one, o, k 1. Next row—slip one, knit all plain, up to last two, then o two, purl two together. Next row—O two, purl two together, k two, n, o, n, o, k three, o, k one. Next, and each alternate row—the same as second row. Fifth row—O two, purl two together, k one, n, o, n, o, k five, knit one. Seventh row—O two, purl two together, k three, o, n, o, n, knit one, n, o, n. Ninth row—O two, purl two together, knit four, o, n, o, n two together, o, n. Next row—O two, purl two together, knit five, o, n three together, o, n. Form the last row, the same as the second.

Saw-Teeth-Edging.—Form thirteen stitches, knitting across plain. Knit one, narrow, throw thread over twice, narrow, knit one, slip one, knit one, throw thread over, narrow, thread over, narrow, throw thread over, knit one. Next row—knit across plain to within six stitches, knit two, purl one, knit three, Next row—knit one, narrow, throw thread over twice, narrow, knit one, slip one, knit two, throw thread over, narrow, throw thread over, narrow, throw thread over, knit one. Next row—knit all but six stitches plain, knit two, purl one, knit three. Next row—knit one, narrow, throw thread over, narrow, knit one, slip one, knit three, throw thread over, narrow, throw thread over, knit one. Next row—knit all but six stitches plain, knit two, purl one, knit three. Next row—knit one, narrow, throw thread over twice, narrow, knit one, slip one, knit four, throw thread over, narrow, throw thread over, knit one. Next row—knit all but six stitches plain, knit two, purl one, knit three. Next row—knit one, narrow, throw thread over twice, narrow, knit one, knit plain across. Last row—slip eight, bind four, knit all but six stitches plain, knit two, purl one, knit three.

Double Oak-Leaf-Lace.—Make a foundation of seventeen stitches, by plain stitches across. Then proceed as follows, viz: S1, k1, tot, ptt, k2, tot, n, tot, n, k2, tot, n, tot, n, k1. Second: S1, k2, p1, k2, p1, k2, n, p1, n, p1, k2, tot, ptt, k2. Third—S1, k1, tot, pttt, k3, t o, n, to, n, k3, to, n, to, n, k1. Fourth—S1, k2, p1, k2, p1, k3, n, p, n, k3, tot, ptt, k2. Fifth—S1, k 1, to t, p t t, k 4, to t, n, to t, n, k4, to t, to t, n, k 1. Sixth—S1, k 2, p1, k 2, p 1, k 4, n, p 1, n, p 1, k 4, t o t, p t t, k 2.. Seventh—S 1, k 1, t o t, p t t, k 5, t o t, n, k 12. Eighth—k 12, n, p 1, k 5, t o t, p t t, k2, k 17, n. Ninth—B,o 5, k 12, t o t, p t t, k 2. This completes the process.

Clydesdale-Lace.—Form a foundation of fifteen stitches, knitting across plain. First—k 2, t o t, s t t, k 1, t o t, n, k 8. Second—k 9, k 1 l, s 1 l, k 1, t o t, s t t, k 2. Third—k 2, t o t, s t t, k 12. Fourth—k 12, t o t, s t t, k 2. Fifth—k 2, t o t, s t t, k 1, t o t, n, t o t, n, k 7. Sixth—k 8, k 1 l. s 1 l, k 1, k 1 l, k 1, t o t, s t t, k 2. Seventh—k 2, t o t, to 2, s t t, k 14. Eighth—k 14, t o t, s t t, k 2. Ninth—k 2, t o t, s tt, k 1, t o t, n, t o t, n, t o t, n, k 7. Tenth—k 8, k 1 l, s1l, k 1, k 1 l, s 1 l, k 1, k 1 l, s 1 l, k 1, t o t, s t t, k 2. Eleventh—k 2, t o t, s t t, k 17. Twelfth—Bo, all but 14 on left-hand needle, k 10, t o t, s t t, k 2. Begin at first row, throw thread over twice before seaming two, each time. This completes the entire process. This nicely executed in Andalusian-wool is charmingly beautiful.

Wide Normandy-Lace.—Form a foundation of thirty-one stitches, knitting plain across. First—k 8, n, o, k 3, o, n, k 9, n, o, k 3, o, k 2. Second—k 2, o, k 5, o, n, k 7, n, o, k5, o, n, k 7. Third—k 6, n, o, k 7, o, n, k 5, n, o, k 1, o, n, k 1, o, n, k 1, o, k 2. Fourth—k 2, o, k 1, n, o, k 3, o, n, k 1, o, n, k 3, n, o, k 9, o, n, k 5. Fifth—k 4, n, o, k 11, o, n, k 1, n, o, k 1, n, o, k5, o, n, k 1, o, k 2. Sixth—k 2, o, k 1, n, o, k 3, o, n, k 2, o, n, k 1, o, k 3 t, o, k 13, o, n, k 3. Seventh—k 5, o, n, k 9, n, o, k 3, o, n, k 1, o, n, k 3, n, o, k 1, n, o, k 1, n. Eighth—Throw off 1, k 1, o, n, k 1, o, n, k 1, n, o, k 1, n, o, k 5, o, n, k 7, n, o, k 6. Ninth—k 7, o, n, k 5, n, o, k 7, o, n, k 1, o, s 1 n, transfer slip-stitch over, then o, k 1, n, o, k 1, n. Tenth—Throw off 1, k 1, o, n, k 3, n, o, k 9, o, n, k 3, n, o, k 8. Eleventh—k 9, o, n, k 1, n, o, k 11, o, n, k 1, n, o. k 1, n. Twelfth—throw off 1, k 1, o, k 3 t, o, k 13, o, k 3 t, o, k 10. Repeat the process to the desired number of times. It will be found to be very pretty.

Collar in Waved-Braid.—The materials required are "waved-braid;" cotton Nos. 20 and 30.

Proceed as follows, viz: Having transferred the fit, and the design upon paper, unite it by means of mucilage on the back, with a piece of muslin; then stitch the braid firmly on the patterns. If the design is in diamonds, rosettes of point d'Angleterre, can be worked in the same. This is done by beginning the first stitch in the diamond, at the point where the braid crosses to the opposite side.

The thread should be twisted seven-times around the first thread. Then the needle should be turned along the back of the braid, in such a manner as to divide the space octagonal, or in eight parts. The last, should be finished in the centre, leaving a half-thread single. Now secure the work by forming one stitch in the centre, and working around it by inserting the needle around one thread and under the next; then around the second, under the third, etc. This should be repeated eight-times; and the work is then completed by folding the cord twice around the single thread, and securing it. The spaces at the edge should be filled closely with "embroidery-work," as fancy or design may dictate. Bars of thick "buttonhole-stitch," should connect the spaces. For the border, form a large "buttonhole-stitch," working into it four closer stitches. Repeat this, entirely around the collar. Now cutting the stitches, joining the design and collar, carefully remove the same; and you will then have a handsome "Waved Braid-Collar."

Worsted-Work.

Mosaic-Knitting :—Suppose a mat, in two distinct colors of wool, deep garnet and old-gold; and that when completed it is finished by a border of cloth, of the darkest shade of wool used; which, should also harmonize with the hangings of the apartment. Proceed as follows, viz : Form a foundation of twelve stitches, and purl one row. Form from these, four designs; each design will, necessarily, have three stitches. Now for the first row, use deep garnet wool; work the wool forward; knit 2† (or jointly) * working wool forward, and knitting three plain; draw the former of the three plain, over the two latter; repeat this process, from * to close : this now leaves one stitch; now bringing wool toward the front, proceed—knitting. You now have thirteen loops, on the hook. For the second row, use old-gold, attaching it, and purling the entire row. Third row : Bring the wool toward the front; knit 2†; then, not bringing the thread forward, knit three plain, drawing the third loop (toward the back) over the latter two; * wool toward front; knit three plain; draw first loop over the latter two; and then repeat from * to the close of this row.

Fourth : Purl the entire row—using the deep-garnet wool. Any design, accurately calculated and based upon any fixed size, can be easily, quickly, and beautifully formed according to the above rules and principles.

Berlin-Work.—This popular method of "worsted-work," admits of great latitude, both in the materials used, and also in the designs. It is worked either in "double," "treble," "quadruple," or otherwise — in "single"-wool. The most popular designs are executed in seven shades of the same color. The stitches are laid over from two to eight threads of the canvas, in a diagonal form; the shades, beginning with black and closing with white. Systems of smaller stitches are arranged and filled-in, at the upper parts of the points. Crossing the centres of the various patterns, long stitches of fine chenille, crochet-silk, beads or gold and silver cords, may be laid, with good taste and with artistic effect.

The Violet.—The usual materials used, are the single threads of fine "Berlin," or "Germantown-wool," in shades green and purple, along with steel or gold beads. Form all the petals separately—as you would the petals of the rose. Do not use the "card-board-process." Proceed as follows, viz : Taking the purple thread, arrange it in the natural "oval-form," by "turning" it a few times; and, having performed this part of the process, lay the fine purple thread across it, each way; thus, crossing it neatly, in each direction. Form the "heart," by threading a very fine gold bead upon a fine wire; and then, arranging a tiny tuft of the green silk or wool underneath said bead, secure your work by bending the wire and neatly twisting and hiding it, underneath. By careful stitches, attach the five petals first made, to said tuft—winding the green material around the ends of the wire, used—which forms the stem, and completes the work.

Fancy-Mitten.—Materials—Four needles, No. 20; and one skein of Saxony-wool. Form foundation of 24 stitches upon each needle. k 2 plain, p 2, k 2, p 2; and repeat this, until five rounds are completed. K 1, to, k 2, s 1, k 2 t, then slip the slipped-stitch over the pair knit together, k 2, t o, and repeat until all are taken up ; 2 rounds plain, 4 rounds as first, next two plain, seventh as first, next two plain; then repeat to desired length of wrist. Carrying four strips up the back, k 4 p between fancy part and thumb. If mitten covers little finger, knit fancy part thus : k 2, t o, k 1, s 1, k 2 t, slip slipped stitch over, k 1, t o, k 1, and repeat over fancy portion. Form next two rounds plain. Fourth : k 3, t o, s 1, k 2 t, slip slipped stitch over, t o, k 2, and repeat. Knit remainder of mitten plain. Narrow on the inner two needles after leaving the palm of the hand, each alternate sixth round, until you drop the fancy part ; then each alternate round is narrowed, until all the stitches are consumed.

French-Open-Hose.—The design is in quadruple, and finished in four rounds. For instep, take number of stitches representing a multiple of four, knitting the rest plain until you reach the instep-needle. First : knit 1, n, tto, and knit one ; repeat the same.

Second : Plain around. Third: narrow, t t o, knit 2 ; and repeat as before. Fourth : Plain, entirely around. Now begin again from first. It can be varied, by knitting either the entire hose open, or only to the instep.

Saxony-Edging.—Form 25 stitches. First: k 2 plain, t o t, n back, *, o, n back, *, repeat from * to *, t o, knit plain to close. Second : Take up first, wind around twice, k, *, putting needle through next stitch wind thrice and knit, *, repeat from * to * for next 15 successive stitches, and k 9 plain. Third : k 2, t o t, n back, *, t o, n back, *, repeat from * to *, k, plain, transfer 8 upon right-needle, slip one back upon left-hand needle, bringing fifth loop over the first four—knit, next three the same, k 4, and remaining stitches, like the first. Knit fourth row entirely plain.

Forest-Moss-Border.—The materials are, all kinds of odds and ends of the work-basket. The pieces are selected by chance, and sewn upon a made-foundation, after being thoroughly steamed. Form a foundation of stitches, double the desired width ; and in knitting, draw the work very closely, in plain stitches. Knit in successive rows, to the desired length.

After completion and steaming, press it thoroughly with a hot iron. The "mossy-effect" is produced by the strip being cut through the centre, the wool raveled open, and then being neatly brushed into curling forms—beautifully simulating the "mountain-moss."

Spanish-Fringe.—Twisted yarn, No. 9 ; tidy needles : Foundation, 12 stitches ; knit plain, entirely across. First: t o, n, t o, n, t o, n, t o, n, t o, n, t o, n, k 1 p. Repeat this, each way, to the desired length. Bind off all—lacking four ; drop those, and ravel back to the place you began.

Ladies' French-Hood. — Requisites : Ivory needles, medium size ; delicate shade of "split-zephyr." Foundation—60 stitches. Knit 15 inches in very easy and plain "garter-stitch." Slipping off half of the stitches and securing them on a "bent-needle," knit alternately backwards and forwards, until the length of 20 inches is produced. Then knit the "reserved-stitches" to the same length, and bind off the work. A fancy border is now crotcheted, around the entire edge ; and then, "gauging" or gathering the extreme lower ends, form a tassel on each. After gathering the top, garnish the same with a handsome "bow of ribbon," of the same shade, upon said gathers. It is jaunty, and very pretty.

Gent's English-Scarf.—Requisites: The desired contrasting shades of " Berlin " or " Germantown-yarn;" 5 skeins of the principal color, and three skeins of the shade for border. Two large needles. Foundation—70 stitches in dark shade. Knit easily across, in plain-stitch. First: * 5 p, seam 5, *.

Repeat process from * to *, entirely across ; and continue thus, until sixth row. Seam 5, and 5 plain, to eleventh row. Then reverse backwards, forming the checked-design. Check 4 rows, and then finish with border, adjusting a neat fringe in each end.

Fancy-Wristlets.—Form 30 stitches on each of two needles, and 31 on the other, and knit around plain. S 1, n and t slipped-stitch over; k 3 p, form 1, k 1, form 1, k 3 p; then repeat to proper length. The last stitch on the needle is always the one having been slipped, which should not be knit, but thrown over those narrowed at the beginning of each needle. Any number of colors and any design, can be adopted.

Persian-Afghan.—Requisites: The desired shade of "quadruple-saxony." Some prefer courser grades. Fine ivory, or coarse steel needles. Form seven stripes: Three of one, and four of the contrasting color. The outside two should be of the same color. Form a ground-work of 24 stitches.

Always knit the first and last three of each row plain.

(1)—* 2 p, s 1, k 1, then draw the "slipped-stitch" over this last "plain-stitch," k 2 p, t o 1, k 1 p, t o 1, *, repeat from * to * to close.

(2)—Seam all in this row, and also all other "even-rows."

(3)—k 1 p, * n, n, k 2 p, t o 1, k 1 p, t o 1, k 2 p, * repeat from star to star.

(5)—* n, n, k 1 p, t o 1, k 2 p; now k 1 p and one "seam-stitch" in the next. T o 1, 1 p, *, and repeat from star to star.

(7)—k 1 p, * narrow, t o 1, k 6 p, t o 1, n, * repeat from star to star.

(9)—* n, t o 1, k 2 p, n, n, k 2 p, t o 1, * repeat from star to star.

(11)—k 1 p, * t o 1, k 2 p, n, n, k 2 p, t o 1, k 1 p, * repeat from star to star.

(13)—k 2 p, * t o 1, k 1 p, n, n, n, k 1 p, t o 1, k 2 p; now form one "plain" and one "seam-stitch" out of the next; * repeat from star to star.

(15)—k 3 p, * t o 1, n, n, t o 1, k 6 p, * repeat from star to star.

(17)—* n, k 2 p, t o 1, n, t o 1, k 2 p, n, * t o 1, k 2 p, n,* repeat from star to star.

The bottom-edge can be finished either by attaching a "knitting-fringe," by means of the needle, or otherwise it can be crochet upon the same; and also to any desired or chosen design and suitable shade of color.

French Half-Hose.—Requisites: 2½ ozs. best "knitting-silk;" four needles, No. 18. Form 113 stitches on three needles. Knit entirely around, plain. Then, k 4 p, seam 2; and repeat this "ribbing" for 72 successive rounds. Then knit six inches, plain. Begin the heel by throwing 57 stitches on one needle, * seaming across and knitting back in plain-stitch. Repeat this process from star to star for 57 successive rows, slipping first stitch of each row, and keeping the seam-stitch at the side back of the heel. Begin to narrow at 58th row, and proceed as follows, viz:—K 13 p, s 1, k 1 p, draw "slipped-stitch" over, k 10 p, n, k 3 p, s 1, k 1 p, draw "slipped-stitch" over, k 10 p, n, k 13 p. 59th row: seam entire row. 60th row: k 13 p, s and b, k 8 p, n, k 3 p, s and b, k 8 p, n, k 13 p. 61st row: seam all. 62d row: k 13 p, s and b, k 6 p, n, k 3 p, s and b, k 6 p, n, k 13 p. Seam all in next row. 64th row: k 13 p, s and b, k 4 p, n, k 3 p, s and b, k 4 p, n, k 13 p. Seam all in next row. 66th row: k 13 p, s and b, k 2 p, n, k 3 p, s and b, k 2 p, n, k 13 p. Seam all in next row. 68th row: k 13 p, s and b, n, k 3 p, s and b, n, k 13 p. Seam all in next row. 70th row: k 17 p, fold needles together reversing heel, s 1, n, draw "slipped-stitch" over, continuing until but one is left. Bring right side of heel out, take up side-stitches of heel toward the left hand. k across "instep-needle;" and then take up stitches of opposite side of heel. This forms the first round of the foot-part.

Form extra stitches in next round, at sides of heel only, in every fourth stitch; and subtract or decrease two stitches, by narrowing at "right-hand point" and slipping and binding at "left-hand point" near instep. Decrease two in next two rounds in the same manner; and then decrease two in each alternate round, down to 112 stitches. Now finish the desired length of the foot-part, plain. Arrange so that an equal number of stitches are on each needle, in narrowing at the toe-part. Beginning at centre of instep-needle, knit all except three; then s and b, k 1 p on next needle, k p until three remain, s and b, k 1 p; k 1 p on next needle, n, k p until all but three, s and b 1 p; this finishes last row of toe-part. Knit three rounds in plain, not narrowing; knit succeeding row like first of toe-part. Two rounds plain; then, as first of toe-part. Two rounds plain; sueceeding row like first of toe-part. Proceed thus, to its completion.

Fancy and Artistic Needle-Work.

Under this general division we embrace and include what is given in separate form, under the various legitimate departments in its province, under "Seaming," "Stitching," "Button-holes," "Tailor-Finishing," "Embroidery," etc.; to which we refer the student.

Hair-Work.

Human hair, can be artistically and beautifully wrought into wreathes, chains, shrubbery, flowers, branches and buds, by a little thought and care. Nearly all the "loopings" and "stitches" given for other fancy work, can be applied in this art. To form chains or braids, many of the "crochet-stitches" come in place. In other forms of working, you will want a simple arrangement in the form of a hand-loom, in order to hold the strands. Anything will do, that will secure them at the ends; and so it will be accessible to the application of the mesh, or the needle, whichever you please to term it. For "nets," you can apply the "Grecian-stitch," as given for "netting." Select your pattern in guards or chains, and produce the different parts by any of the given stitches that may be required; make it in "sections," and then let the junctions come together within gold mountings, secured thoroughly with glue; and having joined the sections, apply the mountings at each end in the same manner. In representing trees, leaves, wreathes, and flowers, etc., use fine "French-wire," to form the stamens and centres of support; upon which frame-work (the wire being bent into the desired shapes), place the hair in such "loopings" and "stitches," as will best answer your purpose; and work on a similar principle with that used in making flowers and shrubbery, in "worsted-work." Try a simple leaf, first; then twigs and branches; then perhaps a rosebush complete; or, out of the emblems and tokens of sorrow and grief, perhaps the "weeping-willow."

Feather-Work.

Art "feather-work" is growing intensely popular. Geese and duck-feathers are made to bud and blossom as the rose. With fine-pointed and very sharp scissors, notches, edgings, and leaves, are formed. Begin at the lower part of the leaf, notching one side first, then the leaf. Petals of any flower can be cut, according to nature. The most delicate feathers, are utilized for curled-flowers. Never paint these. Draw the scissors quickly over their under-side, to curl them. Leaves are painted with dry paints. Paris-green is much used. Balsam of fir reduced with alcohol, forms a good medium, A feather can be converted into a brush. When dry, they can be arranged and mounted on soft, fine wire; and these again, covered with green tissue-paper, for stems. The main colors used are rose-madder, for pink; ultramarine, for blue; gamboge, for yellow; carmine, for scarlet, etc. By means of wax, secured within bent wire of the proper shape, the flowers and petals are fixed and retained in position.

Brown and gray feathers of the duck, mingled with both the painted and natural, form fine combinations. Ingenious methods of arranging sprays, among flowers, buds, vines, etc., are shown with great credit of artistic effect.

Lime-water, cleanses feathers. Heating them and drawing a sharp instrument over them, makes them curl.

Indigo in boiling water, gives shades of blue; verditer, verdigris, and gum-water, shades of green; cudbear in boiling water, lilac; Brazil-wood, vermillion, alum and vinegar, red; tumeric, yellow; cochineal, cream-of-tartar, and muriate of tin, scarlet, etc. All the colors that have naturally fallen upon the wings of birds and the fair forms of flowers, can be placed upon feathers, by the trained and skilled hand of art.

Leather-Work.

The Russian-Cleansing-Bath.—Uncolored leather, is cleansed by first washing with benzine; and applying with a sponge, a solution of oxalic-acid dissolved in warm water.

German Leather-Cement.—Take ½lb. gutta-percha; 2 ozs. india-rubber; 1 oz. pitch; ½ oz. shellac; 1 oz. "boiled-oil." Melt, mix, and use hot.

French Water-proof-Leather-Cement.—One pint boiled linseed-oil; 1 lb. mutton-suet; ¾ lb. beeswax; ½ lb. resin. Melt, mix, and apply warm.

"American-Process:"—Equal parts of gutta-percha and india-rubber, dissolved in wood naptha.

French Stamping-Process.—Wash right side, in weak solution of oxalic-acid and warm water. Placed upon a plain surface, the heated stamp (engraved upon steel), is heavily pressed upon the same. Let remain, till dry. Finish with "leather-varnish," when removed. Leaves, flowers, buds, blossoms and vines, can thus be transferred, and made to show exquisitely perfect and beautiful. In its perfectly cleansed and dry form, leather is cut, by various means and instruments, into "shavings," "strips,," "stalks," "branches," etc.; and is then, upon the principle employed in making paper and wax-work, converted into all the diversified forms of nature's beauties. And thus, the material out of which were made the ever memorable garments called "coats of skins," is made "to bud and blossom as the rose."

Bead-Work.

French Card-Case.—Conceive any design; and as materials, select a piece of the best kid, steel beads, and gold-"thread." Outline, in the gold-"thread;" and fill-in the interior parts, with very delicate "cut-steel-beads." Pass the end of the "thread" through to the under side of kid, when commencing the outlining-work; and repeat the same, on returning and concluding. Form the border, by "loopings" of the gold-cord, ornamenting each loop at the centre, with one bead. Bronze is a good color, for the kid. Complete, by mounting the same upon a card-board-"form;" decorating the inside with silk, and the edges with beads; which will hide the stitches, as well as ornament the work. Only one end, is left open to receive the cards.

Bracelets and Chains.—Any desired number of rows, may be adopted. When jet, or "immitation-pearl" is used for the bracelet, it should be repeated in the chain. Suppose three strings: thread 1 bead for the centre, larger than those outside. Thread two on right-hand, and four on left hand. In doing this, pass needle of right-hand through two beads on left-hand; and leave the centre-thread, underneath. Again, two on left-hand-side, passing needle from right-hand-side through beads on left-hand-side: and so arranging, as to have the centre-thread pass over the top. By these alternations, the centre-thread and beads, are firmly secured ; this is repeated to any desired length ; and is then finished with a jet-ornament, of the desired or fancied pattern.

The American Toilet-Cushion.—Requisites: ¼ yd. bright-colored "silk-velvet;" ⅜ lb. "alabaster-beads;" 2½ ozs. "chalk-beads;" 2½ ozs. "crystal-beads;" 4¼ yds. silk-"Russia-braid" in white ; fine "purse-silk," in black.

Conceive any fancy design. Arrange artistically, for a centre, squares of the braid ; and secure the same, by the "cross-stitch" in the "purse-silk." Form a circle out of card-board, and secure it upon the velvet, covering the ends of the braid. Cover this, by the bead-work, using the chalk-beads, threading and laying them upon it in an angular form ; and threading a sufficient number, at one time, to form a row.

The arrangement of the leaves, flowers, buds, etc., are necessarily, a matter of taste. For the leaves, you might take equal parts of the alabaster and chalk-beads ; for the flowers, crystal and alabaster ; and then form the tendrils, in chalk ; and the fringe and border, in alabaster; or according to taste and fancy. Never work the last row of loops forming the fringe, until after the cushion is filled and mounted. Thus, to the careful and thoughtful student, innumerable designs and methods will present themselves. And by a little practice, coupled with good taste, the whole art will soon become exceedingly simple, as well as interesting ; and beads will fall into positions of grace and beauty, as naturally as dew-drops fall upon opening buds, and leaves and flowers.

Wax-Flowers and Wax-Work.

Almost, every school-girl knows the secrets of this beautiful art. We will simply say, whatever you do be true to nature. To cultivate accuracy in this, do not trust your memory; at least, until after long practice. Secure the best "French-wax-sheet-paper," and the finest and best quality of wire ; and always work in a room of the proper degree of temperature, suitable to the material. If it be a rosebud, or flower, take it apart carefully; and very accurately trace a pattern of each and every part of the same, in the wax ; and then rebuild it, from the beginning, accurately securing it upon the "frame-work" of wire; or in some work, upon its own sufficient strength. Let the fastening of the leaves and stamens, and the petals, all be done with exceeding neatness ; but with sufficient firmness, to give it strength. In the shading of colors, study the "apple-blossom," or that of the "peach;" rather than what any one might do, or say, or write, upon the same. It is one of those arts which comes easy to all, after schooling one's self, to accuracy of observation and work ; and which, to those endowed with a true appreciation and conception of the beautiful, is exceedingly simple; and when mastered, is an art of both exquisite pleasure and charming beauty.

Shell-Flowers and Shell-Work.

The formation of "shells" by means of "embroidery," "crocheting," etc., we have treated elsewhere. We now refer to actual shells, found along the sea-shore and elsewhere ; and to the art of transferring upon them the highest human conceptions of that which is beautiful and decorative, according to correct art. As we look upon those exquisitely beautiful things called "shells," it would seem almost, that to touch them by decoration, would only be to mar their perfect pristine beauty. But in arranging them, taste is certainly allowable, and also valuable ; and as to decoration, we offer a few suggestions, viz.:

"Outline-patterns" of vines, wreaths, buds, blossoms, leaves and flowers, can all be nicely outlined upon shells, by means of compasses and the graver.

An application of nitrate of silver into the incisions, exposed to the sunshine, fixes the same. Repeat, until sufficiently deep and clear. Polish with emery-paper. The incisions, can be inlaid with gold; or any other method of decoration can be adopted. By boiling shells in various solutions, we stamp upon them the decoration of color. Bi-chromate of potash gives brown; lime-water and Brazil-wood, red; increasing the amount of alum in the latter solution, gives a brilliant scarlet; verdigris and vinegar, green; after being green, dipped into a strong and boiling solution of pearl-ash, we get blue; first boiled in alum, and afterward in a solution of French-berries and pearl-ash, brings green, etc.

Abundant room is here presented to the artist, of both form and color. The decoration first described, admits of an endless variety in design, method, and effect. If the work is artistically executed, any of the many naturally beautiful shells, can still be made more beautiful; and can also be made to reflect an almost matchless degree of culture in artistic taste, as well as being at the same time precious gems of adornment in our earthly homes—"while gathering shells upon the sea-shore."

Transparencies.

A beautiful art of copying the various designs and outlines, we find scattered throughout books, papers, fashion journals, etc., consists in the following, viz: Get the finest French "transparent-copying-paper;" fasten it securely upon the sheet, on which appears the given design. Then, with a very finely pointed pencil, indicate by delicate points or lines, the exact image or form of the outline, while the copying-paper is pressed firmly upon the same.

On removing the paper, you will have a true copy of the given design; which, is then easily transferred or reproduced upon any "canvas-work;" or, which can be formed in wax, or in many different ways. By a little practice, all the shadings and colorings can be carried over, in the same manner.

Transfer-Work.

By means of a fine and sharply-pointed tracing-wheel, puncture small holes or "openings" through the paper, very accurately around the design, to be transferred. Otherwise, unthread the needle of the sewing-machine, and run the needle carefully around it; thus leaving regularly graduated "holes," to outline the design. Then finely pulverize a mixture of charcoal and pipe-clay, or any plain or colored "French-chalk;" place it in a coarse white cloth, and secure it from loss, by tying the gathered ends by means of a cord or string. Now securely fasten the punctured paper, upon which the desired design appears, upon the material upon which you wish it to be transferred; and then rub over the punctures, with this cloth of chalk; until all the "openings," are well filled. Carefully remove the paper with the design, and "set" the outlines with white paint mixed with gum. Any design, can thus be beautifully and artistically transferred.

Leaf-Work.

Leaf-Impressions.—To transfer an impression of some choice leaves upon paper, proceed as follows, viz.: Cover a sheet of white paper with a thin coating of glycerine; smoke the same side, with "lamp-black;" placing the leaf upon this sheet, fold it in such a manner that the smoked paper will touch both sides of the leaf; remove the leaf, placing it upon a clean sheet; now, folding the leaf within the paper, press it for a time, very heavily under a weight.

When removed, both sides of the leaf are transferred upon said paper; which will be found to be almost equal to the best pencil or pen-drawing; besides, being much more true and accurate. Any leaf, or similar prize, can thus be transferred and kept.

Oil and India-Ink Leaf-Work.—Place the leaf between two sheets of "French Impression-Paper," securing each sheet, against a substantial plain white surface of card-board. Press this very heavily between two plain hard surfaces, for one hour. then carefully remove the same, and touch-up the outlines according to the rules and principles of painting; either in oil, or otherwise in India-ink, as may be desired; and as the natural leaf shows, and appears. The most exact and perfect reproductions, can thus be given very quickly, and very prettily.

Skeleton-Leaf-Work.—To preserve the "skeleton" of a leaf, proceed as follows, viz.: Form a solution of ½ gallon of water; ¼ pound common potash; ⅛ pound quick-lime; boil this, and place the leaves carefully within it, boiling them thoroughly. After removing and cleansing them, all the cellular matter will come from off the outline or "skeleton" of the leaf; which will then show very beautifully, and can then be kept in this form.

And thus, the leaf—the very emblem of the transitoriness of all earthly things, may be made the enduring monument or "remembrance-token" of memories of persons, places, things, or circumstances; which perhaps we cherish, and might wish to perpetuate.

Millinery.

The term "millinery," has baffled the skill and wisdom of the best lexicographers. The exact limit of its provincial meaning, is yet to be decided. The waters that wash upon its beach seem a shoreless ocean, whose waves roll in all directions of both art and science. Suffice it to say for all practical purposes, its province lies over the head of woman! Its generally accepted legislation has to do with the natural and artificial coverings of woman's head—the hair, and the hat or bonnet. Naturally, it is divided into two great general departments, viz: the scientific and the artistic. Science has fixed the perspective of the artist in painting, and has also legislated upon the same in millinery. It is true in a certain sense, that "fashion rules the world;" and as the greater includes the lesser, we might say, fashion rules millinery! But the term "fashion" expresses a variable and changeable quantity; and its boundaries and limitations are both uncertain and illy-defined. Neither does the term "fashion," always express correct artistic taste; but correct artistic taste is always good fashion! Neither is correct artistic taste, as seen and displayed upon any certain and particular head, necessarily any longer the same, when this same given medium is dropped upon a different head. Differently-shaped heads and faces, require different methods of arranging the hair and the hat or bonnet, if we wish to keep in the line of correct art and true taste. To detect the various deficiencies and necessities, under all exigencies and emergencies, and to promptly supply the same, is the province and work of the true milliner. Art, holds the scepter in the province of garnishing; but outside of her legitimate territory she is powerless, and must appeal to science. Millinery is therefore, both an art and a science. Mathematics legislates upon, and regulates the lines and angles for the proper and symmetric arrangement of the hair, for each particular head and face; and also the proportional size, shape, height and width of the proper hat or bonnet, most suitable for each; while it is the province of art to select the materials, and to garnish it with correct taste and beauty in displaying harmony in materials, colors, and complexion; and to also make it, as a whole, both pleasing and restful to the eye.

Science also controls the processes of manufacturing the materials, and governs the chemistry in the solutions used for the processes of bleaching, dyeing, etc. Science dyes the ribbons, feathers, etc., and art selects, associates, and decides upon their locations, etc.

In the perfect-shaped head and face, a horizontal line drawn through the centre of the eyes, divides the face and head into two equal parts, on a perpendicular line from the under-side of the chin to the horizontal line touching the top of the head; the upper-half of the head and face is again equally divided, by the horizontal line touching the upper part of the forehead, at the junction of the hair with the forehead; and the lower-half is again divided into equal parts, by the horizontal line touching the upper lip, equidistant from its lower-edge and its junction with the lower part of the nose.

This divides the perfect-shaped head and face into four equal parts, horizontally. Two perpendicular lines equidistant from the centre between the eyes, and the first touching the inside of the right eye and the right side of the mouth; and the second touching the inside of the left eye and the left side of the mouth; these said lines will then include between them the nose and mouth, and represent one-fifth of the width of the head and face. Now by two more perpendicular lines, one at the outside of each eye, we divide the width of the perfect face and head into five equal parts; one-fifth included within and between said lines, respectively; and two-fifths outside of each respective outside line. Upon the divisions included within these said five horizontal lines and six perpendicular lines, all mathematical symmetries in the science of millinery depend; at least, so far as the front of the head or the face is concerned, as well as the back part of the head. A similar division of the side of the head might be founded upon the width of the ear; having twice its width back of the ear, and four times its width in front of the ear; thus dividing the entire width of the head into seven equal parts; having two-sevenths back of the ear, and four-sevenths in front of the ear; and then founding or basing all symmetries in the side-view, upon said mathematical divisions.

Where these symmetries, whether as tested from the front, side, or back-view, are naturally deficient, it is the province of the millinery-scientist to detect and supply them, in the arranging of the coiffure; and also, in proportioning the size, shape, height, width, and position of the hat or bonnet, so as to correct them; and certainly, where said symmetries exist, and are naturally perfect, not to destroy them by the same, in a mere blind devotion to some stereotyped and falsely so-called "fashion!" This is, the science of millinery. The art of millinery is, to select the appropriate materials, their associations, their colors and blendings, etc.; and also to stamp the work with appropriateness as to complexion, color of the eyes, hair, etc.; and finally, to make it one grand harmonious whole, according to the principles of what is universally recognized as correct art. To be in fashion, does not necessarily mean to be in sameness! To be in the line of elegance and good taste in the choice of materials, coupled with mathematical symmetry and appropriateness, is to be in the very height of fashion—regardless of ignorant and false criticism to the contrary. The true artist will not be the servant of fashion, but will make fashion her slave! Furthermore, such artists will naturally have such a degree of influence over their customers, as to voluntarily compel the latter, to submit both the selection of the materials and their arrangement; which, will secure the artist against the adverse criticisms, so very often and so very unjustly heaped upon the artists head on account of some inappropriate and ridiculous outfit, which was produced in answer to the demand of an ignorant conceit and false taste in the customer. If this would seem to open the door to an opportunity for imposition, on the part of the artist, we would simply remark that the general public will very quickly detect the degree of honesty or dishonesty, with which said service is rendered, as well as decide upon its actual worth. First, decide upon the arrangement of the coiffure; then, upon the hat or bonnet. A well-shaped head, forbids a large accumulation in arranging the hair, on the centre-line at the back of the head; while a depressed-head at the back, requires the deficiency to be supplied in the same. High-cheeked faces, demand broad foundations in the arrangement of the hair, at the top of the front-side of the head; depressed cheek-bones, want the opposite.

Angular-faces, call for a close and plain coiffure; curving features, waves and curls. Long-necked customers, are improved by wide-curls at the back; Shorter necks, demand a higher and closer coiffure, at the back, etc.

Again, broad-faced persons need narrowness, in the appearances of the hat or bonnet; while narrow faces can be helped by broad-rimmed hats, or broader bonnets.

Short faces with low foreheads, can be improved and helped by a high-coiffure at the front-top of the head; and also by "throwing-up" the hat or bonnet, at the front; so as to apparently furnish through art, that which is lacking in nature. These will serve as general indices, to the thoughtful student. The manufacturing department, pertains to the preparation of material, etc.. As things are arranged in the commercial world, this department is of but little interest to the practical student; since, they largely come ready-prepared. The great work of the modern student, necessarily lies in the direction of the scientific and artistic departments.

The mechanical methods employed in the artistic department are exceedingly simple, and naturally come to any one, by practice.

The particular method, has no special value; only, so that the result is truly obtained and thoroughly secured. Having a few of the leading sizes and shapes, in "blocks," the "home-milliner" can do her own re-shaping; and thus, "old hats" often become "new!" For after all, the hat itself is but little effected by style, from season to season.

Fashion lives more particularly in the land of laces, ribbons, velvets, silks, and plushes; and in the world of plumes, tips, leaves, vines, and flowers; and in the gorgeous plumage of the beautiful birds, which are so wickedly and relentlessly sacrificed upon her consuming altar. For the benefit of the mere beginner, we would simply suggest the following, viz.: Frames are generally covered either with the same material as that of the dress, or otherwise, with some other material properly shading with the same. A "shirring" is put on the inside; about the junction of the rim and crown. Artistic bows of ribbons, generally grace one side; while either a long plume falls about the crown, or a cluster of tips droop over the other side.

The range in stitchery, in the methods of adjusting and forming trimmings, etc., is largely at the option of the artist; the same rules to be observed only, as given under said department, elsewhere.

If, in all this world of flowers, ribbons, laces, jets, silver, steel, and gold ornaments, fashion can not be satisfied; and the sweet and innocent birds must, additionally thereto, be wickedly and cruelly slain—as an offering of life and blood, upon her altar! Then, their wings should be first carefully secured together, by stitching entirely through them, hiding the stitches underneath the feathers; and afterwards, carefully, securely and artistically attached, as parts of the trimmings. In trimming, the knots of the thread should be arranged on the outside, of the hat or bonnet; which will then, naturally fall underneath the trimmings. A "draw-string," is tastefully arranged in the lining; which however, should not be drawn until the entire completion of the hat or bonnet; so as to prevent it from being caught in the stitches. Odds-and-ends of silks, can often be used to great advantage, and with artistic effect; by utilizing them as binding, shirring, etc.

Small tufts of lace and ribbon, otherwise considered useless, often become elements of the principal graces in a hat or bonnet; when arranged by the deft and trained hand of art. In millinery, art flows from the brain through the fingers; and that which once seemed a mere homely thing, is often changed as if by magic, into matchless grace and beauty, at the mere touch of the artistic hand. This power is a natural endowment, in some; in others, while not impossible, is only attained by hard study, severe training, and long practice. It was very natural for one, to have placed one tip upon another, and thereby to have created a "French-tip!" Others would not have thought of it. Such things come, as painting comes! All can paint, if they but tried; but it is given to some, to excel. Steel-cut-ornaments, such as arrows, pins, buckles, etc., often can be set as brilliant stars in an open "firmament;" where naturally otherwise, a vacancy would appear.

The very carelessness in the method of trimming, often seems to make it appear all the more beautiful; and lends it a peculiar and lustrous richness, and gives it that peculiar charm of grace that both captivates and rests the eye.

When the "tips" and "plumes" have lost their curl, place them over a current of heat, for a time; and afterwards carefully, and lightly draw their edges between some dull instrument of steel, or the back of a knife and your thumb or finger; this will re-curl them, beautifully. The finest effects do not, as a general thing, fall in the line of flowers. Let the trimming have a proper "ground-work" of some subdued color, and then decorate by a few simple but rich and striking ornaments; to brilliantly contrast, and brighten the same. Nature, has often enhanced the charm of the opening bud by half concealing it, underneath a cluster of leaves and a shower of dew-drops. The very carelessness of the arrangement, pleases and rests the eye. In this art above all others, will be needed a correct and thorough knowledge of the laws which govern the shadings, and the blendings of color. Leghorn and straw can be beautifully bleached, by first cleansing them in a bath of borax soap; and when yet moist, covering them with "sulphur-dust." Let the sulphur remain a sufficient number of hours upon the same, to complete the work. Or, it can also be done by the confined "sulphur-vapor-process." A strong solution of soda and glutten, also cleanses straw hats or bonnets; and by letting them remain in said solution for a day, and afterwards dipping them into a solution of logwood, acetate of iron, and bichromate of potash, they come out a beautiful black. Thus, the "home-milliner" can change a light and delicate shade, in the summer-hat, into the darker and more becoming shades for the autumn-season. Bathing plumes, tips, or ribbons, in a solution of alum, and dipping them afterwards, into a solution of archil and indigo, brings them out a beautiful blue. Potash in solution with ground annotta, makes them buff. Logwood and copperas brings black. A solution of sulphuric-acid, glauber-salts, violet-liquor, magenta-liquor, and water, brings plumb-color. Sulphuric-acid, imperial violet-liquor, and water, gives a light mauvé. Magenta-liquor, and water, gives magenta and magenta-pink. Paste cochineal, cream of tartar, and nitrate of tin, gives crimson. All the analine-colors can also be used, to great advantage in securing the various shades and colors. Here, as in all other departments of the work, much that goes toward securing success, depends on the natural taste, education, and habits of the artist. Study the mathematical principles governing symmetries, position, shape and proportion. Study correct taste in the combination and display of color and beauty, as seen in nature; and above all, strive to be independent, natural, and true to your own highest conceptions, as to what is beautiful and artistic; and you will certainly reach a high standard of success. The finest productions of the world's choicest looms, are laid upon the table of the milliner; and the navies of the world bring to her, their richest treasures from every nation and people; and it is her work, as the artist, to select out of all these the most befitting; and to arrange the same, in the most appropriate and becoming manner for a drapery and covering for the head—the royal palace and chamber of the spirit.

Harmony of Complexion and Color.

There is charming music or harmony, in color; and the opposite, is discord! Discord in sounds, grates no more harshly upon the educated ear, than discord in color, grates upon the cultivated eye.

We have room for but a few thoughts or hints, to the inexperienced, viz.:

Violet and purple are beautiful to look upon; but bring only discord, when draped about the fairest faces; and as do also, the softer shades of ecru, yellow, or grayish-blues. The darker shades in red and garnet, like the rose, reflect rich lustre and lend strength to beauty, everywhere.

Dark brown on wool, is a general friend; but when placed on silk or satin-threads and bordering on the golden-shade, is at constant variance with all blondes;

while orange and bright yellows, almost annihilate their existence! The blonde with golden hair, would do well to seek the friendship of the softer shades, in blue.
Where whiteness in the complexion is desired, seek the help of violet. Dark myrtle greens are friendly to all, and cling to their brilliant contrast, in red; but blondes alone, should venture upon the lighter "cresson-shades."
A complexion having a redish-tint will find a contrast in the softer shades of blue, lemon, green, gold, or pearl. White, is kind to all complexions; and is also, the friend of all colors. Green, finds a contrast either in its own lighter-shades; or, otherwise, in yellow, stone, gold, lemon, flesh, purple, dove, or pink. Yellow loves the lighter-shades of all colors, but is the special friend of orange; also contrasts white, and does not shrink from purple. Gold, drab, salmon, buff, and yellow, are the servants of blue; while red, makes music with crimson, and beautifully contrasts with green. In a canopy of white, we might set stars of brown, blue, green, purple, and violet; while light colors, bring harmony with gold; which, again is contrasted by all the shades in dark. Black, is at peace when wedded to deep colors; and it also finds contrast, in those of a paler shades.
We term the colors from yellow to purple, "cold-colors;" combinations in these alone, are in bad taste. Red is at constant war with violet, and has but little affection for ultramarine; but, is a warm and true friend to blue. Colors have the power, in certain combinations, to reflect both "coldness" and harshness;" either of which effects, should be avoided. Warm colors should be mixed with cold colors, and *vice-versa*. The ground-work of the color of the heavens and the earth, are vast oceans of low-colors; while brilliant colors are only seen, in suns and stars and planets, in the heavens; and only appear in the leaves and flowers, on earth.
Hence, low-colors should form our "ground-works;" and out of brilliant colors, we should mainly form our "decorations." It is the province of the true artist, to bring this music out of color; and to so touch the entire "key-board" of color in material, eyes, hair, and complexion, as to bring harmony and cadence in all their shadings and blendings; and so to minister at the altar of good taste, as to approximately at least, imitate nature in her bountiful and perfect ministrations of constant, satisfying, and enduring pleasure and restfulness to the eye.

Fancy Dyeing.

Black.—A solution of equal parts of sumac, galls, logwood, and copperas.
Blue.—Bathe in a solution of alum; then in a solution of indigo; and then in a solution of Archil.
Brown.—A solution of equal parts of Persian Berry liquor, saffron, and bichromate of potash.
Green.—A solution of equal parts Persian Berry liquor, extract of indigo, quercitron-bark, and alum.
Gray.—A beautiful gray is obtained by simply reducing or diluting the solution for dyeing black; as given above.
Crimson-Dye for Silk.—For each lb. of silk, dissolve in boiling water 2½ oz. of alum, 1½ ozs. of white tartar. Boil two minutes; put in the silk and boil 1½ hours. Remove silk, and add fresh water; when it boils add 1½ ozs. of cochineal powdered; boil 1 hour. The shade is made lighter, by using less of the ingredients.
Golden-Yellow-Dye.—Take of "shaggy spunk" or *Boletus hirsutus of Linneaus*, a species of mushroom on apple or walnut-trees, and pound it in a mortar. Mixing with water, boil ¼ hour. 1 oz. will tinge six quarts of fluid. Immersing the goods in it, boil 20 minutes. Wash silk or goods then, in soft soap-water.
"Harmless"-Black.— Equal parts of potash, and vitriol of copper, separately dissolved; gradually, mix the solutions. Then combine, with decoctions of logwood. This gives a beautiful glossy-black, and is entirely uninjurious to goods of any kind; which, can not be said of any "verdigris-blacks."

Lilac.—Logwood, first dipping the article in a weak solution of alum; or use archil.

Pink.—From safflower and alum; or brazil wood, and quercitron bark, or Persian berry liquor.

Purple.—From archil, or logwood and roche alum.

Scarlet.—Cochineal and Persian berry liquor. Or dye in a soap lather with annotta, boiling hot, for an orange bottom. Wash the article; then dye with cochineal and a little nitrate of tin.

Red.—From Brazil-wood and madder.

Nankeen—From annotta and pearl-ash.

Orange.—From Brazil-wood and quercitron bark, or young fustic, or Persian berry liquor, and alum.

Lavender.—Extract of indigo, alum, a little plum liquor. The shade may be altered by decreasing the indigo.

Blue-Black.—Steep in nitrate of iron for about one hour; then wash, make a soap lather, and add logwood-liquor; dip the article into the liquor.

Black, for Worsted or Woolen.—Water, 3 gallons; bi-chromate of potash, ¾ oz. Boil the goods in this 40 minutes; then wash in cold water. Then take 3 gallons of water, add 9 ozs. of logwood, 3 ozs. of fustic, and one or two drops of Double Oil of Vitriol; boil the goods 40 minutes, and wash out in cold water. This will dye from 1 to 2 lbs. of cloth, or a lady's dress, if of a dark color, as brown, claret, &c.

All colored dresses with cotton warps should be previously steeped one hour in sumach liquor; and then saddened in 3 gallons of clean water, with one cupful of nitrate of iron for 30 minutes, then it must be well washed and dyed as first stated.

Black, for Silk.—Dye the same as black for worsted; but previously steep the silk in the following liquor: Scald 4 ozs. of logwood, and ¼ oz. of tumeric in a pint of boiling water. Then add 7 pints of cold water. Steep 30 or 40 minutes; take out, and add 1 oz. of sulphate of iron, (or copperas) dissolved in hot water; steep the silk 30 minutes longer.

Brown, for Worsted or Wool.—Water, 3 gallons; bi-chromate of potash, ¾ oz. Boil the goods in this 40 minutes. Wash out in cold water. Then take water, 3 gallons; 6 ozs. of peachwood, and 2 ozs. of tumeric. Boil the goods in this 40 minutes. Wash out.

Imperial-Blue, for Silk, Wool, and Worsted.—Water, 1 gallon; sulphuric acid, a wine-glassful; imperial blue, 1 tablespoonful, or more, according to the shade required. Put in the silk, worsted, or wool, and boil 10 minutes. Wash in a weak solution of soap lather.

Sky-Blue, for Worsted and Woolen.—Water, 1 gallon; sulphuric acid, a wineglassful; glauber salts, or crystals, two tablespoonfuls; liquid extract of indigo, a teaspoonful; boil the goods about 15 minutes. Rinse in cold water.

Claret, for Wool or Worsted.—Water, 3 gallons; cudbear, 12 ozs.; logwood, 4 ozs.; old fustic, 4 ozs.; alum, ½ oz. Boil the goods in it one hour. Wash. This will dye from 1 to 2 lbs. of material.

Crimson, for Worsted or Wool.—Water, 3 gallons; paste cochineal, 1 oz.; cream of tartar, 1 oz.; nitrate of tin, a wine-glassful. Boil your goods in this one hour. Wash out in cold water. Then in another vessel with three gallons of warm water, a cupful of ammonia, the whole well mixed. Put in the goods, and work well 15 minutes. For a bluer shade, add more ammonia. Then wash out.

Fawn-Drab, for Silk.—Hot water, 1 gallon; annotta liquor, a wine-glassful; 2 ozs. each of sumach and fustic. Add copperas liquor, according to the required shade. Wash out.

It is best to have the copperas liquor in another vessel.

A dark-drab, may be obtained by using a little archil, and extract of indigo.

Flesh-Color, for Dyeing Silk.—Boiling water, 1 gallon; put in one ounce of white soap, and one ounce of pearlash. Mix well; then add a cupful of annotta liquor. Put the silk through several times, and proportion the liquor till you obtain the required shade.

A salmon-color may be obtained, by first passing through the above liquor, and then through diluted muriate of tin.

Magenta for Silk, Wool, or Worsted.—Water, 1 gallon, heated up to 180 degrees; add magenta liquor, 1 tablespoonful; stir it up well. This will dye a broad ribbon, 4 yards long; or a pair of small stockings. To dye a large quantity of material, add more magenta liquor and water. The shade of color may be easily regulated by using more or less. Magenta-pink may be obtained by increased dilution.

Mauvé for Silk, Wool, or Worsted.—Water, 1 gallon; add one tablespoonful of sulphuric acid; then heat to boiling point. For a very light mauvé, add one teaspoonful of imperial violet liquor; boil the same amount of material, as stated under magenta, about 10 minutes. Rinse in cold water. If the color be too deep, use a little soap in rinsing, using warm water.

A violet color may be produced, by using a tablespoonful of violet liquor instead of a teaspoonful.

Pea-Green for Silk.—To one quart of water, put half a teaspoonful of picric acid, and rather more than half a wine-glassful of sulphuric acid, and a teaspoonful of paste extract of indigo; boil about five minutes; then add water to cool it down to blood heat, or 100 degrees. Put in the silk, and work about 20 minutes. The shade may be varied by adding more or less of the picric acid, or extract of indigo; if more of either be added, boil separately in a little water, and add to the previous liquor.

Pea-Green for Worsted.—Use the same materials as the aforesaid; but boil all the time in 1 gallon of water for about 20 or 30 minutes.

A darker green may be obtained by using a larger quantity of material.

Plum-Color for Worsted, Silk, or Cotton.—Water, 1 gallon; sulphuric acid, a teaspoonful; glauber salts, or common dyers' crystals, 2 tablespoonfuls; violet liquor, a tablespoonful; magenta liquor, half a tablespoonful. Boil the article (silk, wool, or worsted,) about 10 minutes.

Cotton should be dyed the above colors separately, and by first running them through weak gall liquor, and weak double muriate of tin. Then wash well, and work in the aforesaid liquor, according to color and shade. The liquor should be cold for cotton.

Scarlet on Worsted or Wool.—Water, 3 gallons; 2 ozs. of dry cochineal, 1 oz. of cream of tartar, nitrate of tin, a wine-glassful; boil the goods 1 hour. To give the goods a yellow hue, and a little young fustic, wash out as before.

Yellow, for Dyeing Silk.—Proceed the same in dyeing as pea-green, omitting the extract of indigo, and using oxalic tin instead of sulphuric acid.

Water-Color Painting.

Painting in water-colors, is a very simple process, and an art very easily and quickly attained. But like oil-painting, it admits of unlimited progress, in its study and practice. The requisites for the beginner, are an out-fit of all the principle water-color-paints, either in the form of cakes or tubes; and a first class out-fit in sable-pencils or brushes, of the proper sizes. Now having a smooth surface, upon which to touch your brush and bring it to a point after supplying it with the color or shades, and any legitimate and proper surface that admits of and serves for the purpose of water-color-painting, you are ready to commence the work. The paints are, of course, simply mixed with water. When applied, they become dry and will "set." Begin on some easy and simple object, first; and as you learn by practice, gradually take up the more difficult and complicated paintings. The art of shading, toning, etc., will soon come by practice. Sounds must be heard by the pupil, before the student in music can even conceive of either discord or harmony, much less understand their science; and in like-manner, the effect and results of the various shadings and blendings of color must be practically seen, and must be self-created by the student, before their science can be either properly conceived of or understood. But it comes by practice, as if by "inspiration!"

The art of photography, has opened up new methods and branches of study, in this beautiful art.

If, it is desired to paint a photograph in water-colors, proceed as follows, viz: Procure a second copy, to guide you as to the exact original truth. The photograph is then, first washed in a delicate tint of raw sienna and rose-madder, forming a basis or ground-work for showing complexion. Let the original guide you. It is deepened, by repetition. "Tone," with carnation. "Shade," with tints of raw sienna and rose-madder; with white and vermilion, for fair faces; with Roman ochre for dark complexions. Use a positive blue for the white of the eyes, in children; the light shades of pearl in a yellowish tint, for the same in old age.

For light, black, and brown eyes, use chinese-white, and purple and white, respectively. Lake and vermillion, properly shade the upper-lip; vermilion and carmine, the lower. The iris is "laid in" with transparent color, and "toned" with Chinese-white. Use a heavy tone of carnation about the mouth, temples, and eyes. Yellow helps it. Dark colors, garnish the pupil of the eye.

For white hair, we use Roman ochre, carmine, and sepia; and sometimes purple and Chinese-white. Use the shadings from Roman ochre, for light auburn hair; darker auburn, is best attained by using burnt amber. For gray hair, a compound of indigo and cobalt shaded with sepia, is best. Burnt amber and lake, gives chestnut-color; and gamboge, lake, and indigo, bring black hair. Let shadows fall, upon your painting first; then, the higher lights. A beautiful glaze is attained by pearly-shadings. The shadings for the draperies can be read by the novice, from those given above for shading the hair. Blue and spirit of carmine, bring shades of purple and lilac. Draperies are painted best, by drawing the pencil-brush in the same direction with that of the fold. Even the most inexperienced, will find the whole process very simple, and of easy attainment; and by a few careful, thoughtful, reflective, and patient efforts, will find himself or herself rapidly mastering an art of fascinating entertainment and pleasure, as well as one of utility and profit.

Water-Color Lace-Painting.

A charming result in the reproduction of lace in water-colors, is as follows, viz: Stretch over the "drawing-board" several thicknesses of good strong muslin; stretching the lace, secure it firmly to said muslin by means of pins. Now, using a fine camel's-hair-brush, and adding a small quantity of mucilage to the water when mixing the colors, proceed to paint the design of the lace, as in ordinary "water-color-painting." Blue mixed with lake, will be best to produce the various shades of purple; Chinese-white with new-blue, for the delicate shades of blue. Mix carmine and lake, for roses, etc. And in the main, observe all the rules given elsewhere, for water-color-painting. After completion, let the work dry thoroughly; and then, carefully removing it, fold it carefully. If carefully and artistically executed, it makes a charmingly beautiful work of decoration.

Painted-Embroidery.

A very attractive and effective style of decoration is as follows, viz.: Take a heavy double-ply of white or toned canvas, muslin, linen, or otherwise, velvet, or silk, etc., and stretch the same firmly over the "drawing-board;" as in lace-painting. Upon this, place the stencils of the desired styles or patterns; and proceed to produce the design in either "water-color-dyes," or in oil-paints. You will find the dyes preferable. When completed and dry, give the work a fine finish by re-touching it. Then with either silk or wool, as the circumstances may demand, outline the entire work; adapting and arranging the color of the same, to the colors used in painting. It gives it a brilliant and tapestry-like appearance, which will be found to be very elegant, when carefully and tastefully performed.

China and Pottery-Painting.

First, learn accurate "drawing." In correctly drawing the outlines and designs, lies the secret of much pertaining to success, in this beautiful art.

Any one can learn "to draw;" it only needs the elementary instructions and principles; and a good supply of patience, to practice them. This is the first step. Then be original, above all things; and true to the conceptions and natural instincts with which you are endowed.

As an instrument with which to draw, use either pencil, pen and India-ink, or the color itself.

The mineral paints to be used come in small tubes; and are sometimes called "mineral tube-paints." The brush should be simply a small camel's-hair; or otherwise the sable brush, used for oil-painting. A tile, of either earthen or otherwise, chinaware, will be required; and some turpentine, with which to mix the paints; which are not so moist as oil-colors. Some artists, also use glycerine; any oil, usually acknowledged as a good "mixer," will answer the purpose. Neatness, and an exceedingly accurate cleanliness is all-important, in china-painting. Begin with the most simple leaf or flower, in but one color at first. Have it "fired," and see what the result is. Do not grow impatient or discouraged; but "try, try again;" until little by little you can extend your combinations, both in design and colors; and you will find yourself surprised, at the degree of perfection you are thus practically attaining. The touch in application, is similar to that in water-colors.

The heat, of course, entirely changes the original appearance of the colors; which, sometimes confuses the beginner. Experience will correct all mistakes here. A "key" can also be made by any beginner, by painting all the colors on a white plate, and having it fired. Thus, showing the correct transformations. Even a common coal-fire can be utilized as a "kiln," by the amateur.

The ingenious wife, daughter, or sister, who has learned to paint on china, will scatter beautiful buds, sprigs, blossoms, berries, leaves, cherries, fruits, etc., over the entire set. And it enhances their charm and beauty, to have no two alike.

In embossed pottery, paints or bronze powders may be used.

In using tube-paints, the pieces should be glazed-over with "amber enamel," or "shellac varnish." Apply the paints with a brush, directly from the tubes; mixing the shades as in other painting, and according to fancy and good taste. A coat of silicate or amber enamel should be first applied, when using bronze powders; which serve as a "back-ground," when carefully applied with a "camel's-hair brush."

Brush the loose and superfluous powder off, before giving the second coat of silicate to flowers, after the first is thoroughly dry; and sift upon it the "metalics," or "flitters." Polish thoroughly, before applying the varnish. Out of practice, will grow all the rarest and choicest vines that fancy and good taste may dictate; and, under the sunshine of experience, they will bud and blossom into the many and various-tinted flowers and fruits of both forest, field, and garden; and in many ways will reflect the culture and refinement, of the hand and mind of the artist, that painted them.

Grecian Oil-Painting.

No one need despair of learning to paint in oil. There are however few artists, in oil-painting. Painting, is like music. All could learn to play, if they but persistently and patiently tried; but the true artist in the harmony of sounds, has "music in his soul!" It seems natural, to some; almost impossible to others. The rudiments and elementary principles of oil-painting, are exceedingly simple and easy of attainment; but the sea that washes upon its shores has depths, that the accumulated wisdom and experience of all the centuries past, have as yet been unable to fathom. Oil-painting seems, the great art-ocean into which all the other streams of the entire province of art pour their crystal waters. No art has had worse specimens of taste and skill in the artist, than oil-painting; but it has had many, which never were equaled in any other department of art. Some paintings call forth both our contempt and pity, for the artist; but there are others, in whose crystal waters we see the reflection of worlds of beauty, and which apparently entrance the soul. But if you cannot be a Raphæl, you can at least, be true to yourself and your own highest and best conceptions of the beautiful, the true, and the good. Having your canvas, paints, and brushes, simply form a groundwork; fix upon some simple scene or design, and begin to paint! The science and theory of mixing and blending colors, etc., will come more quickly and more perfectly from practice than from books. The laws of shading and blending of colors, etc., are unchangeable; and ever remain the same, in all departments of painting. What you have found true in other departments as pertaining to color, will be of service here. Study nature; and learn correct principles of color and taste, as well as size, shape, and form, from her truthful and unerring lips.

Paint the water as you would the sky, guided by nature and taste in tints, reflections, etc. Let the ripples appear distinctly, yet so as to blend carefully and softly. Water, should not always have the same hue as the sky above it; different scenes will explain this principle, if carefully observed. The shades and varied hues of green can be beautifully shown in landscapes having hills, ravines, etc.; by bringing each portion, with its own tint, into distinctness and clearness, in the painting.

Animals, houses, buildings, etc., should be so touched up as to be most true to natures own best choice in colors. Foliage should appear in colors natural to the season, and to its kind.

Each season has its tints and shades, peculiar to itself. These should appear correctly and true to the same. Autumn-foliage, is beautifully shown in Grecian oil. The little things, such as the edges of leaves, etc., should be carefully shown in their true form, only omitting this where they become a thick mass, and forbid the same. The trunks of trees, etc., should be in the color that nature has dressed them. And each species, as well as the season in which it appears, should be carefully considered.

Let the foliage of the painting dry, before you throw the lights and shadows in among the branches of the trees, etc. Let the color of the sky govern the hue of the lights and shadows. A mild and mellowish tint of gray, merging into blue, should tinge the distant mountains and hills; and let the shade of green increase, in the ratio of the nearness of their approach. Let all departments of the applications dry thoroughly, before beginning another. Above all, let nature be truly shown in the appearance of flowers, leaves, etc. Much depends upon the condition of the atmosphere, and the degree of warmth, etc., in the place of painting; in order to have the oil and colors work, properly. Having observed all of the above, paint your thoughts and conceptions naturally; and as they flow from the brain and heart through the brush upon the canvas. Be true to yourself, to your highest conceptions of truth and beauty, and true to nature.

For the benefit of the beginner we submit the following, in conclusion, viz:

For instance, suppose a cluster of natural flowers to be painted. First, study them as they truly are, in nature. Now forming your ground-work, place the flowers in a good light, having an appropriate back-ground; and carefully observe the relations of lights and shades, the half-tints, fourth-tints, reflected lights, etc. These make up a large part of the sum of values, in oil-painting.

Take for instance the "Jonquils;" in forming the shades of creamy-white, use the combinations of silver-white, raw-umber, vermilion, yellow value, along with light tints of burnt sienna, permanent blue, and ivory black. For painting the "Narcissus" for instance, use raw umber, cadmium, and white; with light shades of ivory-black, permanent blue, and burnt sienna, for bringing out the deeper tones. In shadow-painting, always select one of the darker shades, by which to measure or upon which to base the degree, of all other lights and shadows. In painting the sky of winter-scenes where it is overcast, arrange so that the gray clouds almost touch the horizon. A clear sky of sunset blending in reds and yellows at the horizon, will form a rich and effective back-ground for trees and foliage, in the distance and the foreground. For the deep-tones of gray clouds, use sky-blue, black, and a small quantity of brown-green. This tone, often reaches down over one half of the entire surface of the sky. Use carnation shadings along the edges of clouds. A large flat brush is most appropriate, for painting the clouds. Below the line of clouds use yellow, deepening it with carnation; until a deep red is thrown over the horizon, in painting sunset-scenes. In painting the portraits of elderly persons, pale complexions, and gray hair, etc., use a back-ground lending warmth, to the surface. Perhaps a rich hue of deep dark-crimson, would not be amiss; or otherwise, a tone of dark blue-gray lights, thrown over the upper part of the canvas; and having rich and warmer shadows over the lower part, thus, forming a correct and legitimate ground-work; and yet, having it tinged with a sufficient degree of variety, to make it pleasing and restful to the eye.

No arbitrary rules of course, can possibly be framed that will cover the entire ground. Better often, that the student had never heard of rules, in oil-painting; for "rules," have often become the chains and manacles, that have bound and fettered the hands and hearts and spirits of many, who ultimately proved failures, that might otherwise have shone as brilliant stars, in the heavens of the artistic world. If you have poetry and song within your heart, paint it! If you have by nature, the endowment of the perception, comprehension, and appreciation of harmony, beauty, melody, pathos, ideality, picturesqueness, proportion, order, fitness, tenderness, irascibility, power, and sublimity, paint them upon canvas!

This, is Grecian Oil-Painting.

Tapestry-Painting.

All kinds and styles of furniture coverings, decorations for the walls, screens, curtains, etc., can be beautifully and artistically decorated, by means of "tapestry-painting." "Tapestry-painting" is done in the following manner, viz.:

Get a supply of liquid "canvas-colors," in black, cadmium, turquois, vegetal green, crimson, vermillion, cobalt, sepia, Prussian-blue, etc.; and a canvas of medium weight and texture, and wide enough without a seam, for curtains. Stretch the canvas thoroughly, as is done for oil-painting. Use some stiff brush for applying the main colors; and the softer-grade, for finishing the painting.

It is best to have a brush for each color. By means of charcoal, outline the design carefully upon the canvas, using it sparingly. Then proceed to sketch with soft colors, as in oil-painting. Moistening the back of the canvas, aids it in absorbing the colors. Re-paint it, after successive dryings; until the tints and shades are sufficiently heavy to suit the design, and according to correct taste. The darker shades of the same color, should be used in the first applications.

Several shades should be in immediate readiness, to properly blend the colors; for it dries rapidly. Use as few colors as is possible, to produce the effect; which is worth remembering in any kind of painting. Always study to do the greatest amount of shading and blending, with the least possible number of colors. For flesh tints, use rose pink, and a light shade of chrome; for the summer-clouds, French ultramarine, or cobalt and carmine. An application of cadmium, washed over the entire surface, gives orange-tint; for the leaves and grasses, mix yellow with blue; and vandyck brown, mixed with yellow ocre and shaded with sepia, gives the proper color for the trunks of trees, etc. The painting should be gone over again, and again; until every object is brought out, in true harmony with nature and truth. Warmth and coldness, can be reflected from the sky and rock and from the depth of the foliage and branches of trees and groves, as well as proportional size ; by the correct and proper choice of the proper color and shading, at each particular place. This can easily be done by even the amateur, after a little experience, study, thought, observation, and reflection ; and any common tapestry, under such hands, will quickly turn from the blank and bleak appearance of a cold and barren winter, into the joyful aspect and blooming freshness and beauty of Spring and Summer.

Painting in Gold and Silver.

An art of exquisite beauty is now becoming intensely popular, by which the same results are obtained in the colors of gold and silver as are obtained in water-color and oil-painting. Terracotta, glass, and many other hard, smooth surfaces, admit of this decoration ; but wood, is generally considered the best medium for the background. The brush should be a good quality of "camel's-hair." Apply the gold and silver-liquid over the entire surface, as a back-ground. Upon this back-ground use the more subdued colors on the gold, and the brighter and more positive colors on silver. Any artistic and expressive design may thus, easily be painted upon said gold or silver; either in oil, or otherwise in water-colors. Richly toned, in the proper shades, all possible artistic designs may be brought out ; and, in exquisitely beautiful effects ; and for an endless variety of occasions and purposes, in "Decorative Art."

Oil Photograph-Painting.

This art, comes to us under the various names of "Oleography," "Cameo-Oil-Painting," "Photographic-Illumination," Crystal-Painting," etc.; all of which, represent one and the same principle, of reproducing the photograph in oil; having only fanciful differences in shades of variation, and in effects and results. By "oil-photograph-painting," a photograph can be reproduced in as permanent a form, as that of the steel engraving. The method is brief and simple, viz: Subject the card upon which the photograph appears, to the process of boiling in clean water. On taking it out, after sufficient boiling, remove the thin paper upon which the photograph appears, from off the card; and then place it between two sheets of blotting paper; and thus, dry it by absorbtion. Now carefully and evenly paste this within the concave-side of an oval-shaped and "prepared-glass;" using equal parts of fine starch, nitrate of strontia, and gelatine; mixed and boiled in a sufficient amount of water to make it the proper consistency, as to its degree of thickness. See that no air is confined between this sheet and the glass. It can be removed, by pressing from the centre toward the edges, with a clean linen cloth. Let this dry thoroughly. Then apply a solution of 1 part of turpentine, 2 parts of balsam of fir, and 2 parts of poppy-oil; spreading it evenly over the entire surface, on the back of said paper on said glass, by means of an ivory "spreading-knife;" or some similar-shaped and clean instrument. This will, after sufficient length of time, make its appearance perfectly transparent. Cleanse from all surplus matter when dry, and varnish it with 2 parts of balsam of fir, and 1 part of turpentine, thoroughly mixed. By means of thin and narrow strips of card-board or heavy paper placed along the edges of the same, so as to protect the glass from actual contact with the former glass and picture, cover it with another and similarly-shaped glass; securing them firmly to each other at the edges, by means of tough thin paper and mucilage.

When dry, holding the picture toward a strong light and with the back toward you, apply the necessary and desired colors, by the regular brushes used in oil-painting; and in like-manner. The work should be done accurately and carefully; and in such a manner, as to enhance its appearance, and yet be true to the original. For the complexion mix red, silver, and yellow ochre; or tinge with silver white, vermilion, and Naples yellow. Ivory black and chinese, is used for blue eyes; and carmine and vermilion for the lips. For black hair, use burnt sienna and chrome-yellow. Ivory black shows black eyes to the best advantage; and Naples yellow and Vandyke brown, best bring out brown hair. A tinge of vermilion, brings out the ruddy complexion of the young. The white of the eyes is best shown by the use of white, slightly tinged with yellow. Turpentine removes mistaken applications. Ochre shows jewelry. The darker complexions are brought out by Vandyke-shadings. Ultramarine blue with Naples yellow, and ivory black, show gray hair to the best advantage. These paints should all be applied on the back of the second glass; as placing them on the first, gives a harsh and rough appearance to the completed work. The laws of mixing and shading, being the same as in general oil-painting, it is deemed unnecessary to go into further details. Any one, can readily and rapidly learn to become quite an expert in the art; which is intensely popular, and has recently become a favorite medium among many for embalming the features of both the living and the dead, in an almost imperishable form.

Lustre-Painting.

By means of "bronze-powders" instead of paints, an exquisitely beautiful result in decorative-art is obtained, upon materials that can be embroidered. From its brilliant and lustrous appearance, it derived the name—"Lustre-Painting." We have this art brought to us, under the various and somewhat fanciful names of "Lustro," "Bronze," "Prasmatine," "Metallic," and "Iridescent"-Painting, etc. The elements and principles of "Lustre-Painting," embrace all of these. It is the parent-stem, out of which all the above species grow; differing nothing in principle, only as fancy has created them, in their names; having only a mere shade of variation, in effect. These same shades of fanciful differences, might with equal good authority be carried to an indefinite extent; and thus, the subject could be transformed into a "world of mysteries." It is a very plain and simple process, and is easily and quickly attained by any ordinarily intelligent mind.

All the general rules of painting, are helpful aids in this province. It doubtless had its origin, in the conception to simulate or imitate the ancient embroideries, in silver and golden threads, or tinsel.

All materials admitting of embroidery, are legitimate ground-works for "Lustre-Painting."

It clings however mostly, to bold and striking displays on large curtains, portieres, lamberquins, bed-spreads, etc.; and loves as a foundation, satins, felts, plushes, etc. Good artists however, show flowers and fine work, to great advantage in this art. But in the main, it is best adapted to heavier work, and bold and lustrous decoration. Instead of paints, the artist supplies himself with "metallic-flitters," and "dry-colors;" these are brightened by "bronze-powders" in all the desired colors, being sprinkled upon the fixing-mediums.

A good recipe is, asphaltum varnish and white balsam of fir, in the proportion of ⅜ of spirits of turpentine to ⅝ of asphaltum, or balsam of fir. The dark medium is, of course, used for fixing the dark colors; and the other, for the lighter shades. Shellac varnish is also, sometimes used; but, the former is preferable. The brushes are bristol and sable. Good sizes run from 2 to 12. Procure a drawing board, a palette, the flitters, powders, and turpentine for mixing. A good list of colors would embrace green gold, antique green, maroon, steel, violet, dark-blue, pale blue, carmine, purple, medium-pink, pale pink, rich gold, pale purple, dark green, dark violet, pale gold, brown, brocade silver, lemon, orange, fire, light green, mauvé, imperial blue, crimson, fawn drab, magenta, pea green, plum, yellow, and claret.

The flitters should embrace purple, blue, orange, red, gold, and silver; and in dry colors, green, pea green, orange, dark red, silver, dark blue, and dark green. All

of these are ready prepared. The artist, of course, forming the desired tints and shades by mixing them. Some differences exist in the rules of shading and blending these, from those in other departments of painting,
But experience will soon correct all mistakes, and will set the student right. Dry powders, tone and subdue brighter shades. From these, we get the "lustre." Use a different brush for each color. Use the flat-side of the brush, in applying the paint, drawing it firmly in the direction of the nap of the goods. The design is carefully stamped first, as in embroidery. Attend to leaves and petals, first; then form the edges toward the heavier and more central parts of the work. Lights, of course, should fall from the upper left-hand-corner. Shadows should be kept underneath the leaves and flowers, etc.; and the laws of light, in nature, should be carefully observed.

A good recipe for stamping, in "Lustre-painting," is as follows, viz: One-fifth blueing-power, and four-fifths pulverized rosin. But very little is required at a time. This is "fixed," by covering it with tissue-paper and passing a hot iron over it. A tube-paint or white powder, can be utilized for stamping on dark goods. Choose an easy design. Stamp it, or have it stamped upon the material. Procure all the above requisites; and then boldly, and with a determination to succeed, proceed to paint. And you will be surprised how rapidly you will attain and master the art, in all its details. It is intensely popular; and, if artistically designed and executed, is exceedingly handsome and very "lustrous,"

Lincrusta-Painting.

This modern and very popular art, does not derive its name either from the method, or the result; but from the material, upon which it is executed. "Lincrusta," is the name of a very recent and peculiar material, composed of paper or a pulpy substance, and of canvas; having a "paper-side" and a "canvas-side." It is a special production, for a special purpose; and comes in the various shades of ecru, in its original form. It is comparatively thick, and ribbed in appearance; averaging 18 inches in width, and having an average cost at present of seventy-five cents per yard. It is used for various purposes and in various ways, for decorating; but it is especially adapted for "pannel-work;" and is also, a very popular "vestibule-decoration." Landscape-scenes, and all designs having objects in bold-relief, are very beautifully and artistically executed upon the Lincrusta-surface. The Lincrusta, is often prepared and pressed into the desired shapes and forms of pannels, plaques, etc.; and all possible styles of relief-work, are thus beautifully arranged and shown, in the completed paintings. The painting is executed in oil. And, of course, all the general principles of "oil-painting" apply to Lincrusta-painting; only, that the work is decorated by means of the various "bronzes," giving it a charming "lustre," and a very striking and beautiful effect. The ordinary colors of flake white, permanent white, chrome yellows, yellow ochre, raw sienna, Roman ochre, gamboge, yellow lake, brown ochre, burnt umber, Vandyke brown, raw umber, emerald green, crimson lake, purple lake, Venetian red, Prussian blue, Chinese blue, permanent blue, burnt Roman ochre, ivory black, blue black, vermilion, violet, carmine, etc., are used, and will be sufficient for the amateur for beginning, in ordinary work. Elegant landscape scenes, showing sky and mountains blending in the distance, are shown with fine artistic effect, in careful and accurate shadings and tonings. The design is sometimes stamped upon the material, as in "Lustre-painting," Kensington-painting," and in embroidery; but it is best adapted to the process of "drawing" the design, accurately upon the surface. After this is properly executed, proceed to paint; observing all the ordinary laws of "oil-painting;" using the bronzes however, for decorating the same. Here as elsewhere in painting, much depends upon the genius, education, and skill, of the artist who executes the work.

Outside of this, perhaps the result depends mainly, on the proper management of the colors. And therefore, for the benefit of the mere beginner, we call attention to the following, viz: Color, is only another term to express the degree of "absorption of light." An object reflecting all the colored rays, we call white; while a body absorbing all the rays, is called black. A ray of light, when decomposed by a prism, or as it is seen in the rainbow, consists of six different colors, viz: blue, red, yellow, violet, green, and orange. The first three are termed "primaries," the last three "secondaries."

The violet is formed, by mixing blue with red; green, by mixing blue with yellow; orange, by mixing yellow with red.

These six colors are termed the "pure colors;" white, in reality being not a color, but the complete reflection of all colors. When three primaries are mixed, we get gray. Colors obtained in a similar way, we call "broken"-colors; which are also sometimes called "tertiary"-colors. The tertiary colors, are colored grays and and browns. Pure white and black give the "normal"-gray; and the shades deepen or grow light, in the exact ratio of increase or decrease of either preponderance. From this principle, the amateur can secure the key to all the secrets of "shading" and "toning;" if, it is properly understood. To get a "tone," we mix a normal color with white or black; and in the varied and desired quantities.

Color can be seen and analyzed in its absolute purity, in the prismatic spectrum. We can only approximate the colors as seen in the spectrum; for there is scarcely a pigment (which is but an opaque medium used to reflect light) that reflects only one kind of rays. To detect the effects of this aberation from truth, and to properly correct it, requires considerable thought, education, and experience. But, it can be thoroughly mastered, by implicitly trusting nature to guide us in the tones and shadings; and then comparing the mediums used to represent color, by the absolute standard as revealed in the spectrum.

"Lincrusta-painting," furnishes a high plane for the display of good taste and skill, in the art of painting. Again we say, study nature! Learn to love the leaves and flowers, and the earth and sky, because of their beauty and perfection. And when you sit down to paint, let the inspiration of truth and love govern the head and heart, and allow nature to guide the hand; and experience will do the rest.

Be true to the above, to yourself, and to nature; and you will be happily surprised at the measure and degree of success attained; and, the rapidity with which you will be made to realize the glad fruition of your highest ambitions and most cherished desires, in this entrancing and beautiful art.

Kensington-Painting.

"Kensington-Painting" and "Kensington-Embroidery" have a common parentage, and are therefore sisters. The brush and paint are to the former, what the needle and silk are to the latter. In "Kensington-painting" we simply aim to imitate or simulate in paint, the peculiarity of the position and effect of the silk, in "Kensington-embroidery." This is Kensington-painting, in a nut-shell! The Kensington principles of stitchery, are the Kensington principles of applying and shading in paint, and *vice versa*. Like "Lustre-Painting," the scope in materials or surfaces, is as extensive and diversified as the materials for embroidery. But in the main, silks, plushes, and nap and pile-goods, are the favorites. Stamp the material, as for embroidery; only observing that the process used, is such as will not show the outlines, after the painting is completed. Much of course, rests upon the good sense, education, refinement, taste, and culture of the artist performing the work. The elementary principles however, are very simple and easy of attainment. The requisites are, a hard wood or steel instrument similar to a "stub-pen," with which to apply the paints. The "stub-pen" is often used, and will answer all purposes for a "brush!" And a good supply of the various "tube-paints." These are not reduced, either with oil or spirits of turpentine. Also, a palette is needed— otherwise, a "dish" of some kind, upon which or in which to mix the paints. A "palette-knife" is used, as the means for mixing the paints. The process of painting, is simply taking up some paint on the pen, by the process of "dipping" it up; the lower part of the same should be cleansed, and then the pen or instrument holding the paint, drawn over the goods; having the cleansed-side next the goods, and turning the instrument slightly, so as to leave a deposit of the paint. Attend to the edges of the outlines first. In careful work here, lies the sovereign secret of beauty in this art. The inside of the leaves and petals are covered, and then completed by the "drawing-process" of the pen; thus imitating the threads and stitches of "Kensington-embroidery." A bodkin, or any instrument by means of which the stitches can be best imitated in the paint, can be used as legitimate instruments. The end, if correctly attained, will justify almost any means! A delicate pen-knife, is often of great value to the artist in this province. The paints should be applied profusely, in this art; and a sufficient length of time should be allowed for it to become perfectly dry. Megilp, mixed with the paints, hastens the process of drying. The general principles of mixing, for shades and tones as taught in "Grecian-oil-painting," apply also, to "Kensington-painting." For the benefit of the beginner, we would simply add that rosemadder and vermilion, give deep red; crimson lake, is used for red. Mixing these with Caledonian brown, gives deep shades; with German yellow, bright lights. For leaf-painting, mix dark and light German greens; for white flowers, light German yellow with flake-white. Flake-white with cork black, forms the shadings. Shadows in general, come beautifully from German-greens and Vandyke-brown. German-greens and white, produce the higher lights. The Chrome yellows mixed with brown, give deep shadows, while silver-white produces high lights. For purple, use mauvé; the shades of which, are produced by mixing with white and light German yellow.

Rose-madder and crimson, are used for roses, etc. Antwerp, or Prussian blue, give blue; and are shaded with flake-white. A sufficient stock of colors would be as follows, viz: Prussian blue, rose madder, Indian red, vermilion, crimson lake, flake white, cork black, mauvé, silver white, caledonian brown, dark chrome green, medium chrome green, Vandyke brown, light chrome green, Antwerp blue, light German yellow, dark chrome yellow, helle jaune brilliant, medium chrome yellow, etc. These will suffice for all ordinary work.

In using transparent colors, first paint in white; and, after drying, apply the desired shades and colors. Hare, as in all decorative arts, a wide scope is allowed for the play of the imagination and the display of good taste in the artist; and those who most ardently love and study nature, and whose pencils and canvas become the reflecting mirrors of her matchless perfection and beauty, will inevitably shine as the most brilliant stars, in the canopy of the heavens bending over its coveted honors.

UTILITO---DECORATIVE ARTS.

Under this general division we embrace all the various processes, methods, and results in art and science, which have the property of utility or usefulness, as well as that of decoration or beauty. Primarily therefore, and in its broadest sense, it embraces the entire scope of subjects treated in this volume; but in a special sense, it refers to the departments outside of the mathematical solutions given, touching the problems in the art and science of garment-drafting; and to which, we refer the student. We therefore close, by simply throwing in a few more buds and blossoms as a free-will-offering.

Painting on Glass.

A process of charming beauty has recently been, through the triumph of science, brought to great perfection and simplicity. The art of painting on glass, is thereby placed within the power of any one, possessing even a moderate degree of the knowledge of drawing. The process, is performed in water-colors; and, of course, the ordinary palette, brushes, etc., used in "water-color painting," will be required. First, carefully outline and draw the design upon the surface of the glass. To accomplish this, use intense brown, diluted with sufficient water to make it flow easily and freely from the pencil. Let this dry, thoroughly. Now, having secured the prepared "glass-varnish," apply it over said drawing while warm, so as to keep it from chilling; this, will insure the drawing from being disturbed by the applications of the paint. As soon as said varnish is dry, which will take but a very short time, proceed to lay on the paints, according to the rules of "water-color painting;" without mixing the different colors more than is positively necessary. After the paints are laid upon the work, re-touch it with the varnish; and then, give it the finishing-tones and shadings, necessary to bring out the most artistic effects. Now, complete by another coat of varnish. The following colors will be needed, and come as "metalic-tubes," viz: scarlet, intense brown, crimson, warm brown, orange, light green, blue, dark green, and opaque black. By the above process, the beauty, richness, and transparent clearness of the painting is greatly enhanced, over the more tedious and imperfect processes formerly used; and the result has a charming freshness and beauty, that is both pleasing and restful.

Electro Glass-Engraving.

A very useful and decorative process of imprinting decorations upon glass, by means of electricity, is as follows, viz:

First, the surface of the glass is thoroughly covered with a solution of nitrate of potash. This is best accomplished by arranging so it will not flow over the edges, and then pouring the solution upon the glass, evenly. A platinum wire, is placed along the edges of glass; in such a manner as to have it immersed in the solution. This is brought into contact with either the positive or negative pole of a secondary battery; and thus, with the electric current. To the opposite pole, a second wire of the same metal is attached; which, is insulated and inclosed in such a manner, that only the point is exposed and left clear. By means of said wire and the current brought to bear upon it, any design or engraving can be beautifully executed upon said glass.

Spatter-Work.

A very simple but beautiful art is called "Spatter-work." It is both entertaining and pleasing, if carefully executed; and, it also has its side of utility. The process is so well understood, that we deem it unnecessary to go into a detailed description of the same. Suffice it to say in a general way, that your designs, or leaves, etc., should be carefully and artistically arranged upon a surface of "Bristol-board." The means of "spattering" are, a tooth-brush, a fine wire-sieve, and India-ink. The ground-work, lying outside of the design, should be thoroughly "spattered" or covered; so as to show the outlines, very accurately and distinctly, when the object is removed. After being thoroughly dry, it is nicely varnished. Thus, a world of beauty in combinations, is often arranged as one general design; showing both order, beauty, and fine conceptions of art, in those who so deftly and artistically executed the work.

Transfering Engravings.

Engravings can be transferred upon paper, as follows, viz:

Place the engravings a few seconds over the vapor of iodine. Dip a sheet of white paper in a weak solution of starch, and, when dry, in a weak solution of oil of vitriol. When again dry, lay a sheet upon the engraving, and place both for a few minutes under the press. The engraving will thus be reproduced in all its delicacy and finish. The iodine has the property of fixing the black parts of the ink upon the engraving, and not upon the white.

Ivory and Bone-Decoration.

Ivory and bone admit of being beautified, in many ways, by the decorating hand of art. We submit the following, viz.:

To Polish Ivory.—Ivory is polished with putty and water, by means of a rubber made of hat; which, in a short time produces a fine gloss.

To Stain Ivory and Bone—Black.—Rub over with diluted oil of vitriol; wash, and then steep in nitrate of silver and good ink.

To Stain Ivory and Bone—Blue.—Steep in a strong solution of extract of indigo and a little potash.

To Stain Ivory and Bone—Green.—Dissolve copper in nitric acid, and steep the ivory in it. Steep in oxalic tin, and then in a strong decoction of Brazil-wood, or lac dye, and alum.

To Stain Ivory and Bone—Purple.—Nitric acid, 2 parts; sal ammoniac, 1 part; mix, and steep the ivory in it.

Gold-Liquid.

A beautiful liquid in gold is made as follows, viz.: Take equal parts of honey and gold-leaf. Grind them thoroughly and finely, mixing them intimately. Now agitate the same with thirty-times its own weght in very hot water. Let the solution settle, and pour off; now repeat the agitation. Dry the gold; and then mix with gum-water, for use. You will then have a beautiful gold-wash, for all kinds of decoration.

Wood-Decoration.

Mahogany-Stain.—Put 2 ozs. of dragon's-blood, in small pieces, into 1 quart of spirits of wine; let remain within a bottle for several days, in a warm place, agitating it freely. Apply with brush.

Black-Stain.—Pour 1 quart of boiling water on ½ oz. of commercial extract ot logwood. When thoroughly dissolved and mixed, add ½ drachm of yellow chromate of potash. Mix thoroughly, and apply with brush.

Marble-Stain.—Two parts alum, one part best glue; mix these with a quantity of whiting, equal to that of the glue; and with three-times as much plaster of Paris, as glue. It can be colored, by first staining the water. It "sets" very hard.

South American-Mahogany-Stain.—Logwood, 1 part; water, 8 parts. Make a decoction and apply the same to the wood, and let dry. Give it three successive coats of varnish, letting it dry each time. For varnish, mix dragons's-blood, 1 part; spirits of wine, 20 parts. It is very fine; and beautifully imitates the real mahogany.

Fixing Pencil and Chalk-Drawings.

Pencil-Drawings.—Dissolve white resin in spirits of wine; lay the pencil-drawing on its face, upon a sheet of clean paper; and brush the back of the drawing, with the solution. This penetrates through the paper, and as the spirit evaporates, the resin is deposited as a varnish on the drawing. It does not cockle the paper, which watery solutions will do; and as the brush only passes over the back of the drawing, none of the pencil marks are in any degree removed.

Chalk-Drawings.—Immerse the drawing in a weak solution of isinglass; allowing no part of the drawing to remain without the isinglass passing over it, or it will look spotty. Drain it, by holding the drawing up by one end, over a plate.

Decorative Table-Spread.

Use any desired color of fine plush, old-gold, or otherwise, crimson. Crimson will do for the ground-work, and old-gold or olive for the decoration. This can be cut into some fanciful design, down toward the centre from the sides. Between these cuttings, form diamonds of silver beads intersecting squares of gold beads; and decorate the centre and lower point, with some spray or flower; having the foliage in silver beads, and using gold for the flowers. Or it could be stamped so as to define a rich design, in the border, and left plain; and simply caught down on each side by small cords of gold or silver, with nicely embroidered squares in each corner. This looks admirably, if cream is used on fawn-colored plush.

Hanging Pictures.

Pictures are made to be seen. Let them not be hung too high! Pictures in detail, should be hung nearer the eye than those in bold outline and relief. Let family-pictures be placed within family-rooms! Light, should fall upon a picture from the same side as the light within the picture. Pictures on the plane of the eye, should hang flat upon the wall; those above that line should droop forward, in the ratio of said height. Large and heavy paintings, should be thoroughly secured. There should be harmony, in groups of pictures; both as to subject, style, and size. Great care should be exercised in choosing befitting subjects, for pictures in the various departments. Death-bed-scenes, are not just exactly the proper thing, for the dining-room! Masonic-pictures, Diplomas, etc., are not strictly pictures, for decorating a home. Study fitness, order, size, shape, appropriateness, etc., in hanging and arranging pictures, as you would in the art of producing them. Hanging pictures, is an art.

Japan-Work.

The Japanese, are noted for their originality and their methods of designing and decoration. Commerce furnishes many minor Japanese articles, ready-prepared; by the help of which, we can admirably simulate the Japanese idea, in house-furnishing and decoration. Especially, where rattan and bamboo furniture are used, in conjunction with these. The circumstances of the student, will necessarily regulate the quality and expense of said Japanese decorations; since, they come in all grades and at all prices, according to the cost in time and labor expended in their manufacture. Beautiful small Japanese pictures, in bright colors and odd figures, can now be procured in the general market. They come in paper-form, and long enough to form a frieze. Various lacquered decorations in the form of boxes, trays, brackets, fans, etc., in innumerable colors and designs, can be had at small cost. Having laid in a supply of said stock, according to a fixed design and purpose in house-furnishing, let us suppose an airy and pleasant summer-room, which we wish to stamp with the Japanese idea, in its peculiarity of decoration, etc., viz: Either paint the bare floor plain, or otherwise in fanciful figures, squares, combinations of diamonds, etc.; or let a cool and fancy-colored matting cover it. Either paint the walls in pale-shaded yellow, olive, or straw-color, etc.; or select a fanciful design in paper, of similar color and shades and tones. Let a frieze, in a striking design of bright paintings, supply the place of a paper border. Let the ceiling have, a finish in a small but similar-shaded design. Blue Japanese "crêpe-cloth," relieved by quaint designs in reds, straw-colors, and olives, can be utilized as easy and full hangings; headed with dado-mouldings. On the walls, you can place a few placid water-color paintings of Japanese conceptions ; joined perhaps, by a pair of narrow-framed "etchings." Japanese chintz, may form the curtains. A screen of bamboo in relief-embroidery, in silk, will hide the grate. Lacquered ornaments, and small vases of blue, olive, or yellow, and darker scrolls of Japanese designs, can be utilized as ornaments of decoration upon small stands, centre-tables, brackets, etc.; and over which, in the corners, chintz-curtains suspended from polished brass rods are quaintly arranged, half-concealing the same. Let the tables be very neat tripple-crossed supports of rattan, in Japanese designs. Furthermore, let the couch or tete, chairs, easy rockers, etc., all be composed of rattan. These may be cushioned with cretonne or "Indian-Pongee," and ornamented with silk embroidered buds, blossoms, vines or tiny leaves of green.

A bamboo case for books and music, a Japanese-painted piano with an embroidered "Indian-pongee" cover; a small centre-table, supported by tripple-crossed brass rods, supporting a quaint card-case along with a Japanese glass-bowl full of leafy crimson buds and roses; now, having these furnishings, arranged in artistic form, we would have, at not a extravagant cost, a very beautiful Japanese-summer-parlor; which, would at one and the same time, combine the properties of healthfulness, restfulness, and pleasure.

Card-Decoration.

"Card-decoration," is growing intensely popular. Secure a delicate shade of any choice color of silk, the ordinary size of which is usually seven inches by five. Fringe it nicely, to a proper depth. Stamp a chosen and befitting design upon it, or draw it upon the same; then paint it artistically, and attach it to a fine bristol-board-card, somewhat larger than the silk. This is done by means of ribbons, taken through holes in the card and quaintly arranged in tasteful tiny bows, on the right-side. Two bows, in opposite corners, will suffice. Or, take as a foundation, a cream-white card. Arrange a cluster of "golden-rods" on the right-hand side, secured in place by a narrow ribbon coming from the back of the card over the stems; and which, is then taken back through a second hole, and is secured at the back. Taking a very fine-pointed brush, paint the desired motto, in but one color. Sprigs of evergreen, showing interlacings of tiny vines with berries intermingled, would help to decorate the same; or, perhaps ferns and pansies, beautifully arranged, might suit another's choice and taste. These tiny, yet often impressive and appropriate decorations, will forever endear to the recipient, the gift of an Easter or a Christmas-Card; which, might otherwise be lightly appreciated.

Art Paper-Flowers.

Subjected to the skilled artistic hand, paper can be transformed into flowers, wreaths, buds, vines, etc., with such effective beauty as to almost equal natural flowers. Those having made "wax-flowers," certainly should be able to make "paper-flowers," and *vice-versa*. Take the natural rose, or any flower, which you wish to reproduce in paper, apart, very carefully; now, cut a pattern out of paper of each separate petal; representing every petal by two thicknesses of the paper, which should be the best quality of "French tissue-paper."

The edges are carefully trimmed with sharp scissors, and carefully and daintily curled and crimped. The scissors can be utilized as the "joining-instrument" of the leaves, petals, etc. By means of tiny wires and a little mucilage, the petals are joined and permanently secured together, in the proper form of the calyx.

The leaves, of course, are cut from the green paper, and should be left a little larger than the natural leaves; which are afterwards reduced, by the process of crimping them with the fingers. The best method of obtaining the natural "centre-rib," and the fine "veinings," is obtained by first crimping them very finely crosswise; and afterwards, in a lengthwise manner, imitating nature.

The wires, are all nicely covered with paper of the same color seen in nature, which serve for stems and stalks; and the whole work is then mounted, and artistically arranged, to imitate the natural sprays and clusters.

Thus, out of the barren wilderness of plain colors and large sheets of paper, a world of beauty can be made to spring, as if by magic.

Lamp-Shade-Decoration.

Lamps are made to give light; and shades are intended to soften the light, and not to intercept it. Dense and ornate lamp-decorations, are incongruous inconsistencies. There is a "fitness of things," even in decorating lamp-shades.

As a general example of correct principles, suppose the following, viz.: Select as the ground-work, a lamp having a glass-globe crowned with an arabasque in a metal that matches the pedestal, upon which it rests; otherwise, a soft porcelain or crystal shade. Now, upon said foundation, draw or transfer some tasteful design, in either landscape or water scenes, etc. By means of the process as given for "engraving on glass," engrave the same beautifully; or otherwise paint it artistically, in such colors and in such a design, that the shades and tones, when the shade is placed over the light, will blend with the greatest degree of beauty and truth, to the original.

Thus, the refreshing scenes of landscapes, clusters of buds and leaves and flowers, or that of the placid and entrancing mountain-lake-scene, can be suspended in exquisite transparent beauty, over the parlor-lamp; or otherwise, placed upon the library-lamp, on which the toiling student can, occasionally at least, rest and refresh his weary eyes, while bending over his midnight-task.

Preservation of Natural-Flowers

Flowers are our close companions in health, happiness, sickness, sorrow, and death; and loving them, we grieve to see them die and pass into oblivion. A good method for preserving then is, to immerse them in melted paraffine. The process of immersing them and withdrawing them, should be quickly performed; and the melted paraffine should only be hot enough, to maintain its fluidity. Holding the flowers by the stalks, dip them thus—one at a time.

They should be instantly moved about and retained just long enough in the paraffine, to relieve them of the bubbles of air. Let the flowers be fresh-cut and free from moisture, when wishing to preserve them. To improve the color of roses, while growing them, place charcoal around the roots of the rosebush. It lends richness of lustre to all flowers.

Glass-Decoration.

A beautiful and very popular method of ornamenting or etching glass, is as follows, viz:

Taking the sheet of glass which is to be decorated, heat or warm it moderately, so as to receive the composition with which it is to be covered; and at the same time make provision, by means of a moistened strip of muslin and wax around the edges, to keep said fluid on the glass. Now coat the glass thoroughly with fluid-wax. By means of the etching-needle, draw any desired design upon this ground-work. Then apply hydrofluoric acid. The length of time for it to remain, depends on the length of time required to make the proper and necessary impressions of the design. The glass is then again heated, and the composition is removed by means of turpentine.

Painting on Ivory.

The design is carefullly and directly drawn upon the ivory, by the experienced artist; but as this is rather difficult for the beginner, and as the correcting of mistakes soils and unfits the surface for the completion of the painting, we do not advise it. Ivory being semi-transparent, a more safe process for the beginner is, to draw a correct and satisfactory design on a white sheet of paper, the size of the ivory upon which it is intended to paint the design. This, when pressed and secured between a pane of glass and the ivory, will show through the latter; and in this simple manner, can then be accurately reproduced by the brush, without the ivory surface having to be subjected to erasures and corrections; as is often the case, in directly drawing upon the ivory. Painting on ivory demands a scrupulous cleanliness, and great accuracy in work. The ivory should be secured from soiling, by having clean paper between the drawing-board and the ivory. The requisites will be six or eight brushes, a palette, some refined gum, and several dishes of water, and the paints.

The accuracy of the painting will, of course, largely depend on the accuracy and perfection of the drawing; which is now being reproduced on the surface of the ivory, by the brush and paint. Therefore, it should be very accurately and finely executed. The required colors should be laid on in as small quantities as it is possible, to effect the process. If it is a miniature-portrait, aim to bring the highest possible degree of naturalness and perfection, in the face; the most effective back-grounds for the fair complexion being blue, with a slight tint of gray; and for a dark complexion, use gray, with a slight tint of green.

Yellow ochre, red, and cobalt, form a fine shadow. Brown hair, can best be shown by sepia and vandyke brown; red hair, by ochre and burnt sienna. Chinese white, best touches up the eyes. But, of course, all general rules in mixing, shading, and toning, as used in painting elsewhere, and as taught heretofore, must be observed for the various shades of complexion, eyes, hair, etc. After brief experiment, and a little thought and care, the entire process of this beautiful department of painting, will become exceedingly simple and plain. If stippled too soon, however, it gives the painting a "watery" appearance.

Pen-Drawing.

Although not generally so accepted, the pen and ink are instruments almost rivaling the brush and paint, in art. Pen-drawing has also a commercial value, raising it to sufficient importance to receive our attention. A distinct clear black line on a pure and perfectly white surface, is one of the strongest possible contrasts; and accurate pen-drawing is a most potent means of art-work, if properly executed. In the main, it comes by accurate practice upon reproductions of accurate and perfect copies. The materials are, simple and few: a bottle of India-ink, six graduated sizes of good steel pens, a blotter, an ink-eraser, a fine-pointed penknife, and the best quality of Bristol-board. In drawing, let the lines be very fine but distinct, and in jet black; this latter property is regulated largely, by the quality of the ink. Clearness and distinctness, are prime qualities in pen-drawing. Let the drawing be executed in a size one-third larger than the intended engraving. The reproduction, by the process of photography, will show contact in the lines, often, if executed otherwise. The beginner should outline, by pencil-sketches, the entire work; and thus, corrections on the surface of the Bristol-board are obviated. In representing lights and shadows, draw the darkest shadows first; this is done by drawing a series of apparently wide parallel-lines intersected by another series of parallels, governed as to their direction by the requirement of the form or object illustrated. Shadows are deepened or left light, by means of, and in the ratio of the degree and number of the intersections used. Let the line first appear in its lightest form; the shadow can afterward be regulated to its proper degree, by deepening the same. Begin by practicing on plain, simple work. Learn to outgrow timidity, and to cultivate a trustworthy confidence and boldness in work. Do not tire of practice, and aim high; and pictures of matchless grace and beauty will soon be seen to fall from your pen, in the simple shades and combinations of purest black and white.

Art Metal-Work.

A very beautiful and useful art consists in fixing impressions upon copper and other metals, by means of the chemical action of acids. The metal should be first, thoroughly cleansed, by means of benzine and whiting. The copper plate or metal is then slightly heated and varnished, with a solution of three parts asphaltum, two parts white wax, and one part Burgundy pitch, thoroughly boiled, mixed, and worked. In applying it, melt the wax and varnish the plate by heating it and pouring said solution over it, letting it drain off, evenly. This leaves a plain smooth surface, when it is "set." The metal should then be thoroughly smoked, so as to not be seen through the solution, immediately before the varnish has set. After solidifying, the design is drawn upon this surface in outlines, by means of the drawing-needle; and the copper is laid bare, showing in fine lines, imitating the same correctly. This, requires exceedingly fine and accurate drawing; so as to show fineness of the lines, in the completed work.

The "closing-out-varnish," is composed of asphaltum, with a slightly greater quantity of turpentine. The development of this solution often requires thirty-six hours; it is then of the consistency of syrup. The outline of the design is then varnished with the same. The plate is now, by means of cords, carefully deposited in a bath of equal parts of water and nitric acid, for four minutes; letting the acid flow over it, into the outlines. The bubbles should be kept from the lines, during the process of the chemical action. Now rinse the plate in clean water, and dry by means of a blotting-pad. Where the work is properly completed, the "protective varnish" first used, is placed over the lines, and the remainder is again submitted to a second or third action in the bath; and, until the work is thoroughly completed. Then cleanse the plate with benzine, and dry and polish the same by paper or cloth, rubbing it. Any design can thus be accurately imprinted upon copper or other metals, and an art of charming beauty and great utility is hereby accomplished.

Linen-Decoration.

A beautiful branch of art needle-work is shown in "Linen-Decorations." By means of the hand of art, in this line, tidies grow on backs of chairs as gracefully as apple-blossoms in the orchard; and linen, fringed with insertions and laces, will fall into place upon sofas, as naturally as autumn-leaves fall from trees. It can be made very simple, and yet, admits of great intricacy; thus, suiting all degrees of genius. If edged with lace, the drawing should be made on fine white sateen; and the laces used, should have the quality of enduring washing. Insertion can also be used, to beautify the work. By using fine linen, beautiful toilet-sets are constructed, by the use of lawns or cambric upon which to form the sketch; and in the process of completion, having the mats lined with silk in pink or blue; the color showing sufficiently through the work, to lend it a charming and refreshing effect. In splashers for washstands, "Jack and Jill" bearing the historical "pail of water" and accomplishing extensive "splashing," might serve as an ornamental design.

Pillow-Shams can also, thus, be beautified by sprigs and branches and clusters of various leaves or flowers.

The ingenious housekeeper, will also scatter these decorations over window-shades, screens, hanging panels, and countless other household articles of both beauty and utility; and thus, please and gladden the hearts of those who gather there.

Portiere-Decoration.

The materials are satin, velvet, and plush.

An elegant design, having both the properties of art and utility, can be arranged as follows, viz: In the upper panel, form the mixed shades of dark, red, and brown; choose a ground of a warm shade in green, for the lower; with sage-green, at the sides. Any suitable color may be chosen for the main body; Persian-satin, looks well. The decorative designs, should be wrought in velvet-material. It is best to adopt some plain material in a neutral shade, ornamented with flower-designs attached in neutral colored velvet. This makes an exquisite design, and will serve as a general index, to "Portiere-Decoration."

Future Art.

Correct art, is the most perfect and beautiful expression possible, of absolute truth, goodness, and beauty; as conceived of within the empire of human thought, and expressed in the handiwork of man. This is true, whether as manifested in the great industrial arts of agriculture, navigation, architecture, law, medicine, etc., or in those we term "The Fine Arts;" such as music, poetry, sculpture, painting, embroidery, etc.

Correct art, must ever be the minister to the desire and gratification of all rightful and lawful pleasures. The pleasures of eating and drinking, of exercise and repose, of warmth and coolness, etc., form a class in direct contrast with the pleasures coming through the ordinances of music, painting, and sculpture, or those of the drama; for all of which, the human heart and mind hunger and thirst, as the body craves for food and drink; and to which, it is the province of art, in unison with nature, to minister. Pure art is therefore, an "angel of mercy," that stamps out impurity, injustice, and wrong; and that lifts the empire of mind, thought, and affection, out of the shadows of darkness and the domain of sensualism, into that higher and better sphere of perfect liberty, true happiness, and complete enjoyment, when sanctified by the Great Teacher of both art and science. It is to the heart and mind of humanity, under these circumstances, what the strong oak of the forest is to the vine;—lifting it out of the dense shadows beneath, into the realms of the radiant light of day—making it to "bud and blossom as the rose." This, must forever remain the ideal of the true artist, and the guiding star to an endless and undying fame; and among the many lustrous names that shine from the heavens of the artistic world today, and as seen in the annals of the nations, those having the most brilliant lustre are such as most closely approximated the above, in their ideal.

The bodily functions, while incidentally ministering to our pleasure, are in the main subservient to maintaining our existence; having been created for that special purpose. And when perverted by deviation from their intended purpose, we find them instantly and intimately connected with the production of results in which their short-lived and apparent pleasures turn only to bitter-ashes, both repulsive and loathsome; all of which proves conclusively, that these sources are not fountains of pure, unalloyed, and satisfying pleasure. The same may truthfully be written of the pleasures of power and dignity. The eye, in viewing the landscape and flowers and buds and blossoms of spring; and the ear, in feasting upon the harmony and cadence of music and the song of the lark, drink in unalloyed pleasure; and that, too, from an entirely different fountain, than the source of those sensations—which for want of a better term, we call "pleasure"—coming to us through eating and drinking, etc. The true province of art, therefore, seems to be encompassed by the boundaries of

UTILITY AND DECORATIVE ART. 221

the eye and the ear. At the altars erected within these provinces, the ministrations of art come to us in the various pleasures of harmony, beauty, melody, pathos ideality, picturesqueness, proportion, order, fitness, tenderness, irascibility, power and sublimity. When sweet sounds are harmoniously combined, we have the musical art; the painter has a similar aim, in reference to colors and forms; and through all the "fine arts," this quality is found recurring, as the crowning work of the artistic hand. Nothing is so indisputably included within the circle of the æsthetical or beautiful as these properties, in their elements and principles, separately and collectively.

Color in itself, imparts pleasure, apart from harmonious union.

Story is essential to romance and poetry; and sweetness, in the separate sounds, is essentially requisite for good music. The unity in art, requires that a building, an oratorio, a poem, a history, a dissertation, a painting, a piece of music, or embroidery, etc., should have a discernable principle of order throughout; the discernment of which, yields an artistic pleasure, even in works of pure utility, as well as in works of pure decoration. A blind devotion to art, however, merely for the sake of art, without esteeming it for its utility, aud as the language and expression of truth, purity, and beauty—in their absolute sense, is, something, akin to idolatry; and deserves censure, rather than praise.

The injurious and deliterious effects of such blind idolatry, is clearly manifest in its excessive forms, in the history of the Greeks; who were intoxicated thereby, and too much drawn from the urgencies of practical truth, and real life; rendering them an easy prey to political despots, and foes within and without; as well as, making them indifferent to moral principle; which, finally, proved the fatal quicksand underneath their political and religious temples and altars; and which, finally dissolved the nation. Grecian art proved itself a delusive snare. Correct art will, some day, have a synonymous definition with that of science—when both will be equal to, and express only "The Beautiful, the good, and the true;" when the artist will subject the imagination, strictly to the conditions of truth; when sculpture, will no longer be permitted to perpetuate a lie; when, instead of the tales of "Fairyland," the "Arabian Knights," and the romances of chivalry, the novelist will both entertain and teach truth, through truthful pictures and manners of living men and women; and as they really are to-day. When poetry, shall become true history as well as prophecy; when music and painting, will be the true language of affection, sublimity, gratitude, devotion and purity. And when, like the clear river that reflects the adjacent city with all its domes and turrets and spires, along with the overhanging foliage of trees and flowers, we shall see reflected in the clear and lucid stream of scientific art, as well as in nature and Revelation, the flowers and foilage of a still more beautiful and a better world; and when, looking deeply into its crystal waters, we can discern, through "The Gates Ajar," the outlines of the uplifted domes and turrets and spires, and can hear the distant music coming to greet us, from within the gates of "A city, not made with hands—eternal in the heavens."

PART FOURTH.

Gentlemen's Coat, Pantaloons, Vest,

AND

OVERCOAT-CUTTING.

PREFATORY AND INTRODUCTORY.

It almost seems a mere formality to write a special preface and introduction to each department; and yet, under the existing circumstances of the present confusion and controversial strife prevailing in some communities, touching the science of garment-cutting, caused by the false teachings of self-constituted authorities, who neither have practice nor correct knowledge, we feel the peculiar circumstances demand that something should be said to aid those who are sincerely seeking the truth, and to submit a few thoughts in vindication of the same. We have, however, only room for but a few words. And we refer such persons to the general preface and introduction, as given in the front part of the Encyclopedia. Suffice it to say here, that the elements and principles of measurement and drafting, as given and taught in this department—as everywhere throughout the work, are not simply theories, but practically demonstrated truths; which, while entirely original in their arrangement and construction, as to the methods of reading the indices and interpreting them in the draft—as a whole, they represent the world's best science and practical experience of many centuries combined; and are worthy of the most careful study of all, both experienced and inexperienced.

In conclusion, we only remark as follows, viz:

The nonsensical theory, that it requires a different system or science for cutting ladies' garments, from that of gentlemen's garments, is fortunately exploded. And tailors, in all departments, have outgrown its confusing and blinding teachings. Garment-cutting is a science; and when properly conceived and understood, can no more be affected by the particular form or style of the garment, than the science of geometry can be affected by the form of the object or figure which it measures. The science of cutting is, to get the mathematical mould of the perfect fit; art and good taste, attend to the style and shape—whether it be a pair of pantaloons for a gentleman, or otherwise a ladies' riding-habit, or a basque. The cutter is armed with a system of mathematical elements and principles, in order to obtain the exact mould or fit; and, he makes fashion minister to his wants; and, makes style his slave. As well might we say, the shoe-fitter must have a different science for fitting a child's shoe, from that of a ladies; or otherwise, that of a shoe or boot for a gentleman. Or again, that we needed a different science or system of surveying, each time we found a differently-shaped body of land or water! Or that we needed a different science or system of music and instrument, whenever we wished to write or play a different piece of music! The age for such folly in thought, is past! Any one who cannot from the same simple measures alone, and by means of the same common plain inch-rule and tape-measure alone, open any fashion journal and correctly reproduce any design therein to said measures, for any garment—either for man, woman, or child—does not understand the true and genuine "Scientific Tailor System," or the genuine science of garment-cutting; and, would do well to begin again, at the very beginning, and at the alphabet of said science. In our illustrations we give the "scale-figures," based upon the scale of the mathematical-inch; not because we use or endorse scales in our method, but to accommodate those who are inseparably wedded to the same The illustrations therefore serve the double purpose of illustrating the principles of the work, and that of accommodating the cutter who insists upon the scale-method. To those from whose eyes the scales have fallen, we commit and commend the "actual-measurement-system," as given and taught within the work. We cannot but feel gratified at the high appreciation that has already been expressed for the same, by the profession and general public at large; and to such, both present and future, it is hereby committed as a sacred trust, in the hope that it will accomplish a mission of great usefulness.

THE MEASURES.

1. Measure over a close and neatly fitting coat, otherwise a jacket, made plain and to different sizes for that special purpose, for coat-cutting.
2. Over the waist-coat only, for vests.
3. Over the pantaloons only, for pantaloons.

Location of Points for the Coat-Measures.

1. With the double corner of the Square toward the back, from the form being measured, and the long arm toward the bottom and in a perpendicular position, place the short arm underneath the junction of the body with the under-side of the arm, from behind; in such a manner that said junction of shoulder, arm, and body, will rest firmly upon the top-edge of the short-arm, and at the same time not being raised up out of its natural position ; maintaining this position of the Square, locate with a flattened pencil of crayon or chalk, upon the garment of the form, points 13 and 15, in their respective positions, even with the upper edge of the short-arm—on each side of the arm or shoulder ; as illustrated in Figure LX, of Plate XIX.

2. Locate point 4 at the junction of the under-side of the arm with the body, on a perpendicular line through the centre of the true and natural arm's-eye ; as illustrated in said cut.

3. In the same manner and by the same principles, repeat the location of points 13, 15, and 4, on the opposite side of the person ; as said points are illustrated in said cut, in their first position.

4. With the short arm of the square down, and the long arm in a horizontal position across the chest at the front of the form, and at the same time having the top of said arm of the square even with point 13 on the right-side and point 13 on the left-side, locate points 9 and 22, even with the top edge, in their respective positions ; as shown in said cut: and, repeat point 22 on the opposite-side, in the like position; as it is illustrated in its first position, in said cut.

5. In a similar manner, and by the same rules and principles that you located points 9 and 22 on the front of the form, locate points 5 and 20 on the back of the form, and in the positions respectively assigned to each; and as illustrated in said cut : also, repeat point 20 on the opposite side, by the same principle that you repeated point 22 at the front-side ; and as point 20 is illustrated in its first position, in said cut.

6. Locate point 26 on the form, back from the top of the shoulder, at the desired and proper intersection of the shoulder-seam with that of the sleeve and coat ; as illustrated in said cut by points 26 and 34, which represent one and the same point on the form, when the measures are taken.

7. Locate point 3 on the form, on a perpendicular line through point 4, at the bottom of the natural waist-depth ; as illustrated in said cut.

8. By the same rule and principle repeat point 3, at the opposite side of the form.

9. Locate point 7 at the desired back waist-depth, on a direct perpendicular line through the centre of the form, at the back ; as illustrated in said cut.

10. Locate point 11 at the desired front waist-depth, on a direct perpendicular line through the centre of the form at the front ; as illustrated in said cut.

11. Locate point 8 at the intersection of a direct perpendicular line through the centre of the form, at the back, with a horizontal line marking the proper junction of the collar with the coat, at the back ; as illustrated in said cut.

12. Take an easy neck-circumference measure, around the bare neck, as low down as possible; calling it, to be placed in the measure-book.

13. Locate point 23 on the form, on a line of the proper junction of the collar with the coat, $\frac{1}{8}$ of the neck-circumference from point 8, toward the front ; as illus-

trated in said cut by points 23 and 21, which represent one and the same point on the form, when taking the measures.

Completion.

14. Guided by the points and lines, as illustrated in said cut ; accurately take the following measures in the following order; calling them, to be placed in the measure-book, viz :

Skirt-Depth : From point 3, on a direct perpendicular line to line M^2.
15. *Blade-Width :* From point 4, on line C extended over the back, to point 4, on the opposite side of the form.
16. *Chest-Circumference :* From point 5, on line C. extended as a circumference line around the entire form, to point 5.
17. *Waist-Depth :* From point 4, on line B, to point 3.
18. *Back-Waist-Depth :* From point 5, on line D, to point 7.
19. *Back-Height :* From point 5, on line D, to point 8.
20. *Front-Waist-Depth :* From point 9, on line E, to point 11.
21. *Front-Height :* From point 9, on line E, to point 12.
22. *Arm's-Eye-Circumference :* From point 26, on lines W and Y, to point 34 ; or the exact and entire circumference-measure, around the junction of the arm and shoulder, with the body.
23. *Back-Shoulder-Height :* From point 20, on line I, to point 21.
24. *Front-Shoulder-Height :* From point 22, on line K, to point 23.
25. *Shoulder-Depth :* From point 23, on line O, to point 26 ; or from point 21, on line X, to point 34; which represent one and the same thing, on the form, when taking the measures.
26. *Waist-Circumference :* From point 2, on line A, extended as a circumference, entirely around the form to point 2.

MEASURES FOR PRACTICE.

The Coat.

		Inches.
1.	Neck-Circumference,	14
2.	Skirt-Depth,	16
3.	Blade-Width,	18
4.	Chest-Circumference,	36
5.	Waist-Depth,	8½
6.	Back Waist-Depth,	10
7.	Back-Height,	7¾
8.	Front Waist-Depth,	9½
9.	Front-Height,	5¼
10.	Arm's Eye-Circumference,	16
11.	Back Shoulder-Height,	8¼
12.	Front Shoulder-Height,	8¾
13.	Shoulder-Depth,	6¼
14.	Waist-Circumference,	32

The Sleeve.

Guided by the points and lines, as illustrated in Figure LXI, on Plate XIX, take the following measures in the following order ; calling them, to be placed in the measure-book, viz :

1. *Extreme Depth :* From point 16, on a direct line to a point on line D, equidistant from points 3 and 8 ; and from said point, on a direct line, to point 7.
2. *Arm's-Eye-Circumference :* The same as taken for the body of the coat ; and, unnecessary to be retaken.

3. *Elbow-Circumference:* From point 3, on line D, extended as a circumference line around the elbow, to point 3.
4. *Elbow-Depth:* From point 16, on a direct line to a point on line D, equidistant from points 3 and 8, or the outside point of the elbow.
5. *Hand-Circumference:* From point 7, on line F, extended as a circumference line, entirely around the front of the sleeve, to point 7.
6. *Arm-Circumference:* From point 11, on line J, extended as a circumference line, entirely around the arm, to point 11.
7. *Inside-Height:* From point 6, on line I, to point 9.
8. *Outside-Height:* From a point on line D equidistant from points 3 and 8, or the outside point of the elbow, on a direct line to a point on line L, equidistant from points 12 and 13.

Measures for Practice.

Inches.

The Sleeve :
1. Extreme-Depth, - - - - 23½
2. Arm's-Eye Circumference, - - - 16
3. Elbow-Circumference, - - - 13½
4. Elbow-Depth, - - - - - 14
5. Hand-Circumference, - - - 12
6. Arm-Circumference, - - - - 15
7. Inside-Height, - - - - 8½
8. Outside-Height. - - - - 12

The Measures for Vests.

Explanation.—The measures for waist-coats or vests, are taken the same; only omitting the skirt-depth measure; and taking the measure over the form, disrobed of the coat; or otherwise, the measures can be safely chosen or made from the coat-measures, by the practical cutter. It is well to take, as extra measures, the depth to the opening; and, desired depths of front and back, etc.

The experienced cutter will trace the design sought, directly from the "mould" or fit of the coat-draft, to the desired degree of closeness or ease of fit; shaping the design, entirely, and instantly, at will. Otherwise, draft by scales, as laid down in the cuts; which are based here, as elsewhere, upon the unit of the mathematical-inch, to the measures given.

The Measures for Pantaloons.

The measures for pantaloons, are so simple and easily seen from the cuts, that we deem it entirely unnecessary to go into details, as regards taking them. We will therefore make the following instructions suffice, viz.:

1. *Knee-Depth.*—From the bottom of the waist, at the side; just above the hip-bone, down the outside, to the knee; and the

Outside-Depth.—From the same point where you began the measure, in the above, to the desired depth at the outside.

2. *Front-Depth.*—From a point on a horizontal line through the first point from which you began, at the centre and in front of the form, on a direct line down the leg, to the desired depth.

3. *Inside-Depth.*—From the inside junction of the leg with the body, (holding the tape-measure up firmly, by the square) down the inside leg-seam, to the desired depth.

4. *Waist-Circumference.*—On a circumference line, around the waist.

5. *Hip-Circumference.*—On a circumference line, around the fullest part of the hips.

6. *Thigh-Circumference.*—On a circumference line around the thigh, horizontal with the junction of the inside of the leg with the body.

7. The same—on the opposite leg.

8. *The Diagonal.*—From the exact point you started for the side-seam, down underneath and through the junction of the inside of the legs with the body, up over the hip, to the point from which you started.

9. *Knee-Circumference.*—On a horizontal circumference-line around the knee.

10. *Bottom-Circumference.*—On a horizontal circumference-line around the bottom of the pants.

MEASURES FOR PRACTICE.

Pantaloons.

		Inches.
1.	Knee-Depth, and Outside-Depth,	23½ 42
2.	Front-Depth,	42
3.	Inside-Depth,	32
4.	Waist-Circumference,	32
5.	Hip-Circumference,	38
6.	Thigh-Circumference,	24
7.	Thigh-Circumference,	24
8.	Diaognal,	34
9.	Knee-Circumference,	16
10.	Bottom-Circumference,	18

N. B.—It is not necessary to take double-measures, either in coats, vests, or pantaloons; only where a noticeable difference exists; when both sides of the form must be measured in each department, and drafted and traced separately; and, accurately marked right and left.

The International Method

OF

French, Prussian, English, and American

SCIENTIFIC TAILOR-PRINCIPLES,

FOR ALL KINDS AND STYLES

OF

BODY-DRAFTING,

WITH ANY

SQUARE OR INCH-RULE AND TAPE-MEASURE ALONE.

PLATE XIX.—FIGURE LX.

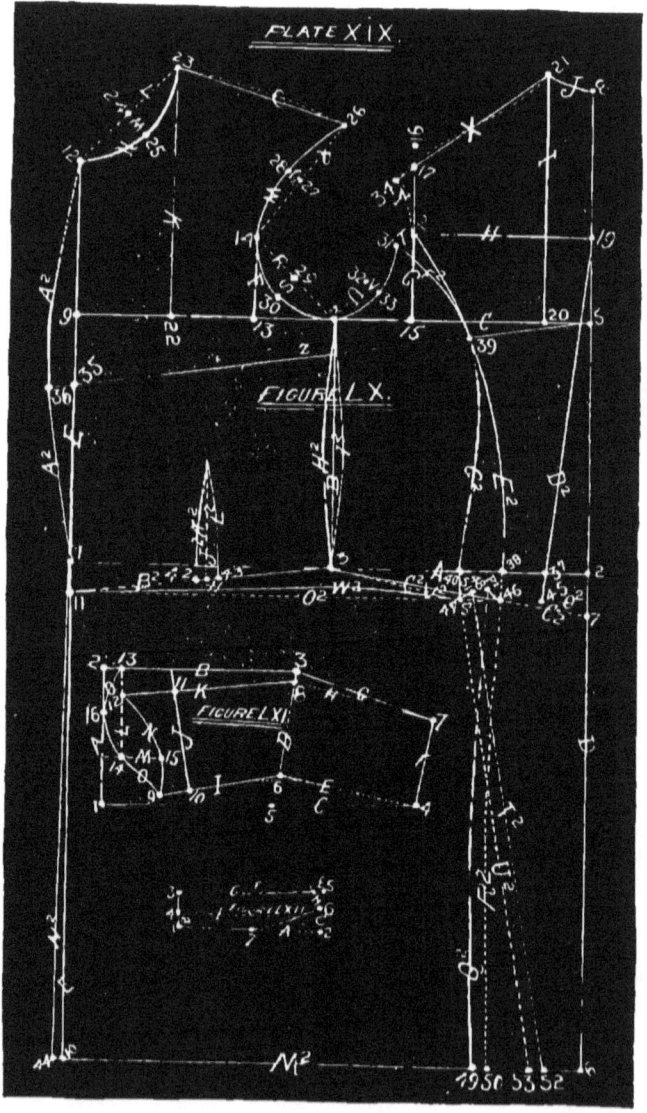

International Coat, Sleeve, and Collar-Plate.

International Coat-Drafting.

PLATE XIX.—FIGURE LX.

INSTRUCTIONS.

1. Add 1½ inches to the skirt-depth: Locate point 1 on paper or material, that distance from the bottom or right-hand edge, and 4 inches from the front edge, toward the opposite edge of the same.
2. Draw line A, at right angles with the front edge, from point 1 toward the opposite edge, to the distance of ½ of the chest-circumference from point 1; at which point, locate point 2.
3. Locate point 3 on line A, the distance of ½ of the blade-width from point 2, toward point 1.
4. Draw line B, at right angles with line A, from point 3 toward the top, to the distance of the waist-depth from point 3; at which point, locate point 4.
5. Draw line C, at right angles with line B, from point 4 toward the opposite edge, to the distance of ½ of the blade-width from point 4; at which point, locate point 5.
6. Find the sum of the waist-depth, plus the skirt-depth: Draw line D, at right angles with line C, from point 5 toward the bottom, to the distance of that sum of depths from point 5; at which point, locate point 6.
7. Locate point 7 on line D, the distance of the back waist-depth from point 5, toward point 6.
8. Extend line D, from point 5 toward the top, to the distance of the back-height from point 5; at which point, locate point 8.
9. Find the difference between ½ of the chest-circumference and ½ of the blade-width:
Extend line C, from point 4 toward the front, to the distance of that difference from point 4; at which point, locate point 9.
10. Find the sum of the waist-depth, plus the skirt-depth: Draw line E, at right angles with line C, from point 9 toward the bottom, to the distance of that sum of depths at which point, locate point 10.
11. Locate point 11 on line E, the distance of the front waist-depth from point 9, toward point 10.
12. Extend line E, from point 9, toward the top, to the distance of the front-height from point 9; at which point, locate point 12.
13. Locate point 13 on line C, at a point which is one-sixth of the arm's-eye-circumference from point 4, toward point 9.
14. Draw line F, at right angles with line C, from point 13 toward the top, to the distance of ⅙ of the arm's-eye-circumference from 13; at which point, locate point 14.
15. Locate point 15 on line C, at a point which is ⅙ of the arm's-eye-circumference from point 4, toward point 5.
16. Draw line G, at right angles with line C, from point 15 toward the top, to the distanc of ⅜ of the arm's-eye-circumference from point 15; at which point, locate point 16.
17. Locate point 17 on line G, the distance of ⅓ of the arm's-eye-circumference from point 15, toward point 16.
18. Locate point 18 on line G, equidistant from points 15 and 16.
19. Draw line H, at right angles with line G, from point 18 to line D; and locate point 19, at the point of intersection of lines H and E.
20. Locate point 20 on line C, the distance of ⅛ of the neck-circumference from point 5; toward point 15.

21. Draw line I, at right angles with line C, from point 20 toward the top, to the distance of the back shoulder-height from point 20; at which point, locate point 21.
22. Draw curved line J, from point 21 to point 8; as illustrated in the cut.
23. Locate point 22 on line C, the distance of ¼ of the neck-circumference from point 9; toward point 13.
24. Draw line K, at right angles with line C, from point 22 toward the top, to the distance of the front shoulder-height from point 22; at which point, locate point 23.
25. Indicate pointed line L, from point 12 to point 23; as illustrated in the cut.
26. Locate point 24 on line L, equidistant from points 12 and 23.
27. Indicate pointed line M, at right angles with line L, from point 24 toward line K, to the distance of one-twelfth of the neck-circumference; at which point, locate point 25.
28. Draw curved line N, from point 23, through point 25, to point 12; as illustrated in the cut.
29. Draw line O, from point 23, toward point 16, to the distance of ¼ inch less than the shoulder-depth; at which point locate point 26.
30. Indicate pointed line P, from point 26 to point 27; as illustrated in the cut.
31. Locate point 27, on line P, equidistant from points 26 and 14.
32. Indicate pointed line Q, at right angles with line P, from point 27, toward line K, to ⅔ of the distance between points 17 and 16 from point 27, over line G; at which point, locate point 28; as illustrated in the cut.
33. Indicate pointed line R, from point 14 to point 4.
34. Locate point 29, on line R, equidistant from points 14 and 4.
35. Indicate pointed line S, at right angles with line R, from point 29 toward point 13, to a point equidistant from points 29 and 13; at which point, locate point 30; as illustrated in the cut.
36. Indicate pointed line T, from point 18 toward point 13, to the same distance, from point 18, as the distance (on line G), between points 16 and 17; at which point, locate point 31; as illustrated in the cut.
37. Indicate pointed line U, from point 31 to point 4.
38. Locate point 32 on line U, equidistant from points 4 and 31.
39. Indicate pointed line V, at right angles with line U, from point 32 toward point 15, to ¾ of the distance between points 18 and 31; at which point, locate point 33; as illustrated in the cut.
40. Draw curved line W, from point 26 through 28, 14, 30, 4, and 33, to point 31; as illustrated in the cut.
41. Draw line X, from point 21, through point 17, to the distance of the shoulder-depth from point 21; at which point, locate point 34.
42. Draw curved line Y, from point 34 to point 18; as illustrated in the cut.
43. Locate point 35 on line E, at a point which is ¼ of the distance between points 11 and 9, from point 9, toward point 11.
44. Draw line Z from point 5, through point 35, to a point which is 1 inch in front of point 35 (or to any desired degree of curve, at the front); at which point, locate point 36.
45. Draw curved line A^2, from point 11, through point 36, to point 12; as illustrated in the cut.
46. Draw line B^2, from point 11 to point 3.
47. Draw line C^2, from point 3 to point 7.
48. Find the difference between the chest-circumference and the waist-circumference: Locate point 37 on line A, ¼ of that difference from point 2, toward point 3.
49. Draw line D^2, from point 19 to point 37.
50. Find ⅛ of the waist-circumference: Locate point 38, on line A, ⅜ of that distance from point 37, toward point 3.
51. With a radius on the tape-measure, of the distance between points 1 and 2, (over line A), from a centre-point in front of line E, common to points 18 and 38, draw curved line E^2, from point 18 to point 38; as illustrated in the cut.

UTILITY AND DECORATIVE ART. 233

52. Locate point 39 at the point of intersection of lines E^2 and Z.
53. With the same radius, and by the same means, rules, and principles, given for drawing line E^2, draw curved line F^2, from point 31 to point 39; as illustrated in the cut.
54. Locate point 40 on line A, the distance from point 38 toward point 3, which is ½ the distance from point 2 to point 37, on line A.
55. With the same radius, and by the same means, rules, and principles, given for drawing lines E^2 and F^2 draw curved line G^2, from point 39 to point 40; as illustrated in the cut.
56. Draw curved line H^2 and I^2, from point 3 to point 4, with an equidistant and symmetrical curve of ⅜ of an inch respectively, at the greatest depth, from line B; and as said lines are illustrated in the cut.
57. Locate point 41 on line B^2, at a point equidistant from points 11 and 3.
58. Locate point 42 on line B^2, ½ of the distance between points 38 and 40 from point 41, toward point 11.
59. Locate point 43 on line B^2, the same distance from point 41 toward point 3; as the distance between points 41 and 42, on the same line.
60. Indicate line J^2, at right angles with line A, from point 41 toward line Z, to the distance of ½ of the waist-depth, from point 41.
61. Draw curved line K^2, from the upper terminus of line J^2, to point 42; as illustrated in the cut.
62. Draw curve line L^2, from the upper terminus of line J^2, to point 43; as illustrated in the cut.
63. Draw line M^2, from point 6, through point 10, to a point in front of point 10, which is the distance of ½ the distance between points 35 and 36, from point 10; at which point, locate point 44.
64. Draw line N^2, from point 11 to point 44.
65. Indicate and draw line O^2, at right angles with line E, from point 11 to line D; as illustrated in the cut.
66. Locate point 45 at the point of intersection of lines O^2 and D^2, when line D^2, is extended toward the bottom, as it should be; and as illustrated in the cut.
67. Locate point 46, at the point of intersection of lines E^2 and O^2, when line E^2 is extended toward the bottom, as it should be; and as illustrated in the cut.
68. Locate point 47 at the point of intersection of lines G^2 and O^2, when line G^2 is extended toward the bottom, as it should be; and as illustrated in the cut.
69. Locate point 48, the distance of ¼ of an inch above line O^2, and ½ inch from line E^2, toward line G^2; as illustrated in the cut.
70. Draw line P^2, from point 48 to point 46.
71. Draw line Q^2, at right angles with line A, from point 48 to line M^2, and locate point 49, at the intersection of lines Q^2 and M^2.
72. Locate point 50 on line M^2, ½ inch from point 49, toward point 6.
73. Indicate pointed line R^2, from point 46 to point 50.
74. Locate point 51 the distance of ¼ inch above line O^2 and ½ inch from line G^2, toward line E^2; as illustrated in the cut.
75. Draw line S^2, from point 47 to point 51.
76. Locate point 52 on line M^2, equidistant from points 49 and 6.
77. Draw line T^2, from point 51 to point 52.
78. Locate point 53 on line M^2, ½ inch from point 52, toward point 50.
79. Indicate line U^2, from point 47 to point 53.
80. Indicate line V^2, from point 3, to point 47.
81. Draw curved line W^2, from point 11, through an imaginary point (on a direct perpendicular line through point 3), equidistant from lines O^2 and A, to point 47, as illustrated in the cut; which process, completes the entire body and skirt of the coat; as the various parts appear and are illustrated, in the cut.   Q. E. D.

N. B.—"The above draft, can, in its various parts, be traced upon the linings; or the parts, cut out and used as patterns; or otherwise, can by the same principles be drafted, in its various parts, directly upon the goods.

2. All kinds and possible styles of body-garments, can instantly be drafted by the above principles for obtaining the "mathematical mould," or fit of the form;—which, when obtained, can be changed, at will (or according to the dictates of fashion), to any desired degree of closeness or ease, in the garment; and the same applies also, to the location, number, and style of the seams; the principles being entirely independent of either, as well as the shape or design of the skirt.

The Sleeve.

PLATE XIX.—FIGURE LXI.

INSTRUCTIONS.

1. Add 1 inch to the extreme-depth:
Locate point 1, that distance from the bottom or right-hand edge, and 1 inch back from the front edge of the paper or material.
2. Draw line A, at right angles with the front edge, to the distance of $\frac{5}{8}$ of the arm's-eye-circumference from point 1, toward the opposite edge; at which point, locate point 2.
3. Draw line B, at right angles with line A, from point 2 toward the right, to the distance of $\frac{1}{4}$ inch less than the elbow depth from point 2; at which point, locate point 3.
4. Find the difference between $\frac{1}{2}$ of the extreme-depth and the elbow-depth; subtract this, from the extreme-depth: Then draw line C, at right angles with line A, from point 1 toward the right, to that distance from point 1; at which point, locate point 4.
5. Locate point 5, on line C, the distance of $\frac{1}{2}$ the extreme depth from point 1; toward point 4.
6. Draw line D, from point 3 toward point 5, to the distance of $\frac{1}{2}$ an inch more than $\frac{1}{2}$ of the elbow circumference from point 3; at which point, locate point 6.
7. Draw line E, from point 6 to point 4.
8. Draw line F, at right angles with line E, from point 4 toward the opposite edge, to the distance of $\frac{1}{2}$ of the hand-circumference, from point 4; at which point, locate point 7.
9. Draw line G, from point 3 to point 7.
10. Locate point 8 on line D, $\frac{1}{2}$ an inch from point 3, toward point 6.
11. Draw line H, from point 8 to point 7.
12. Draw line I, from point 6 toward point 1, to the distance of the inside height from point 6; at which point, locate point 9.
13. Locate point 10 on line I, $\frac{1}{4}$ of the distance between points 9 and 6 (on line I) from point 9, toward point 6.
14. Draw line J, at right angles with line I, from point 10 to line B.
15. Subtract the distance between point 10 and line B—measuring over line J —from the arm-circumference: Locate point 11, on line J, the distance of that difference from point 10.
16. Draw line K, from point 8 through point 11, to the distance of the outside height from point 8; at which point, locate point 12.
17. Indicate line L, at right angles with line B, from point 12 to line B; and locate point 13, at the point of intersection of lines L and B.
18. Extend line L, by indication, from point 12 toward line C, to the distance of $\frac{3}{8}$ of the arm's-eye-circumference from point 13; at which point, locate point 14; as illustrated in the cut.
19. Indicate line M, at right angles with and from line A, through point 14, to the distance of $\frac{1}{4}$ of the arm's-eye-circumference from line A; at which point, locate point 15; as illustrated in the cut.

20. Draw curved line N, from point 12 through point 15, to point 9; as illustrated in the cut.
21. Locate point 16, on line A, ⅓ of the distance between points 1 and 2, from point 2 toward point 1.
22. Draw curved line O, from point 13, through points 16 and 14, to point 9; as illustrated in the cut.

Q. E. D.

N. B.—This draft can, also, be eliminated by tracing its parts and using the patterns; or otherwise, by the thoughtful student, can, by applying the same principles separately to each part, be drafted in each part directly upon the goods; by the same construction lines being used, to each part.

The Collar.

PLATE XIX.—FIGURE LXII.

1. Locate point 1, on the paper or material, ½ inch from the right-hand edge, and ½ inch from the front edge, toward the opposite edge.
2. Draw and indicate line A, at right angles with the front edge, from point 1 toward the opposite edge, to ½ of the desired circumference-depth of the collar; at which point, locate point 2.
3. Draw line B, at right angles with line A, from point 1 toward the top, to the desired full-depth or width of the collar, at the back; at which point, locate point 3.
4. Locate point 4, on line B, the desired distance from point 1, at which you wish the crease-line of the collar.
5. Indicate line C, at right angles with line A, from point 2 toward the top, to the desired full-depth or width of the collar, at the front; at which point, locate point 5; as illustrated in the cut.
6. Locate point 6, on line C, the distance of ⅛ of the neck-circumference from point 2, toward point 5.
7. Locate point 7, on line A, the distance of ⅓ of the neck-circumference from point 1; toward point 2.
8. Draw curved line D, from point 7 to point 6; as illustrated in the cut.
9. Indicate line E, at right angles with line C, to the distance of ½ inch from point 5—toward line B; at which point, locate point 8; as illustrated in the cut.
10. Indicate pointed line F, from point 3 to point 8; as illustrated in the cut.
11. Draw curved line G, from point 3 to point 8, so that its extreme depth of curve, at the centre, will be over a point ¼ inch below line F; as illustrated in the cut.
12. Draw line H, from point 8 to point 6.
13. Indicate line I, at right angles with line B, from point 4 to the point of intersection of lines I and D.

Q. E. D.

N. B.—All kinds and styles of collars may be drafted to measure, from these construction lines and principles; and shaped at will, and in accordance with the demand of fashion. We take the crease-line as a basis, passing at the corner of the shoulder-seam, and reaching the lapel-seam higher or lower, in the proportionate degree that the coat is buttoned higher or lower.

The International Method of French, Prussian, English, and American Scientific Tailor- Principles for all Kinds and Styles of

PANTALOON-DRAFTING

With any Square or Inch-Rule and Tapemeasure Alone.

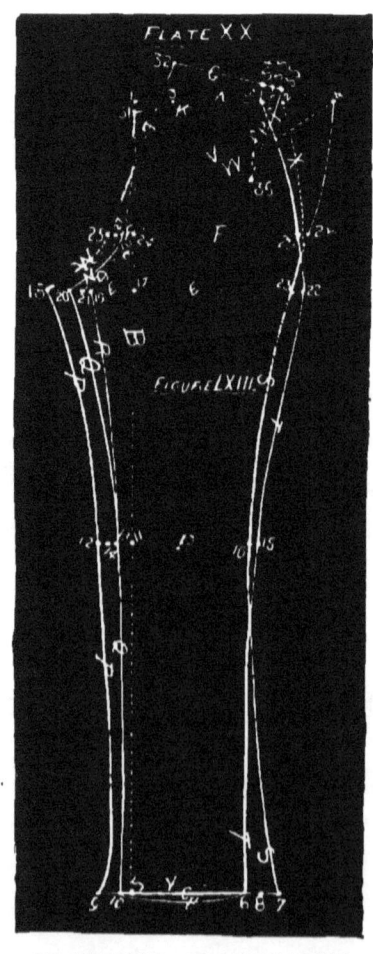

PLATE XX.—FIGURE LXIII.

The International Method of Pantaloon-Drafting.

PLATE XX.—FIGURE LXIII.

INSTRUCTIONS.

1. Add 1 inch to the outside-depth; Locate point 1 on the paper or material, that distance from the bottom or right-hand edge, and the distance of ⅙ of the hip-circumference from the front, toward the opposite edge.
2. Indicate line A, at right angles with the front edge, from point 1 toward the opposite edge, to the distance of ⅛ of the waist-circumference from point 1; at which point, locate point 2.
3. Locate point 3 on line A, ¼ of the distance between points 1 and 2 from point 1, toward point 2.
4. Indicate the extension of line A, from point 2, toward the opposite edge, to 1 inch more from point 2, than the distance between points 1 and 3; at which point or terminus, locate point 4.
5. Indicate line B, at right angles with line A, from point 1 toward the bottom, to the distance of the outside-depth, from point 1; at which terminus, locate point 5.
6. Draw line C, at right angles with line B, from point 5 toward the opposite edge, to the distance of ⅓ of the bottom-circumference; at which terminus, locate point 6.
7. Extend line C, from point 6, toward the opposite edge, to ¼ of the distance between points 5 and 6 from point 6; at which terminus, locate point 7.
8. Locate point 8 on line C, equidistant from points 6 and 7.
9. Extend line C, from point 5 toward the front, to the same distance from point 5, as the distance between points 6 and 7; at which terminus, locate point 9.
10. Locate point 10 on line C, equidistant from points 5 and 9.
11. Locate point 11 on line B, the distance of the knee-depth from point 1, toward point 5.
12. Indicate line D, at right angles with line B, from point 11 toward the front edge, to the distance of ⅛ of the knee-circumference from point 11; at which terminus, locate point 12.
13. Locate point 13 on line D, equidistant from points 11 and 12.
14. Locate point 14 on line D, equidistant from points 12 and 13.
15. Indicate the extension of line D, from point 11 toward the opposite edge, to the distance of ¼ inch more than ½ of the knee-circumference from point 14; at which terminus, locate point 15.
16. Locate point 16 on line D, the same distance from point 15 toward point 11, as the distance between points 12 and 14.
17. Locate point 17 on line B, the distance of the inside-depth from point 5, toward point 1.
18. Indicate line E, at right angles with line B, from point 17, toward the front, to the distance of one-fifth of the thigh-circumference, from point 17; at which terminus, locate point 18.
19. Locate point 19 on line E, equidistant from points 17 and 18.
20. Locate point 20 on line E, equidistant from points 18 and 19.
21. Locate point 21 on line E, equidistant between points 19 and 20.
22. Indicate the extension of line E, in the opposite direction from point 17, to the distance of ½ inch more than ½ of the thigh-circumference, from point 21; at which terminus, locate point 22.
23. Locate point 23 on line E, 1 inch from point 22, toword point 17.

24. Locate point 24 on line B, at a point which is ⅓ of the distance between points 1 and 17 from point 17, toward point 1.

25. Indicate line F, at right angles with line B, from point 24 toward the front, to ½ of the distance between points 17 and 18 from point 24; at which terminus, locate point 25.

26. Locate point 26 on line F, equidistant from points 24 and 25.

27. Locate point 27 on line F, equidistant from points 25 and 26.

28. Indicate the extension of line F, in the opposite direction from line B, to the distance of ½ inch more than ¼ of the hip-circumference from point 27; at which terminus, locate point 28.

29. Locate point 29 on line F, 1 inch from point 28; toward point 24.

30. Locate point 30, on line A, 1 inch from point 2 toward point 1.

31. Locate point 31, on line B, ½ inch from point 1 toward point 24.

32. With the short arm of the square toward the front, and the double-corner on point 4 and the outside edge of the long arm on point 23, draw line G, from point 4 toward the front, to 1 inch more than ¼ of the waist-circumference from point 4; at which terminus, locate point 32.

33. Locate point 33 on line G, ⅓ of the distance between points 4 and 32.

34. Locate points 34 and 35, on line G, ½ inch each respectively, from point 33; and in opposite directions, from point 33.

35. Indicate line H, at right angles with line G, from point 33 toward line F, to ½ the distance between points 1 and 17, from point 33; at which terminus, locate point 36.

36. Now draw the following lines, from and to the following given points, in the following order, and as said lines and points are illustrated in the cut, viz:

Line I, from point 35 to 36; J, from 34 to 36; K, from 31 to 30; L, from 25 to 32; M, from 25 to 18; N, from 31—through 26—to 20; O, from 26 to 19; P, from 18—through 12—to 9; Q, from 20—through 13—to 10; R, from 19 to 13; S, from 4, through 28, 23, and 16, to 7; T, from 30, through 29, 22, and 15, to 6; U, from 19 to 6.

Q. E. D.

N. B.—1. The indicated lines V and W, show the method of applying the proof "Diagonal-measure." Line A, may be used at option, instead of line K. In very full waist-measures, use indicated line X, for that part of line T between points 22 and 30. In very small waists, draw line L through point 3; and arrange for the necessary changes elsewhere, effected thereby. Line Y, shows the "crescent-cut" at the bottom.

2. All possible kinds and styles of pantaloons can quickly and accurately be drafted from these same construction lines, and by the same principles.

3. Many prefer to make the compound-draft, and preserve it;—as is also done in coat-drafting—for the special form to whose measures it is drafted; and simply tracing the parts separately upon pattern-paper, they use said patterns to mark or chalk the fit upon the goods—cutting the patterns the exact fit, and allowing all seams and extensions for finish, everywhere.

4. To draft directly upon the goods—for each part by itself—use the same construction lines to each part separately, as they are used in the compound draft,—locating and numbering all the points, as in the compound draft—and then only using the necessary points for that particular piece, and drawing only the necessary lines through said points; and as each is illustrated, separately and distinctly, from the others, in the compound-draft. All of which also applies to coat, vest, and overcoat-drafting. The careful and thoughtful student will find the application of these principles exceedingly simple, perfect and exhaustive; whether as applied in either compound or direct-drafting.

In the following pages we give a number of styles; in the various departments which, while illustrating styles for the Actual-Measure-method, also are arranged, for those who wish to do so, to cut the same by the Scale-Method.

Overcoats.

Overcoats are drafted precisely by the same principles, as the undercoat; only that the measures are increased in the following ratio, and drafted to the desired style, viz:

	Inches.
Neck-Circumference	Increase ½
Shoulder-Depth	" ½
Arm's-Eye-Circumference	" ¾
Chest-Circumference	" 1 to 2.
Waist-Circumference	" 1 to 2.
Hip-Circumference	" 1 to 2.

Lapels.

Lapels may be cut separate from the forepart, or in one piece with the same. When cut separate, cut the seam joining with the forepart, straight. The width depends entirely upon fashion. Perhaps the best method is, to bring the lapel-seam back to the button line—if fashion does not forbid. From the following plates and styles the thoughtful student will have no trouble in thoroughly mastering the art, in all its details and practical applications.

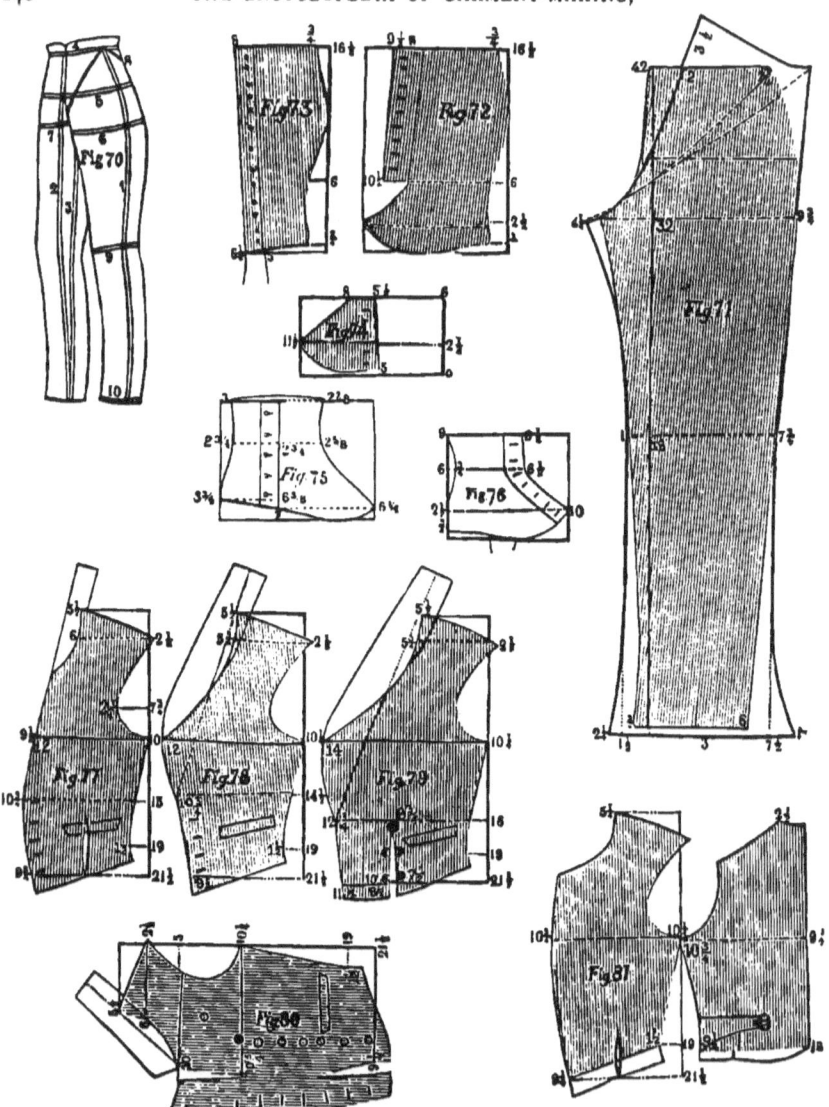

PLATE XXI.—FIGURES 70-81.

EXPLANATION.

Figure 70, represents the Method of Measureing; Figure 71, the Pantaloon-Draft; Figures 72, 73, and 74, the inside, the outside, and the part of the outside containing the buttons, respectively, of a pair of long gaiters; Figures 75 and 76, two different styles of short gaiters, respectively; Figures 77, 78, 79, and 80, four different fronts of the vest, as shown in its draft in Figure 81, showing the back and front. Draft the actual measures, per instructions given for each respective garment; or, in the above ratios, by the "Scale-System," from the respective construction lines for each, as is easily seen; and, as illustrated in the cuts.

PLATE XXII.

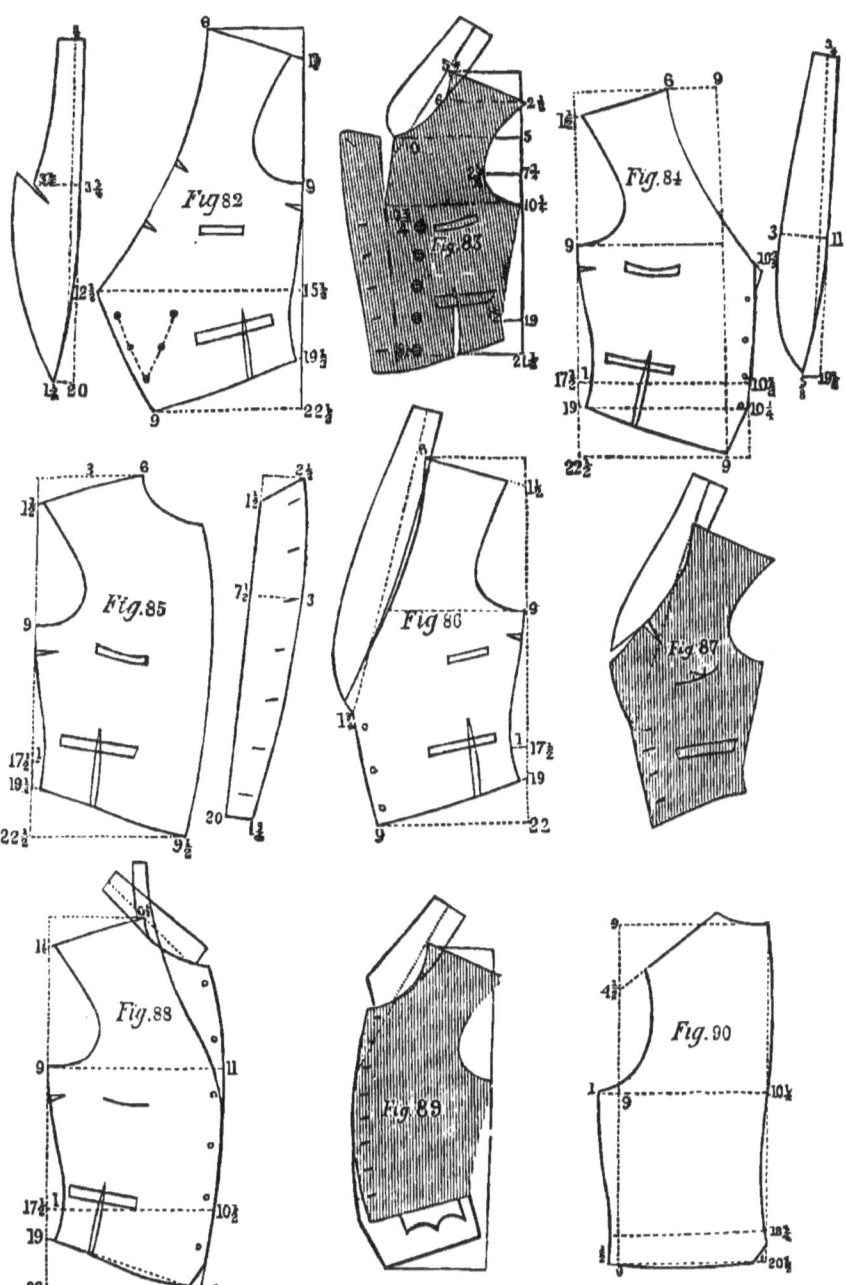

PLATE XXII.—FIGURES 82-90.

EXPLANATION.

Figures 82, 83, 84, 85, 86, 87, 88, and 89, of Plate XXII, show respectively, eight different styles of vest-fronts of one and the same back, as shown in Figure 90. These styles can be drafted from the actual measures, per instructions given for body-drafting; or, can be drafted to the ratios given ner the "Scale-Method."

PLATE XXIII.

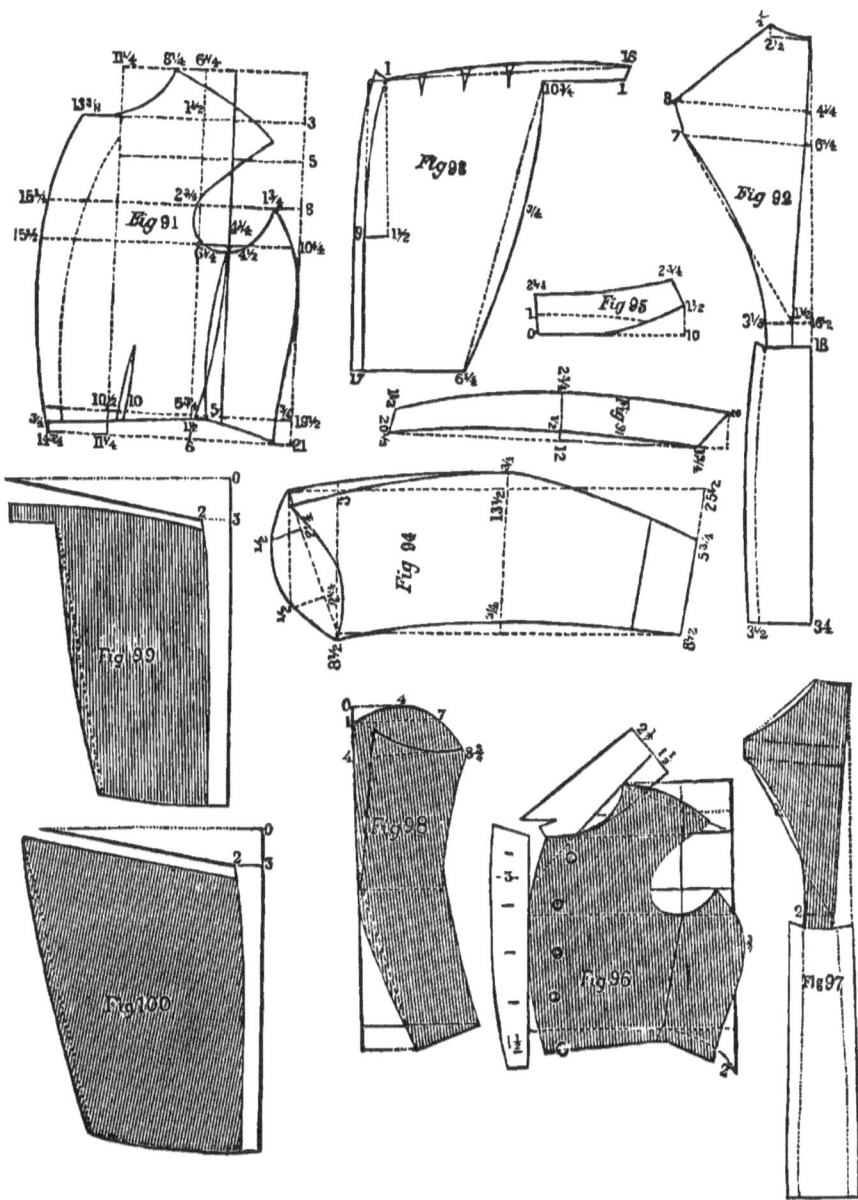

EXPLANATION.

PLATE XXIII.—FIGURES 91-100.

Figures 91, 92, 93, 94, and 95, respectively represent the front, the back, the skirt, the sleeve and the collar, of a stylish dress-coat. And Figures 96, 97, and 98, show still a different style of dress-coat; of which, figures 99 and 100, represent two different styles of skirts.

Plan of Measure-Book for Ladies' Measures.

Name,..
Residence,..
Materials,...
Garments,...
To Fashion-Plates, Nos...............Page............of......................A. D. 188....,............Number.
Time for Completion,...................day,............M.................the................A. D. 188......
Price for Materials Furnished,...$...............
Price for Manufacturing,..$...............

        Total,................$...............

THE BODY-MEASURES.	The Close or Standard Measures	The Increased, or Coat-Measures.		
	For all Kinds of Goods.	Medium Weight Goods.	Heavy Weight Goods.	Very Heavy Weight Goods.
	INCHES.	INCHES.	INCHES.	INCHES.
1. Neck-Circumference,.........				
2. Skirt-Depth,................				
3. Blade-Width,................				
4. Chest-Circumference,........				
5. Waist-Depth,................				
6. Back-Waist-Depth,...........				
7. Back-Height,................				
8. Front-Waist-Depth,..........				
9. Front-Height,...............				
10. Arm's-Eye-Circumference,....				
11. Back-Shoulder-Height,.......				
12. Front-Shoulder-Height,......				
13. Shoulder-Depth,.............				
14. Bust-Circumference,.........				
15. Waist-Circumference,........				
16. Hip-Depth, ...				
17. Hip-Circumference,..........				
18. Sleeve: 1. Extreme-Depth,.........				
2. Arm's-Eye-Circumference,.				
3. Elbow-Circumference,....				
4. Elbow-Depth,............				
5. Hand-Circumference,.....				
6. Inside-Height,..........				
7. Arm-Circumference,......				
8. Outside-Height,.........				

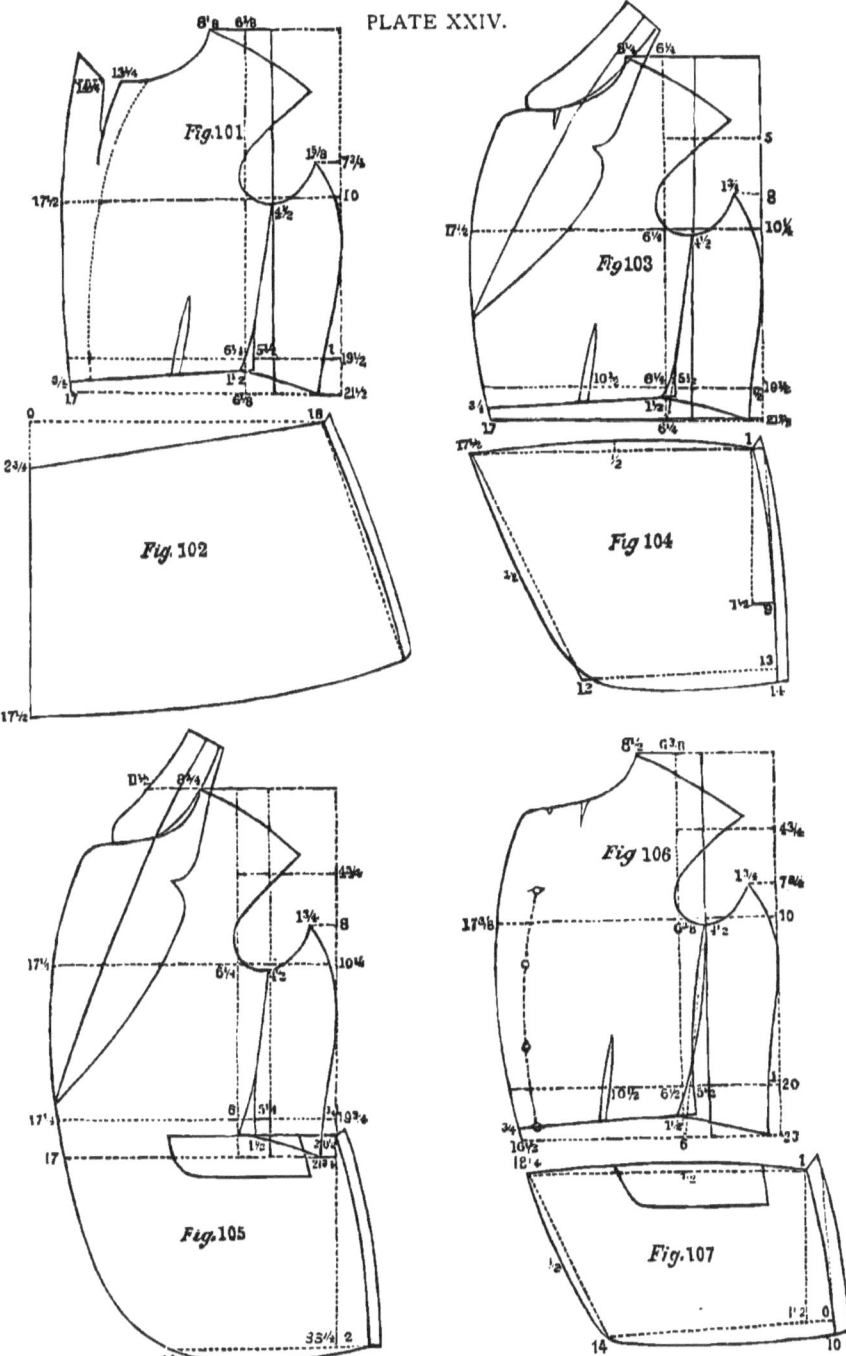

PLATE XXIV.—FIGURES 101-107.

EXPLANATION.

Figures 101 and 102, of Plate XXIV, respectively represent the front and skirt, of a single-breasted Frock-Coat; the back, sleeve, collar, etc., of which are drafted the same, and by the same principles, as the Dress-coat. Figures 103 and 104, respectively represent the front and skirt, of the open walking-coat. The principle and shape of the backs, are all the same—only differing in lengths, to suit the skirts, chosen. Figure 105, shows the form of the front of a neat jacket, of the register style ; the back, is the same as for any other style—only, made suitable in length; which, pertains to all the cuts in the above plates, unless stated otherwise. Figures 106 and 107, represent the front and skirt of the "English-Jacket."

PLATE XXV.

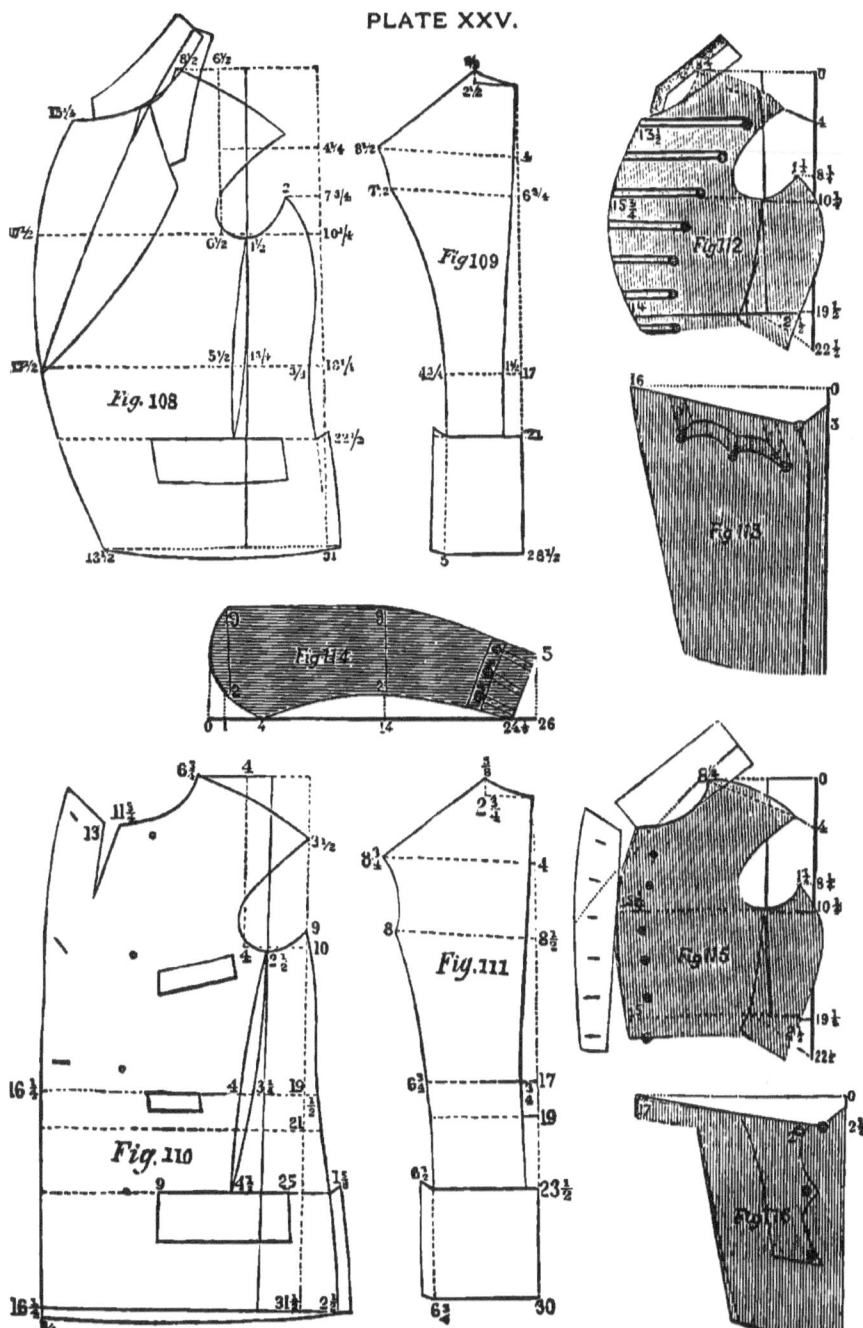

PLATE XXV.—FIGURES 108-116.

EXPLANATION.

Figures 108 and 109 of Plate XXV, represent respectively the front and the back, of the New York-Jacket. Figures 110 and 111, represent the front and back of the Pea-Jacket or "Reefer." Figures 112, 113, and 114 represent the front, skirt, and sleeve of a style of Livery, or Uniform-Coat; as also do Figures 115 and 116.

UTILITY AND DECORATIVE ART. 249

PLATE XXVI.

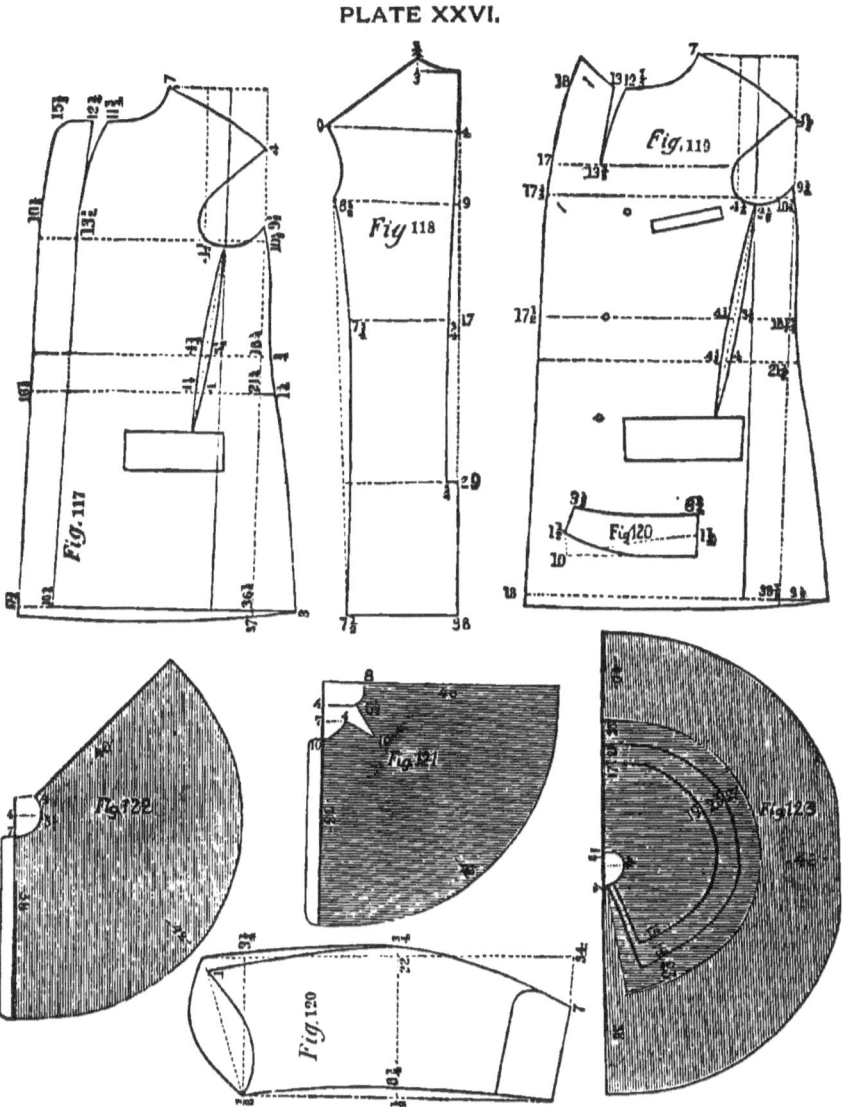

EXPLANATION.

1. Figures 117, 118, 119 and 120, of Plate XXVI, show the front and back of the, single-breasted Sack-Overcoat; and the front and collar of the double-breasted style. Figure 121 shows the Half-Circle–cloak. Figure 122, the Three-Fourth-Circle-Cloak. Figure 123, the Full-Circle or Spanish-Cloak ; as well as the various sizes of the Spanish Cape.

Surtout-Overcoats.

2. Surtout-Overcoats, are cut on the same principle as Frock-coats; only, increasing the measures in the ratios given; and following the styles, outside of the fit, at will, or according to fashion and good taste.

3. In placing the patterns upon the goods, always begin at the bottom and place the largest piece first, then the second; and continue thus, to the end; arranging them, in the most economical way. This rule, knows no exception. It is true of all garments.

Received of M..resident of.............................County of ....State of........................,this...........day of.......................A. D. 188....., the full sum of Five Dollars, for the right of personal possession and individual and personal use, only, of one copy of "THE INTERNATIONAL ENCYCLOPEDIA OF GARMENT-MAKING, ARTISTIC NEEDLE-WORK, UTILITY, AND DECORATIVE ARTS," etc., the receipt of which is hereby acknowledged by the said...

It is furthermore distinctly understood, mutually acknowledged, and declared hereby, that said money is paid and accepted, only, for and in consideration of the possession of said Book and said right; for which this said receipt, as printed within said Book, shall be the mutual and sufficient voucher.

....Author and Pro'p.
　　　　　Per
....Agent.

Plan of Measure-Book for Gentlemen's Measures.

Name,..
Residence,..
Materials,...
Garments,...
To Fashion-Plates, Nos.....................Page..............of.............A. D., 188...,.........Number.
Time for Completion,.........................day,..............M............the..................A. D., 188.....
Price for { Materials Furnished,...$...........
{ Manufacturing,..$...........

Total,....................$

THE COAT AND VEST MEASURES.	The Close or Standard Measures	The Increased Measures for Overcoat Cutting.	
	For all Kinds of Goods.	For Heavy Weight Goods.	For Very Heavy Weight Goods.
	INCHES.	INCHES.	INCHES.
1. Neck-Circumference...			
2. Skirt-Depth,...			
3. Blade-Width,..			
4. Chest-Circumference,...			
5. Waist-Depth...			
6. Back-Waist-Depth,...			
7. Back-Height,..			
8. Front-Waist-Depth,..			
9. Front-Height,...			
10. Arm's-Eye-Circumference,..			
11. Back-Shoulder-Height...			
12. Front-Shoulder-Height,...			
13. Shoulder-Depth,...			
14. Waist-Circumference,...			
15. Sleeve: { 1. Extreme-Depth,.......................................			
{ 2. Arm's-Eye-Circumference,.........................			
{ 3. Elbow-Circumference,..............................			
{ 4. Elbow-Depth,..			
{ 5. Hand-Circumference,...............................			
{ 6. Arm's-Circumference,...............................			
{ 7. Inside-Height,.......................................			
{ 8. Outside-Height,.....................................			
16. Pantaloons: { 1. { a. Knee-Depth...........................			
{ and			
{ b. Outside-Depth,.....................			
{ 2. Front-Depth,..			
{ 3. Inside-Depth,..			
{ 4. Waist-Circumference,..............................			
{ 5. Hip-Circumference,.................................			
{ 6. Thigh-Circumference,...............................			
{ 7. Thigh-Circumference,...............................			
{ 8. Diagonal...			
{ 9. Knee-Circumference,...............................			
{ 10. Bottom-Circumference,............................			

www.ingramcontent.com/pod-product-compliance
Lightning Source LLC
Chambersburg PA
CBHW031730230426
43669CB00007B/303